On-Line LC–NMR and Related Techniques

On-Line LC–NMR and Related Techniques

Edited by

Klaus Albert
Universität Tübingen, Tübingen, Germany

JOHN WILEY & SONS, LTD

Telephone (+44) 1243 779777

Email (for orders and customer service enquiries): cs-books@wiley.co.uk
Visit our Home Page on www.wileyeurope.com or www.wiley.com

Other Wiley Editorial Offices

John Wiley & Sons Inc., 111 River Street, Hoboken, NJ 07030, USA

Jossey-Bass, 989 Market Street, San Francisco, CA 94103-1741, USA

Wiley-VCH Verlag GmbH, Boschstr. 12, D-69469 Weinheim, Germany

John Wiley & Sons Australia Ltd, 33 Park Road, Milton, Queensland 4064, Australia

John Wiley & Sons (Asia) Pte Ltd, 2 Clementi Loop #02-01, Jin Xing Distripark, Singapore 129809

John Wiley & Sons Canada Ltd, 22 Worcester Road, Etobicoke, Ontario, Canada M9W 1L1

Library of Congress Cataloging-in-Publication Data

On-line LC-NMR and related techniques / edited by Klaus Albert
p. cm.
Includes bibliographical references and index.
ISBN 0–471–49649–9
1. Nuclear magnetic resonance spectroscopy. 2. Liquid chromatography. I. Albert, Klaus.

QD96.N8 O5 2002
543′.0877—dc21

2002072626

British Library Cataloguing in Publication Data

A catalogue record for this book is available from the British Library

ISBN 0–471–49649–9

Typeset in 10/12pt Times by Kolam Information Services Pvt. Ltd, Pondicherry, India.
Printed and bound in Great Britain by Biddles Ltd, Guildford, Surrey.
This book is printed on acid-free paper responsibly manufactured from sustainable forestry in which at least two trees are planted for each one used for paper production.

For Heidi

Contents

Contributors

Klaus Albert
Institut für Organische Chemie, Universität Tübingen, Tübingen, Germany

Ulrich Braumann
Bruker BioSpin GmbH, Rheinstetten, Germany

Holger Fischer
Institut für Organische Chemie, Universität Tübingen, Tübingen, Germany

Tobias Glaser
Institut für Organische Chemie, Universität Tübingen, Tübingen, Germany

Markus Godejohann
Bruker BioSpin GmbH, Rheinstetten, Germany

Heidrun Händel
Institut für Organische Chemie, Universität Tübingen, Tübingen, Germany

Michael Lacey
School of Chemical Sciences, University of Illinois at Urbana-Champaign, IL, USA

John C. Lindon
Department of Biological Chemistry, Imperial College of Science, Technology and Medicine, London, UK

Jeremy K. Nicholson
Department of Biological Chemistry, Imperial College of Science, Technology and Medicine, London, UK

Alfred Preiss
Fraunhofer Institut für Toxikologie und Aerosolforschung, Hannover, Germany

Daniel Raftery
Department of Chemistry, Purdue University, IN, USA

Martin Sandvoss
GlaxoSmithKline, Ware, Hertfordshire, UK

Alexandre Bezerra Schefer
Institut für Organische Chemie, Universität Tübingen, Tübingen, Germany

John P. Schockcor
Department of Biological Chemistry, Imperial College of Science, Technology and Medicine, London, UK

Manfred Spraul
Bruker BioSpin GmbH, Rheinstetten, Germany

Jonathan V. Sweedler
School of Chemical Sciences, University of Illinois at Urbana-Champaign, IL, USA

Andrew G. Webb
Department of Electrical and Computer Engineering, University of Illinois at Urbana-Champaign, IL, USA

Ian D. Wilson
AstraZeneca Pharmaceuticals, Alderley Park, Macclesfield, Cheshire, UK

Preface

It has been more than 20 years since the very first HPLC–NMR experiments were conducted, employing iron magnets for the registration of ^{1}H NMR spectra. Since then, an enormous increase in sensitivity has been accomplished by the combined use of cryomagnets, new NMR coil designs and materials, together with effective pulse sequences for solvent suppression. In the late 1970s, only model separations with mg amounts could be performed, but nowadays LC–NMR is an established analytical technique in biomedical, pharmaceutical, environmental, drug metabolism and natural product analysis (see References [1–70] in Chapter 1 of this volume). The newest development of capillary NMR shows detection limits beyond 10 ng.

Whereas LC–NMR was considered to be an exotic technique in the late 1970s, today over 200 LC–NMR systems are installed world-wide. The success of LC–NMR is due to the enthusiastic work of people in both industry and academia, who have combined their skills and efforts to continuously improve the reliability of this coupled technique. Some of the early pioneers of LC–NMR are coauthors of this book and thus ensure a guarantee for competent contributions.

The aim of this text is to introduce the fascinating topic of the hyphenation of chromatographic separation techniques with nuclear magnetic resonance spectroscopy to an interested readership with a background either in organic, pharmaceutical or medical chemistry. The basic principles of NMR spectroscopy, as well as those of separation science, should previously be known to the reader.

The specific constraints and requirements of continuous-flow NMR will be explained in the first chapter, whereas specific applications, such as biomedical and natural product analysis, LC–NMR–MS and LC–NMR in an industrial environment, together with polymer analysis, will be discussed separately. Thus, the reader will obtain a broad overview of the application power of LC–NMR and the benefits of its use. He/She will also be introduced to the pitfalls of this technique. Special attention will be given to the exciting newer coupled techniques such as SFC–NMR and capillary HPLC–NMR. However, new emerging future developments will also be discussed thoroughly.

Finally, I hope that this text will become a standard in all analytical laboratories and I would like to thank all contributors for their ongoing interest in completing this book.

Klaus Albert
Universität Tübingen, Germany

1 LC–NMR: Theory and Experiment

KLAUS ALBERT
Institut für Organische Chemie, Universität Tübingen, Tübingen, Germany

1.1 INTRODUCTION

The conventional way of recording solution-state NMR spectra is by the use of a 5 mm cylindrical NMR tube in which the compound of interest is dissolved in 0.5 ml of a deuterated solvent. The sample is constantly available for an infinite time period for the registration of NMR spectra. With the commonly applied Pulse Fourier-Transform acquisition mode, a gain in signal-to-noise ratio (S/N) of the acquired NMR spectrum can be obtained by co-adding the Free Induction Decays (FIDs) resulting from pulse excitation. The FID is dependent upon the relaxation time T_2, which effects the line shape and the resolution of a spectrum. The recovery of equilibrium magnetization is determined by the spin–lattice relaxation time T_1 (Figure 1.1). After pulse excitation, it takes a time period of three to five times the T_1 to establish the full Boltzmann distribution, together with full magnetization of the nuclei. Then, a new excitation pulse can be applied. Thus, longer T_1 times, e.g. of aromatic moieties of an organic molecules, necessitate longer pulse repetition times. Whereas the S/N value is defined by the square root of the number of transients (NS), the pulse repetition time for a new excitation of fully relaxed nuclei is dependent upon the spin–lattice relaxation time T_1.

1.2 NMR IN A FLOWING LIQUID

In the conventional measuring mode the sample stays in the NMR tube, and thus in the radiofrequency Helmholtz coil all of the time. In the continuous-flow mode it resides within the NMR detection coil only for a distinct time of some few seconds (Figure 1.2). This residence time τ is dependent upon the volume of the detection cell and the employed flow rate (Table 1.1). For example, a detection volume of 120 µl, together with a flow rate of 0.5 ml/min, results in a residence time of 14.4 s, while with a detection volume of 8 µl the residence time is only 0.96 s. A shorter residence time τ within the NMR measuring coil results in a reduction of the effective lifetime of

On-Line LC–NMR and Related Techniques
Edited by K. Albert.

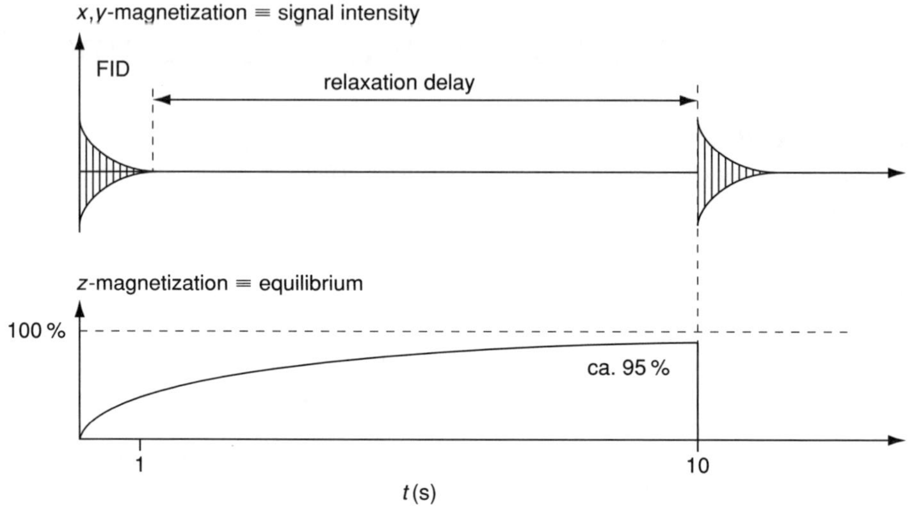

Figure 1.1 Acquistion scheme used for a static NMR experiment

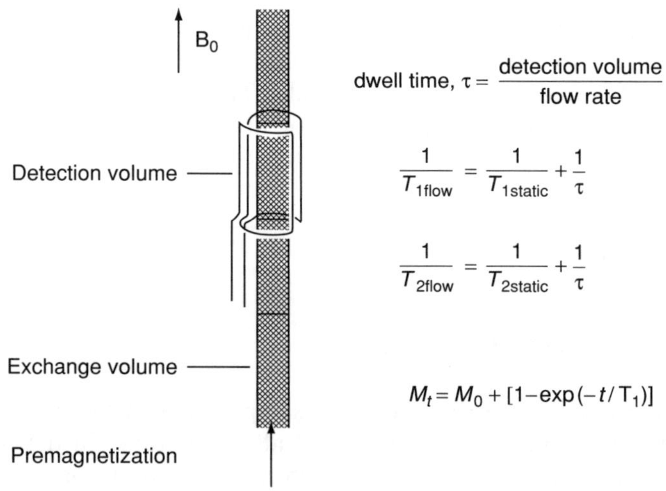

Figure 1.2 Continuous-flow NMR detection principle

the particular spin states. Thus, the effective relaxation rates, $1/T_n$, are increased by $1/\tau$:

$$1/T_{n\ \mathrm{effective}=} \sum 1/T_i + 1/\tau \qquad (1.1)$$

In a flowing system, the reciprocal relaxation rates, the relaxation times $T_{1\mathrm{flow}}$ and $T_{2\mathrm{flow}}$, are reduced according to the following:

Table 1.1 Variation of residence time and line broadening as a function of detection cell volume and flow rate in continuous-flow NMR spectroscopy.

	Dwell time τ (s)		Line broadening $1/\tau$ (Hz)	
Volume (μl)	Flow rate 0.5 ml/min	Flow rate 1.0 ml/min	Flow rate 0.5 ml/min	Flow rate 1.0 ml/min
44.0	5.28	2.64	0.19	0.38
60.0	7.20	3.60	0.14	0.28
120.0	14.40	7.20	0.07	0.14
200.0	24.00	*12.00*	0.04	0.08
Volume (nl)	Flow rate 10 nl/s	Flow rate 20 nl/s	Flow rate 10 nl/s	Flow rate 20 nl/s
5	0.50	0.25	2.00	4.00
50	5.00	2.50	0.20	0.40
500	50.00	25.00	0.02	0.04

$$1/T_{1\text{flow}} = 1/T_{1\text{static}} + 1/\tau \tag{1.2}$$

$$1/T_{2\text{flow}} = 1/T_{2\text{static}} + 1/\tau \tag{1.3}$$

In a net effect, the pulse repetition times in flowing systems can be reduced to the decrease in the apparent spin–lattice relaxation times $T_{1\text{flow}}$, whereas at a given detection volume an increase in flow rate leads to an increase in the signal half-width W due to the decrease of $T_{2\text{flow}}$.

$$W = (1/\pi)T_2 \tag{1.4}$$

$$W_{\text{flow}} = W_{\text{stationary}} + 1/\tau \tag{1.5}$$

Thus, the resolution of a continuous-flow ^{1}H NMR spectrum is strongly dependent upon the flow rate/detection volume ratio.

Given a detection volume of 44 μl, a flow rate of 0.5 ml/min results in a residence time of 5.28 s and a line broadening of 0.19 min, whereas a flow rate of 1.0 ml/min leads to a line broadening of 0.38 Hz (see Table 1.1). This effect can be easily varified from Figure 1.3, which shows the signal half-width of chloroform at different flow rates. The static signal linewidth is 0.55 Hz; at a flow rate of 0.5 ml/min this is increased to 0.75 Hz (theoretical value, 0.74 Hz), and at a flow rate of 1.0 ml/min to 1.05 Hz (theoretical value, 0.93 Hz). To minimize the flow-induced broadening effect, NMR flow cells should provide residence times of the order of 5 s. Thus, the resulting line-broadening values will be about 0.2 Hz, which are tolerable for the acquisition of ^{1}H NMR spectra.

In an on-flow NMR experiment, the excited nuclei leave the flow cell whereas 'fresh' nuclei enter. Due to the decrease of the apparent $T_{1\text{flow}}$ rates, faster pulse repetition rates can be used and more transients in a distinct time-period can be accumulated (Figure 1.4). The theoretical maximum sensitivity is obtained

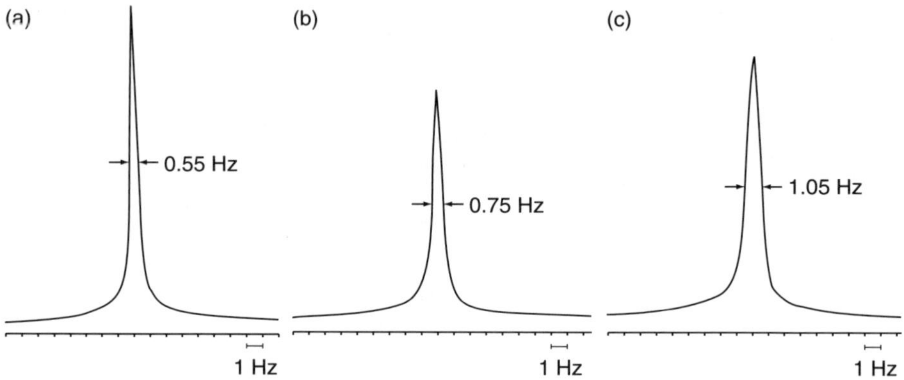

Figure 1.3 ^{1}H NMR spectra (400 MHz) of chloroform recorded at flow rates of (a) 9 (stopped flow), (b) 0.5 and (c) 1.0 ml/min

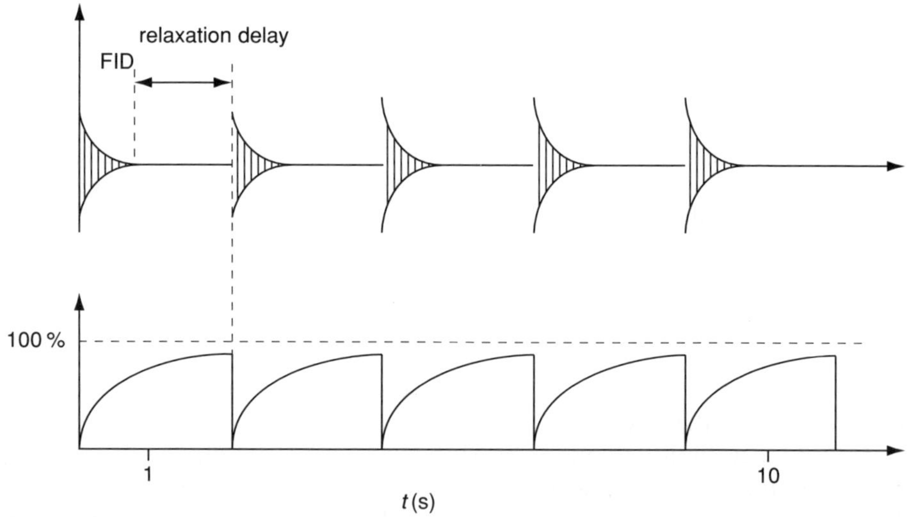

Figure 1.4 Acquisition scheme used for a continuous-flow NMR experiment

when the pulse repetition time, PRT (The sum of the acquisition time, AQ, and relaxation the delay D1) is equal to the residence time τ in the NMR flow cell

$$\text{PRT} = (\text{AQ} + \text{D1})\text{optimal} = \tau \tag{1.6}$$

If the 'fresh' incoming nuclei are fully magnetized upon entering the flow cell, e.g. the Boltzmann distribution is established, an increase in sensitivity can be obtained. This is demonstrated in Figure 1.5, which shows the ^{1}H NMR spectra of the CH_2 protons of ethanol under stopped-flow and continuous-flow (2.4 ml/min) conditions.

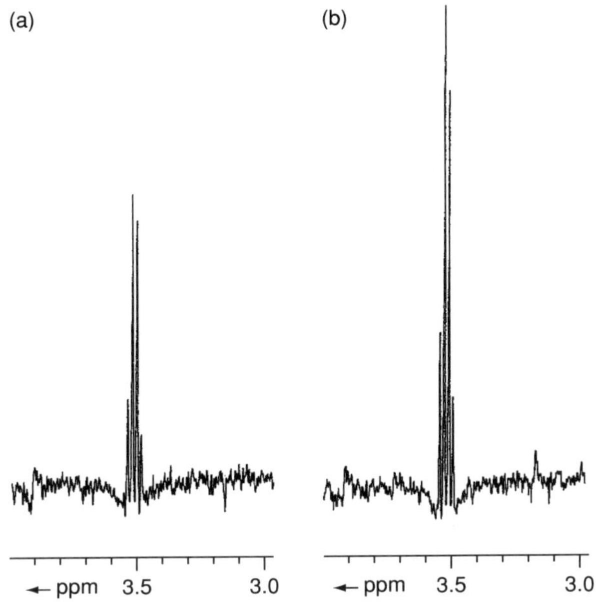

Figure 1.5 ^{1}H NMR spectra (400 MHz) of the CH_2 protons of ethanol recorded under (a) stopped-flow and (b) continuous-flow (2.4 ml/min) conditions

1.3 DESIGN OF CONTINUOUS-FLOW NMR PROBES

The first approach for continuously recording NMR spectra was to use the conventional existing probe for the registration of NMR spectra [1]. The latter are usually recorded under rotation of the NMR tube with a rotational speed of 20 Hz in order to remove magnetic field inhomogeneities. Thus, it should be sufficient to introduce a capillary within a rotating NMR tube and to suck off the effluent (mobile phase) with the help of a second capillary (Figure 1.6). The problem with this design is that no complete transfer of the mobile phase is guaranteed by the employment of the second capillary, and peak mixing, together with memory effects, will occur at the bottom of the rotating NMR tube. Thus, it would be more straightforward to employ a 'bubble cell' design of a widened glass tube. This approach was used for the registration of the first continuous-flow NMR spectra with iron magnets (Figure 1.7) [2-6] and also together with cryomagnets (Figure 1.8) [7–15]. This design, which was already introduced in the early 1980s, is still mostly used today.

Such a design combines the bubble-cell characteristics, together with an U-type design of the glass tube employed as the NMR detector. The glass tube is positioned within a glass Dewar, thus enabling temperature-dependent measurements. Another feature is the direct attachment of the NMR radiofrequency

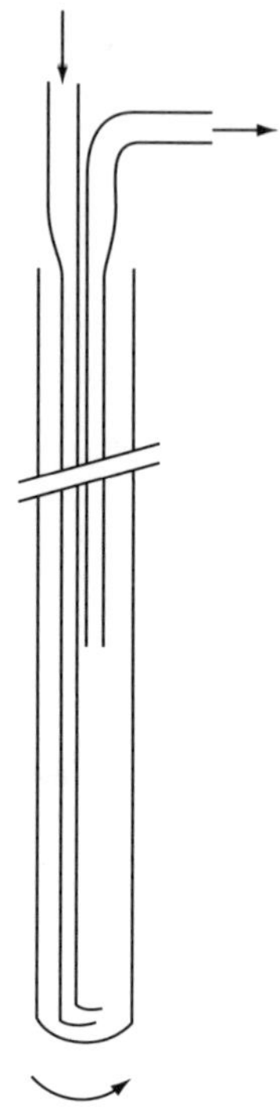

Figure 1.6 System used for recording continuous-flow NMR spectra in a rotating NMR tube

(rf) coil to the glass tube, thus rendering any rotation of the tube impossible. However, in contrast to the conventional NMR probe design, the filling factor (ratio of sample volume to the NMR detection cell volume) is much higher (Figure 1.8). Because both the inlet and outlet of the continuous-flow detection cell are at the bottom of the cylindrical NMR probe, the whole probe body can easily be inserted into the room temperature bore of the cryomagnet. No problems with air bubbles exist because the NMR detection cell is filled from the bottom to the top against the earth's gravity.

Due to the fact that within this design the radiofrequency coil is positioned parallel to the z-direction of the magnetic field of the cryomagnet, magnetic field homogeneity can be readily achieved, because the device for the correction of the magnetic field, the so-called 'shin system', is optimized for correcting inhomogeneities in the z-direction. Thus, the U-type flow cell shows very good NMR characteristics, despite the non-rotation of the cell. The quality of cryomagnets, together with the NMR probes, is checked by determining the signal linewidth of chloroform at the height of the ^{13}C satellites and at 1/5 of this altitude. The obtained values (Figure 1.9) are very close to the conventional ones obtained without any rotation of the NMR tube. The large detection

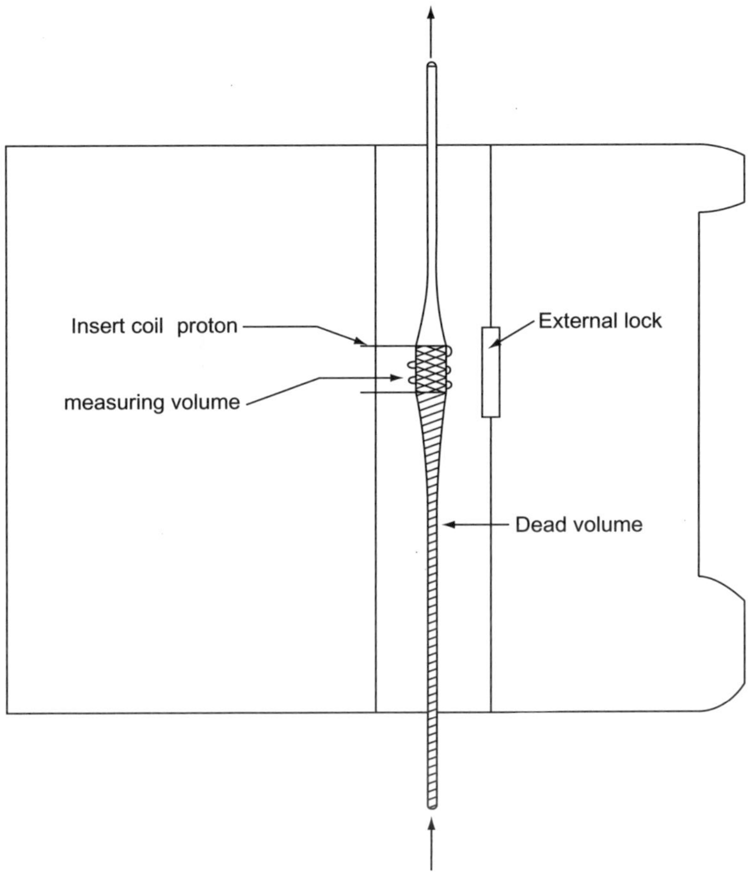

Figure 1.7 Schematic of a continuous-flow probe suitable for iron magnets

volume of 120 μl employed leads to NMR sensitivity levels of about 100 ng for one-dimensional (1D) acquisitions, and of about 1 μg for two-dimensional (2D) NMR spectra. On the other hand, the chromatographic peaks are broadened by approximately 20 %. The peak dispersion effects of several NMR flow cells were measured by directly evaluating chromatographic separations with the help of a modified fluorescence detector. These measurements lead to the conclusion that the plate height is adversely affected for capacity factors below 2.5 [15]. A further evaluation of peak dispersion effects is given later in Chapter 6 of this present book.

Better chromatographic peak performance is obtained with an NMR detection volume of 60 μl, although NMR sensitivity values suffer from the low amount of nuclei in the detection cell. Thus, despite its degraded chromatographic performance the 120 μl flow cell seems to be a good compromise

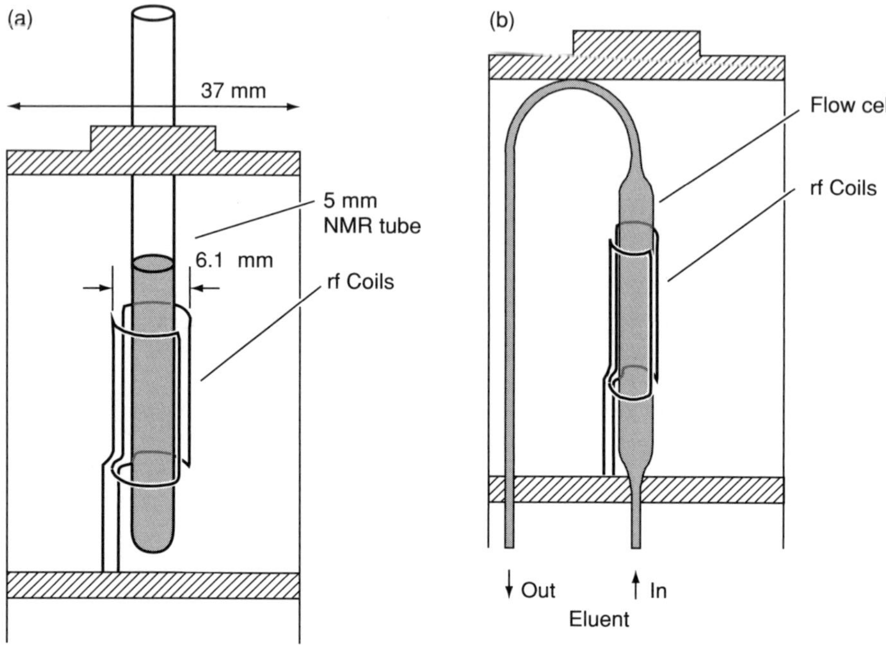

Figure 1.8 Schematics of (a) conventional and (b) continuous-flow NMR probes suitable for cryo magnets

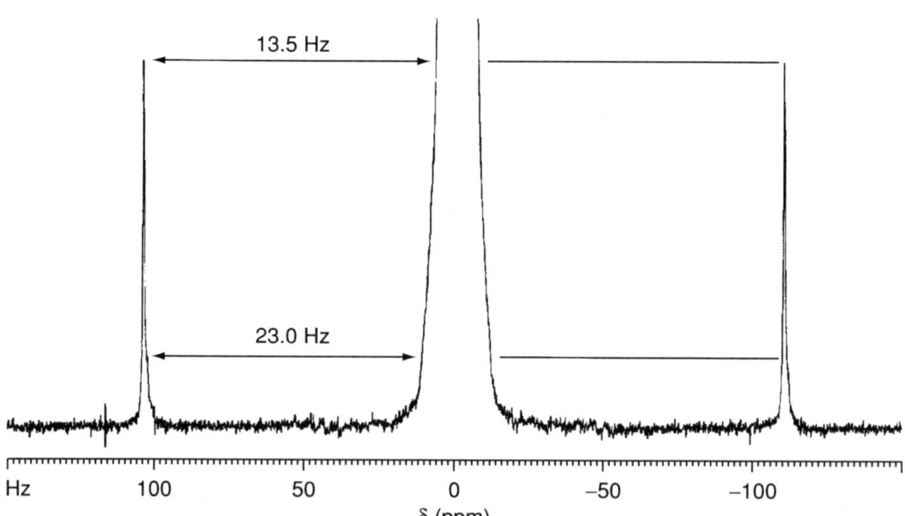

Figure 1.9 ^{1}H NMR signal line shape of chloroform in acetone-d_6 (hump test), measured with a 120 μl continuous-flow probe (600 MHz)

between the NMR and chromatographic requirements. From the NMR viewpoint, it is not a problem to discriminate between the signals obtained from a major compound and a minor component, and thus the chromatographic peak broadening can be tolerated for the gain in NMR sensitivity. Major ongoing improvements in NMR sensitivity with micro-coils will lead to the development of capillary probes with superior characteristics.

1.4 EXPERIMENTAL ARRANGEMENT FOR HPLC–^{1}H NMR COUPLING

The main prerequisite for on-line LC–NMR, besides the NMR and HPLC instrumentation, are the continuous-flow probe and a valve installed before the probe for the registration of either continuous-flow or stopped-flow NMR spectra.

Due to the current development of cryomagnet technology, no benchtop-like cryomagnets will be available with a magnetic field strength between 9.4 and 14 T in the next few years. Therefore, the current available types of cryomagnets have to be used (Figure 1.10). Whereas a proton resonance frequency of 300 MHz is sufficient for GPC–NMR experiments it is advisable to use magnetic field strengths higher than 9.4 T (^{1}H resonance frequency of 400 MHz) for HPLC–NMR coupling. The position of the HPLC instrument is dependent upon the size of the stray magnetic field of the used cryomagnet.

Thus, in conventional installations the HPLC instrument is located at distances between 1.0 and 2.0 m from the cryomagnet, whereas with new available shielded cryomagnets the HPLC instrument can be directly hooked to the cryomagnet.

The analytical NMR flow-cell (see Figure 1.8) was originally developed for continuous-flow NMR acquisition, but the need for full structural assignment of unknown compounds led to major applications in the stopped-flow mode. Here, the benefits of the closed-loop separation-identification circuit, together with the possibilities to use all types of present available 2D and 3D NMR techniques in a fully automated way, has convinced a lot of application chemists [17-70]. A detailed description of the different modes for stopped-flow acquisition (e.g. time-slice mode) is found in Chapters 2 und 3.

Figure 1.10 shows an experimental arrangement of HPLC–NMR coupling which is currently employed in many analytical laboratories. In most laboratories, unshielded magnets are used at the moment, and thus the HPLC instrument, consisting of an injection device, HPLC pumps together with a gradient unit, an HPLC column (4.6 × 250 mm) and UV detector, is located at a distance of 1.5 m from the cryomagnet. The outlet of the UV detector is either connected via a stainless steel capillary (id, 0.25 mm) to a valve or to

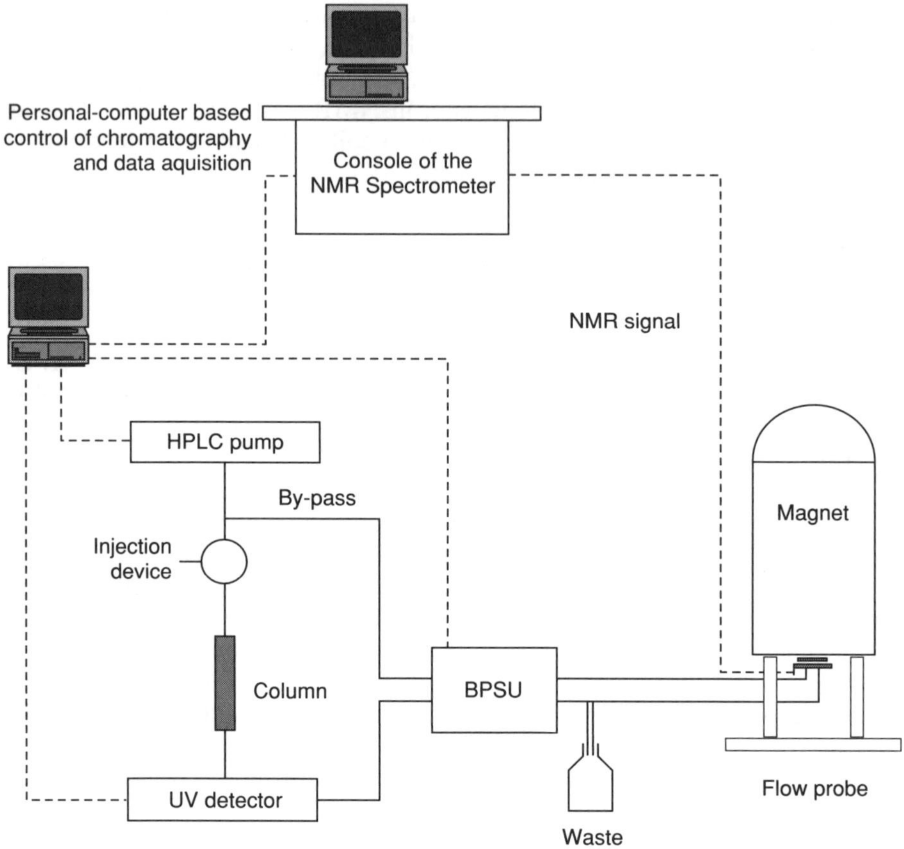

Figure 1.10 Schematic of the experimental set-up used for HPLC–NMR coupling: BPSU, Bruker peak sampler unit; (——) capillary junctions; (- - -) electronic junctions

a peak sampling unit which are connected to the continuous-flow probe in the cryomagnet. With either the valve or the peak sampling unit being under software control, stopped-flow NMR acquisition of all peaks of an HPLC separation can be performed due to the triggering of the UV signal. For a proper timing and recording of chromatographic peaks, the transfer time of peak passage between the UV detector and the NMR flow cell has to be carefully recorded.

This experimental design has the big advantage that it can be easily accomplished. The operation mode of the NMR instrument from routine NMR data acquisition to the LC–NMR mode can be easily changed by removing the routine probe from the room-temperature bore of the cryomagnet and inserting the continuous-flow probe by fixing two screws. The magnetic field homogeneity of the continuous-flow probe can be readily adjusted by using standard reference shim files.

The transfer volume of the capillaries between the HPLC instrument and the NMR probe is about 150 μl. For minimum peak dispersion, the insertion of the HPLC column into the probe body of the continuous-flow probe would be desirable. This experimental arrangement was proposed by Wilkins and co-authors (16) and is used in (supercritical fluid chromatography) (SFC)–NMR employing immobilized free radicals (see Chapter 7.2 below). However, here column exchange is much more demanding than in the design outlined in Figure 1.8. With the increased use of shielded cryomagnets, the distance between the HPLC and NMR instruments will be reduced, thus rendering the need for inserting the HPLC column within the probe body unnecessary.

1.5 PRACTICAL CONSIDERATIONS, SOLVENT SUPPRESSION TECHNIQUES, GRADIENT ELUTION AND PURITY OF HPLC SOLVENTS

In real-life application of HPLC–NMR, three main types of data acquisition have been established, namely continuous-flow acquisition, stopped-flow acquisition, and time-sliced acquisition with the help of storage loops. For all of these acquisition techniques the major prerequisite is an optimized HPLC separation. Because sensitivity is still the crucial point of this coupling technique it is extremely important to develop a chromatographic separation where the quantity of the available separated compound is concentrated in the smallest available elution volume. This need necessitates the development of stationary phases which exhibit optimum separation characteristics, together with the capability to tolerate column overloading. The newly developed C_{30} phases are typical representatives of these types of columns, which is evidenced by the separation of tocopherol isomers. Figure 1.11 clearly shows that it is possible to overload a C_{30} column by a factor of 200 without loosing any chromatographic resolution.

The major amount of HPLC separations is performed with reversed-phase columns employing binary or tertiary solvent mixtures with isocratic or gradient elution. The protons of the solvents of the mobile phase cause severe problems for an adequate NMR registration. The receiver of the NMR instrument (either a 12-bit or a 16-bit analog–digital converter (ADC)) is unable to handle the intense solvent signals and the weak substance signals at the same time.

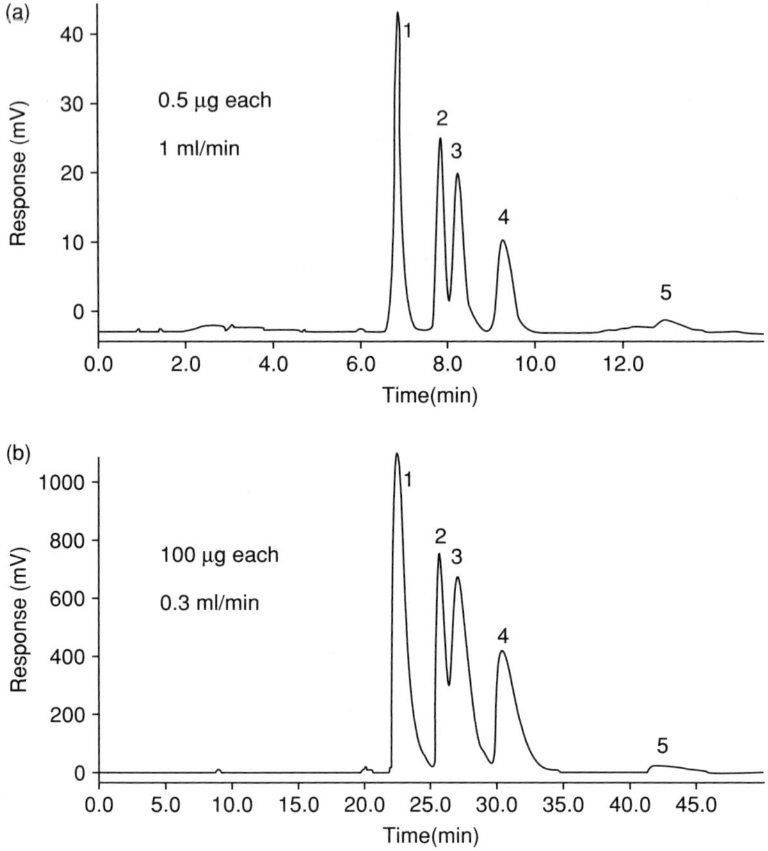

Figure 1.11 UV chromatograms (295 nm) of the separation of tocopherol isomers (1, δ-tocopherol; 2, γ-tocopherol; 3, β-tocopherol; 4, α-tocopherol; 5, α-tocopherol acetate: (a) analytical separation; (b) with a 200-fold amount of sample

Figure 1.12 shows the free induction decay (FID) and the transformed spectrum of a 0.01 % sample of butylbenzylphthalate in acetonitrile (ACN) D_2O (80/20). The FID is dominated by the methyl group signal from the acetonitrile. In order to get an undistorted spectrum, a small receiver gain has to be chosen, leading to a low signal-to-noise (S/N) value for the sample signals. An increase in receiver gain does not lead to the desired result. Figure 1.13 shows the effect of overloading the receiver with solvent signals. By the 'clipping' of the FID, the transformed spectrum is distorted and thus useless for interpretation (see Figure 1.13) and the sensitivity of detection is severely decreased. In order to avoid this problem, the signal intensity of the solvent signals has to be reduced. Now the receiver gain can be increased and

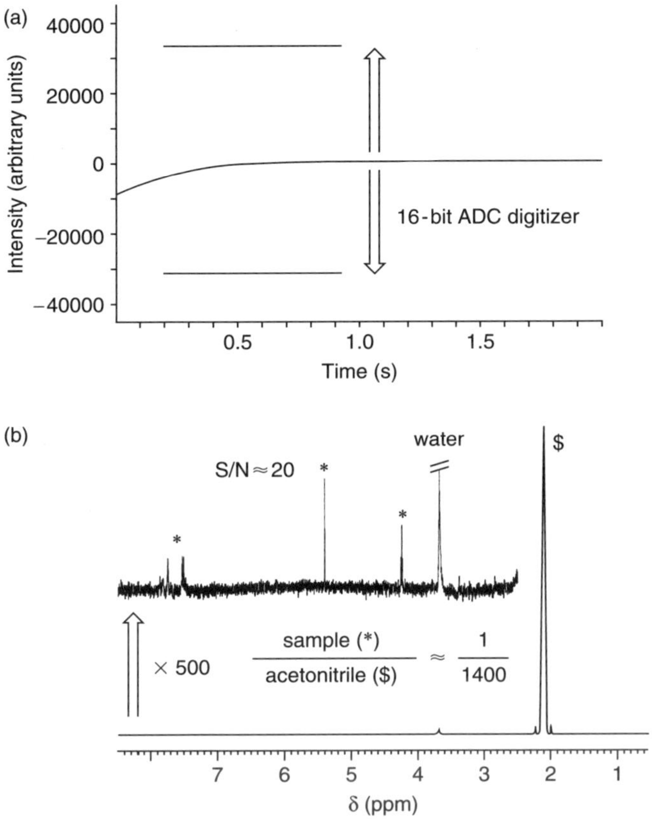

Figure 1.12 Conventional ^{1}H NMR spectrum of a 0.01 % sample of butylbenzyl phthalate in ACN/D_2O: (a) free induction decay; (b) transformed spectrum

adjusted to the smaller FID without any problems. Figure 1.14 shows the FID and the resulting ^{1}H *NMR* spectrum of the same sample after reducing the solvent signal intensity. The remaining signals of the suppressed methyl group resonance of acetonitrile can be seen at 2.1 ppm. The signals of butylbenzylphthalate show a much higher S/N of 320:1. The 16-fold enhancement of the signal-to-noise value corresponds to a saving factor of 256.

1.5.1 SOLVENT SIGNAL SUPPRESSION

Solvent signal suppression is necessary in order to achieve a reduction of the NMR signal entering the receiver for observing small analyte signals in the

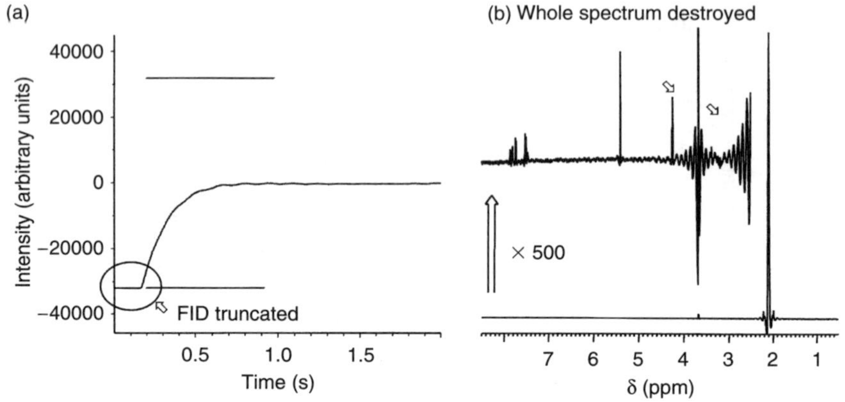

Figure 1.13 Increase in receiver gain without solvent signal suppression: (a) free induction decay; (b) resulting NMR spectrum

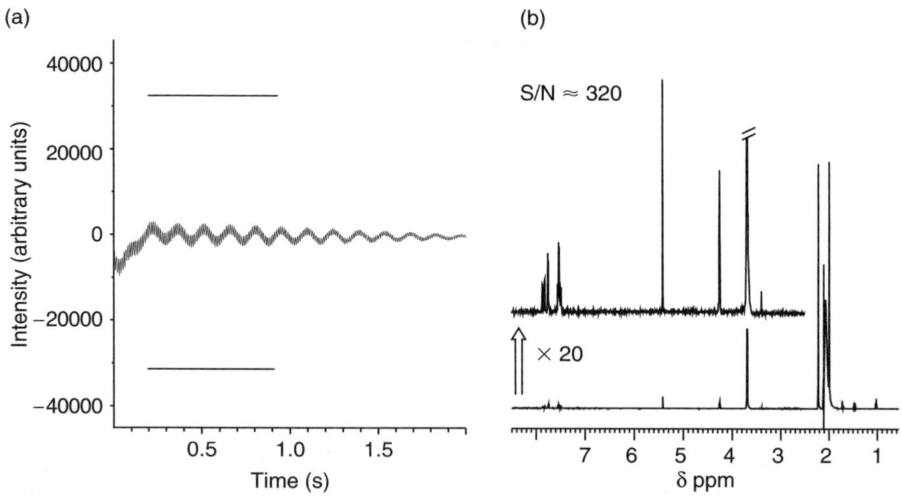

Figure 1.14 Optimized receiver gain with solvent signal suppression: (a) free induction decay; (b) resulting NMR spectrum

presence of much larger signals from the mobile phase. Solvent signal suppression is efficiently performed by using three techniques:

- Presaturation (NOESY presaturation)
- Soft-pulse multiple irradiation
- WET presaturation employing a *z*-gradient.

Presaturation

The principle of presaturation relies on the phenomenon that nuclei which are unable to relax, because their population in the ground state α and the excited state β is the same, do not contribute to the free induction decay after pulse irradiation. Prior to data acquisition, a highly selective low-power pulse irradiates the desired solvent signals for 0.5 to 2 s, thus leading to saturation of the solvent signal frequency. During data acquisition, no irradiation should occur. NOESY-type presaturation is an effective pulse sequence of presaturation. The pulse sequence consits of three 90° pulses (similar to the first increment of a NOESY experiment):

$$\mathrm{RD} - 90^\circ - t_1 - 90^\circ - t_m - 90^\circ - \mathrm{FID} \tag{1.7}$$

where RD is the relaxation delay, and t_1 and t_m are the presaturation times (e.g., 0.6 and 0.08 s). The effect of NOESY-type presaturation is illustrated in Figure 1.15, which shows the ^{1}H *NMR* spectra of 12 μg retinoic acid in ACN/D_2O (60/40), without and with solvent suppression. With the same number of transients as the spectrum without solvent suppression, the S/N value of the olefinic signals is much increased. In addition, the remaining ^{13}C-satellites of the suppressed solvent signal and the baseline distortion around the presaturation frequency can be seen.

Soft Pulse Multiple Irradiation

Here, presaturation is performed with the use of shaped pulses, which have a broader excitation profile. This method is therefore better suitable for the suppression of multiplets. The advantages of this technique are that it is easy to apply, easy to implement within most NMR experiments, and multiple presaturation is possible, and that it is very effective. The disadvantages are that transfer of saturation can occur (in aqueous solutions) to slowly exchanging protons that would be detectable without saturation. Another drawback is that spins with resonances close to the solvent frequency will also be saturated and 2D cross peaks will be absent.

WET Presaturation

The WET sequence (Water Suppression Enhanced through T_1 Effects) uses four solvent selective pulses of variable lengths (Figure 1.16). Each selective rf pulse is followed by a dephasing field gradient pulse. By varying the tip angle of the selective rf pulse, the WET sequence can be obtimized. This approach provides a fast and highly efficient saturation of multiple solvent frequencies. It can be combined with ^{13}C decoupling to remove the ^{13}C satellites of the solvent.

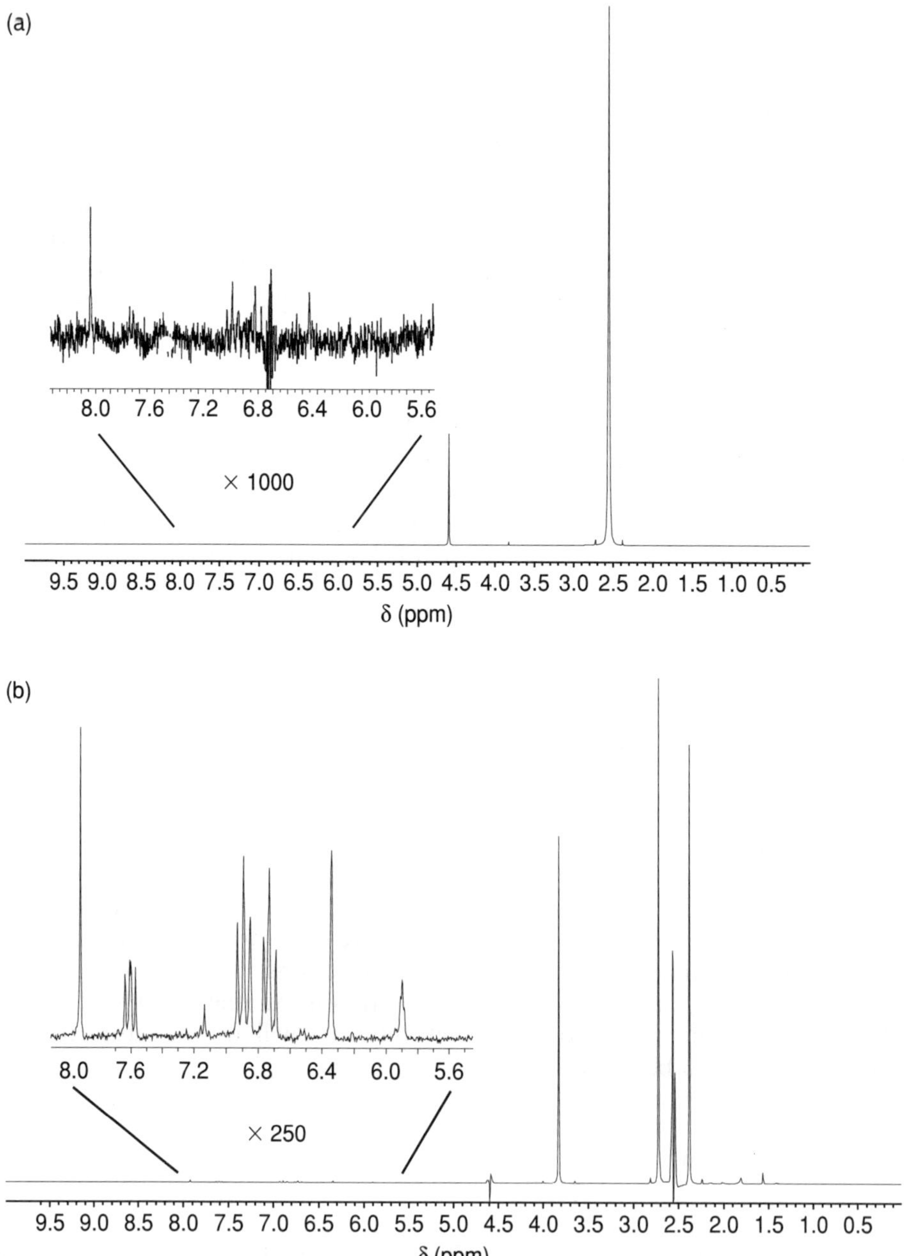

Figure 1.15 The ^{1}H NMR spectra of 12 μg retinoic acid in ACN/D20 (60/40), (a) without and (b) with solvent signal suppression

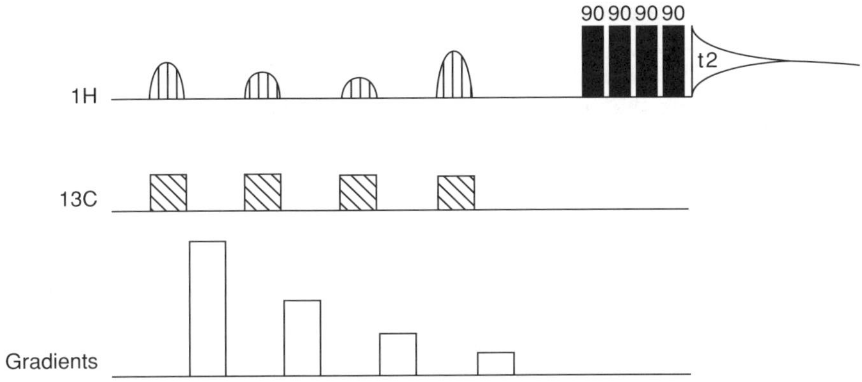

Figure 1.16 Representation of the WET pulse sequence for multiple solvent suppression

Advantages and Disadvantages

The NOESY sequence proved to be very effective for the reduction of one particular signal such as the methyl group of acetonitrile. However, very often the mobile phase has a composition of several solvents, together with up to six solvent signals. Here, the application of the soft pulse multiple solvent suppression technique is advisable.

Both of these techniques, NOESY presaturation as well as soft pulse multiple solvent suppression, lead to a reduction of signal intensity of 1000:1 (see Figure 1.15). Whereas the former solvent signal results in a distortion of the baseline of the NMR spectrum, the ^{13}C satellites which exhibit 0.55 % intensity of the solvent signal, are now the most intense signals of the ^{1}H NMR spectrum. Because the natural abundance of ^{13}C nuclei is 1.1 %, ^{1}H, ^{13}C heteronuclear coupling results in a doublet centered by the singlet of protons adjacent to the NMR-inactive ^{12}C nuclei (natural abundance of 89.9 %). These ^{13}C satellites are only visible at intense signals in a routine ^{1}H NMR spectrum, and it is clear that they contribute to the HPLC–^{1}H NMR spectra. With the application of the WET pulse sequence, these ^{13}C satellites can be eliminated by a combined ^{13}C decoupling together with a proper alignment of the proton magnetization with the help of gradient pulses.

All three suppression techniques can be used either for stopped-flow or continuous-flow acquisition. Presaturation works quite well in the stopped-flow mode, whereas the WET sequence seems to be superior in the continuous-flow mode.

However, all three techniques have the big disadvantage that compound signals lying under the solvent signal are also suppressed. Thus valuable information may have disappeared. This is also the reason why multiple solvent suppression is only useful to a limited extent because too much spectroscopic information may be lost after eliminating too many signals.

Therefore from a practicle viewpoint it is advisable to use only two protonated solvents in a HPLC separation, e.g. H_2O and CH_3CN. To obtain field/frequency stabilization of the cryomagnet it is advisable to exchange H_2O versus D_2O, but even here the intense HDO signal has to be suppressed.

Gradient Elution

In isocratic separations the ^{1}H NMR signals exhibit the same chemical shift, while in gradient separations the changing dielectric constant of the different solvent compositions leads to severe chemical shift alterations. This is outlined in Figure 1.17, which shows the proton NMR spectra of solvent mixtures of acetonitrile and water, from a 100 % concentration of water to a 100 % concentration of acetonitrile. With gradient separations, solvent signal suppression may be carried out by a ‘scout scan’ which detects the effective shift of solvent signals first, and then performing solvent suppression together with the registration of the NMR spectrum in a second step.

Conclusions

The choice of the suppression method to be used depends on both the solvent and sample characteristics. Therefore, HPLC–NMR suppression via presaturation

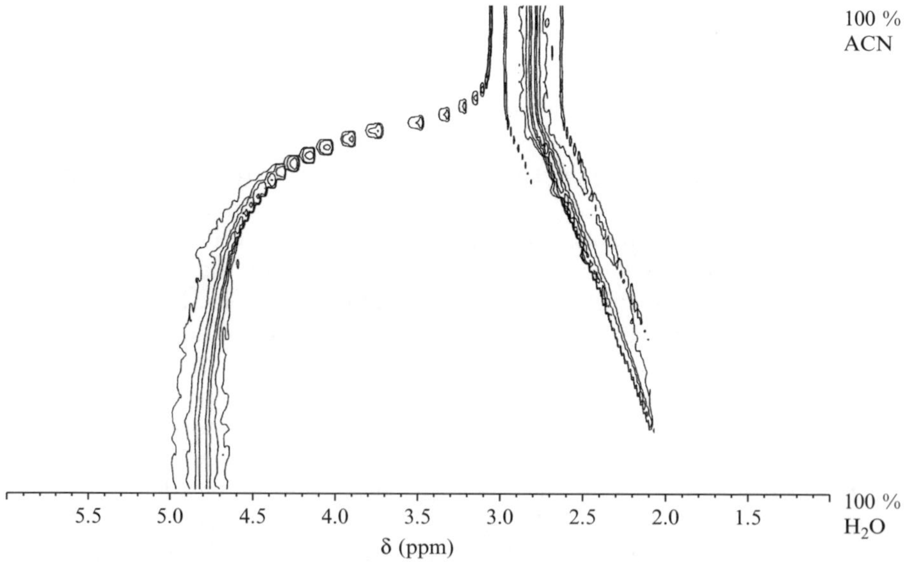

Figure 1.17 ^{1}H NMR contour plot of a solvent mixture over the range from 100 % ACN to 100 % H_2O

or shaped pulses has to be preferred because of the roubustness, simplicity and high suppression ratios, even for multiple solvent signals. Solvent signal suppression should be applied for as short a time as possible and with the lowest power as is necessary. A good line shape (shimming), optimal temperature control and lock stability are a prerequisite for optimal solvent signal suppression.

1.5.2 PURITY OF HPLC-GRADE SOLVENTS

Most solvents contain a small amount of impurities, and often stabilizing chemicals have been added. The HPLC-grade solvents are supposed to be especially pure, although the criteria of purity for these solvents is their interference with the UV adsorption of the solute molecules. NMR detection is much more sensitive to smaller amounts of additional chemicals, especially since the concentration of the sample molecules is often of the order of 0.001 % (m/v).

The mostly employed solvents, such as water/D_2O and acetonitrile, are available with high NMR purity. For all other solvents, the amount of the impurity has to be examined by using a reference spectrum. Figure 1.18 shows a

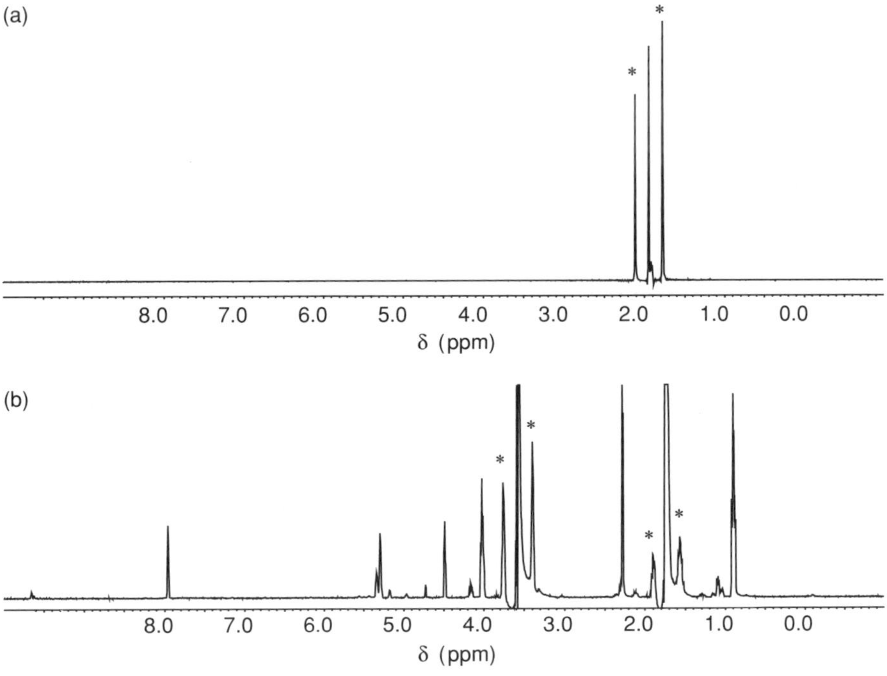

Figure 1.18 ^{1}H NMR spectra of two HPLC-grade solvents, ACN (a) and THF (b). The ^{13}C-satellites of the suppressed signals are marked with asterisks

comparison between the ^{1}H NMR spectra of HPLC-grade pure acetonitrile (a) and tetrahydrofuran (b). Solvent signal suppression was carried out to eliminate the main solvent signals. In this figure, the ^{13}C satellite signals of the suppressed resonances are marked with an asterisk. It can be seen that the spectrum of acetonitrile is free from impurity signals. In the spectrum of tetrahydrofuran, however, several additional signals occur, which are distributed over the whole spectral range. Therefore, this solvent is not advisable for use in GPC–NMR.

REFERENCES

1. Watanabe, N. and Niki, E., *Proc. Jpn. Acad., Ser. B*, 1978, **54**, 194.
2. Bayer, E., Albert, K., Nieder, M., Grom, E. and Keller, T., *J. Chromatogr.*, 1979, **186**, 197.
3. Haw, J. F., Glass, T. E., Hausler, D. W., Motell, E. and Dorn, H. C., *Anal. Chem.*, 1980, **52**, 1135.
4. Bayer, E., Albert, K., Nieder, M. Grom, E. and An, Z., *Fresenius Z. Anal. Chem.*, 1980, **404**, 111.
5. Buddruss, J. and Herzog, H., *Org. Magn. Reson.*, 1980, **13**, 153.
6. Buddrus, J., Herzog, H. and Cooper, J. W., *J. Magn. Reson.*, 1981, **42**, 453.
7. Haw, J. F., Glass, T. E. and Dorn, H. C., *Anal. Chem.*, 1981, **53**, 2327.
8. Haw, J. F., Glass, T. E. and Dorn, H. C., *Anal. Chem.*, 1981, **53**, 2332.
9. Bayer, E., Albert, K., Nieder, M., Grom, E., Wolff, G. and Rindlisbacher, M., *Anal. Chem.*, 1982, **54**, 1747.
10. Haw, J. F., Glass, T. E. and Dorn, H. C., *J. Magn. Reson.*, 1982, **49**, 22.
11. Haw, J. F., Glass, T. E. and Dorn, H. C., *Anal. Chem.*, 1983, **55**, 22.
12. Buddruss, J. and Herzog, H., *Anal. Chem.*, 1983, **55**, 1611.
13. Laude, Jr, D. A. and Wilkins, C. L., *Anal. Chem.*, 1984, **56**, 2471.
14. Dorn, H. C., *Anal. Chem.*, 1984, **56**, 747A.
15. Albert, K., Nieder, M., Bayer, E. and Spraul, M., *J. Chromatogr.*, 1985, **346**, 17.
16. Laude, Jr, D. A., Lee, R. W.-K. and Wilkins, C. L., *Anal. Chem.*, 1985, **57**, 1281.
17. Laude, Jr, D. A., Lee, R. W.-K. and Wilkins, C. L., *Anal. Chem.*, 1985, **57**, 1464.
18. Laude, Jr, D. A. and Wilkins, C. L., *Trends Anal. Chem.*, 1986, **5**, 230.
19. Laude, Jr, D. A. and Wilkins, C. L., *Anal. Chem.*, 1987, **59**, 546.
20. Allen, L. A., Glass, T. E. and Dorn, H. C., *Anal. Chem.*, 1988, **60**, 675.
21. Albert, K., *Habilitationsschrift*, University of Tübingen, Tübingen, 1988.
22. Albert, K. and Bayer, E., *Trends Anal. Chem.*, 1988, **7**, 288.
23. Caswell, K. A., Glass, T. E., Swann, M. and Dorn, H. C., *Anal. Chem.*, 1989, **61**, 206.
24. Ha, S. T. K., Wilkins, C. L. and Abidi, S. L., *Anal. Chem.*, 1989, **61**, 404.
25. Albert, K., Kunst, M., Bayer, E., Spraul, M. and Bermel, W., *J. Chromatogr.*, 1989, **463**, 355.
26. Albert, K., Kunst, M., Bayer, E., de Jong, H. J., Genissel, P., Spraul, M. and Bermel, W., *Anal. Chem.*, 1989, **61**, 772.
27. Spraul, M., Hofmann, M., Glauner, H., Gans, J. and Albert, K., *Bruker* Report, No. 2, 1990, 12.
28. Grenier-Loustalot, M. F., Grenier, P., Bounoure, J., Grall, M. and Panras, R., *Analysis*, 1990, **18**, 200.
29. Görög, S., Balogh, G. and Gazdag, M., *J. Pharm. Biomed. Anal.*, 1991, **9**, 829.
30. Spraul, M., Dvortsak, P., Hofmann, M. and Glauner, H., *Bruker Report*, No. 129, 1993, 23.

31. Albert, K. and Bayer, E., in *HPLC Detection: Newer Methods*, G. Patonay (Ed.), VCH New York, 1992, pp. 197–229.
32. Hofmann, M., Spraul, M., Streck, R., Wilson, I. D. and Rapp, A., *Labor Praxis*, 1993, **10**, 36.
33. Spraul, M., Hofmann, M., Dvortsak, P., Nicholson, J. K. and Wilson, I. D., *J. Pharm. Biomed. Anal.*, 1992, **10**, 601.
34. Spraul, M., Hofmann, H., Dvortsak, P., Nicholson, J. K. and Wilson, I. D., *Anal. Chem.*, 1993, **65**, 327.
35. Spraul, M., Hofmann, M., Wilson, I. D., Lenz, E., Nicholson, J. K. and Lindon, J. C., *J. Pharm. Biomed. Anal.*, 1993, **11**, 1009.
36. Wilson, I. D., Nicholson, J. K., Hofmann, M., Spraul, M. and Lindon, J. C., *J. Chromatogr.*, 1993, **617**, 324.
37. Spraul, M., Hofmann, M., Lindon, J. C., Nicholson, J. K. and Wilson, I. D., *Anal. Proc.*, 1993, **30**, 390.
38. Spraul, M., Hofmann, M., Lindon, J. C., Farrant, R. D., Seddon, M. J., Nicholson, J. K. and Wilson, D., *NMR Biomed.*, 1994, **7**, 295.
39. Roberts, J. K. and Smith, R. J., *J. Chromatogr., A*, 1994, **677**, 385.
40. Johmnson, S., Morgan, E. D., Wilson, I. D., Spraul, M. and Hofmann, M., *J. Chem. Soc., Perkin Trans. 1*, 1994, 1499.
41. Albert, K., Schlotterbeck, G., Braumann, U., Händel, H., Spraul, M. and Krack, G., *Angew. Chem. Int. Ed. Engl.*, 1995, **34**, 1014.
42. Albert, K., *J. Chromatogr., A*, 1995, **703**, 123.
43. Albert, K. and Bayer, E., *Anal. Methods Instrum.*, 1995, **2**, 302.
44. Sidelmann, U. G., Lenz, E. M., Spraul, M., Hofmann, M., Troke, J., Sanderson, P. N., Lindon, C., Wilson, I. D. and Nicholson, J. K., *Anal. Chem.*, 1996, **68**, 106.
45. Korhammer, S. A. and Bernreuther, A., *Fresenius J. Anal. Chem.*, 1996, **354**, 131.
46. Pursch, M. Strohschein, S., Händel, H. and Albert, K., *Anal. Chem.*, 1996, **68**, 386.
47. Albert, K., *Analusis*, 1996, **24**, M17.
48. Höltzel, A., Schlotterbeck, G., Albert, K. and Bayer, E., *Chromatographia*, 1996, **42**, 499.
49. Godejohann, M., Mügge, C., Wünsch, G. and Preiss, A., *Anal. Chem.*, 1997, **69**, 3832.
50. Sidelmann, U. G., Braumann, U., Hofmann, M., Spraul, M., Lindon, J. C., Nicholson, J. K. and Hansen, S. H., *Anal. Chem.*, 1997, **69**, 607.
51. Lindon, J. C., Nicholson, J. K., Sidelmann, U. G. and Wilson, I. D., *Drug Metab. Rev.*, 1997, **29**, 705.
52. Strohschein, S., Pursch, M., Händel, H., and Albert, K., *Fresenius J. Anal. Chem.*, 1997, **357**, 498.
53. Strohschein, S., Schlotterbeck, G., Richter, J., Pursch, M., Tseng, L.-H., Händel, H. and Albert, K., *J. Chromatogr., A*, 1997, **765**, 207.
54. Godejohann, M., Mügge, C. and Preiss, A., *Anal. Chem.*, 1998, **70**, 590.
55. Strohschein, S., Pursch, M., Lubda, D. and Albert, K., *Anal. Chem.*, 1998, **70**, 13.
56. Albert, K., in *Analytiker-Taschenbuch 20*, H. Günzler, (Ed.), Springer-Verlag, Berlin, 1998, pp. 107–139.
57. de Koning, J. A., Hogenboom, A. C., Lacker, T., Strohschein, S., Albert, K. and Brinkman, U. A. Th., *J. Chromatogr., A*, 1998, **813**, 55.
58. Dachler, M., Kohler, K. and Albert, K., *J. Chromatogr., B*, 1998, **720**, 211.
59. Wilson, I. D., Morgan, E. D., Lafont, R., Shockcor, J. P., Lindon, J. C., Nicholson, J. K. and Wright, B., *Chromatographia*, 1999, **49**, 375.
60. Strohschein, S., Rentel, C., Lacker, T., Bayer, E. and Albert, K., *Anal. Chem.*, 1999, **71**, 1780.
61. Albert, K., *J. Chromatogr., A*, 1999, **856**, 199.

62. Albert, K., Dachtler, M., Händel, H., Lacker, T., Schlotterbeck, G., Strohschein, S. Tseng, L.-H. and Braumann, U., *J. High Resol. Chrom.*, 1999, **22**, 135.
63. Albert, K., in *NMR Spectroscopy in Drug Development and Analysis*, U. Holzgrabe, I. Wawer and B. Diel (Eds), Wiley-VCH, Weinheim, 1999, pp. 102–118.
64. Pusecker, K., Albert, K. and Bayer, E., *J. Chromatogr., A*, 1999, **836**, 245.
65. Strohschein, S., Pursch, M. and Albert, K., *J. Pharm. Biomed. Anal.*, 1999, **21**, 669.
66. Dachtler, M., Glaser, T., Händel, H., Lacker, T., Tseng, L.-H. and Albert K., in *Encyclopedia of Separation Science*, Vol. II, Academic Press, London, 2000, pp. 747–760.
67. Vilegas, W., Vilegas, J. H. Y., Dachtler, M., Glaser, T. and Albert, K., *Phytochem. Anal.*, 2000, **11**, 317.
68. Santos, L. C., Dachtler, M., Andrade F. D. P., Albert, K. and Vilegas, W., *Fresenius J. Anal. Chem.*, 2000, **368**, 540.
69. Tseng, L.-H., Braumann, U., Godejohann, M., Lee, S.-S. and Albert, K., *J. Chin. Chem. Soc.*, 2000, **47**, 1231.
70. Dachtler, M., Glaser, T., Kohler, K. and Albert, K., *Anal. Chem.*, 2001, **73**, 667.

2 LC–NMR: Automation

ULRICH BRAUMANN and MANFRED SPRAUL
Bruker BioSpin GmbH, Rheinstetten, Germany

2.1 PRACTICAL USE OF LC–NMR AND LC–NMR/MS

The coupling of LC (liquid chromatography) with NMR (nuclear magnetic resonance) spectroscopy can be considered now to be a standard analytical technique. Today, even more complex systems, which also include mass spectrometry (MS), are used. The question arises as to how such systems are handled efficiently with an increasing cost and a decreasing availability of skilled personal. LC–NMR and LC–NMR/MS combine the well-established techniques of LC, NMR and MS. For each of those techniques, various automation procedures and software packages are available and used in analytical laboratories. However, due to the necessary interfacing of such techniques, completely new demands occur and additional problems have to overcome.

In the first section of this review the possible types of experiments and their application fields will be described, while in the second section the individual steps of those experiments and the possibility of how to perform these tasks automatically will be discussed.

2.2 DIFFERENT WORKING MODES IN LC–NMR

The theoretical basics of LC–NMR coupling have already been discussed in the previous chapter. Except for one type of experiment, the connection of the chromatographic system and the NMR detection cell via a capillary is not sufficient. Most of the experiments require a special interface with switching valves under software control for reliable and reproducible results. The level of equipment and the application field depends on the types of experiment which are being conducted and will be discussed in the following.

The working modes can be first of all differentiated by the status of the sample during the measurement. First, it can be measured while the chromatographic separation is continuing. In this situation, the sample is flowing through the NMR detection cell while the NMR spectra are being acquired ***on-flow***.

On the other hand, the sample can be measured under static conditions. Depending on how the sample is transferred into the NMR detection cell we

On-Line LC–NMR and Related Techniques
Edited by K. Albert.

can further distinguish the experiments. The first possibility is to submit it directly from the chromatographic column. When the sample has arrived in the NMR detection cell, the separation must be interrupted in order to provide static conditions and to thus allow the measurement of further peaks from this separation. Therefore, we call this working mode ***direct stop-flow***.

It is also possible to store fractions of the chromatogram intermediately in sample loops. From these storage loops the samples are transferred into the NMR detection cell, after when the separation has finished. The NMR measurement is then again carried out under static conditions. We refer to the combination of these two procedures as ***loop storage/loop transfer***.

Measurement under dynamic conditions	On-flow
Measurement under static conditions	Direct stop-flow
	Loop storage/loop transfer

A schematic set-up of the minimum requirements for the individual experiments is shown in Figure 2.1.

2.2.1 ON-FLOW

Here, the outlet of the chromatographic system is connected to an NMR detection cell. The NMR spectra are acquired continuously while the sample is flowing through the detection cell. The result is a set of one-dimensional (1D) NMR spectra which cover the whole chromatogram and are typically displayed as a two-dimensional (2D) matrix showing NMR spectrum against retention time, similar to an LC–diode array detection (DAD) plot.

The chromatography and NMR systems perform completely independent by of each other. The only necessary link is the liquid connection between the column and the NMR detection cell. The NMR spectrometer can act as a detector for the chromatographic system, so that even a conventional LC detector in the chromatographic system is not necessary.

The typical peak width with analytical columns of 4.6 mm i.d. and a 1 ml/min flow rate is of the order of 10–30 s. The acquisition of NMR spectra with a short relaxation delay and an acquisition time of below 1 s allows the acquisition of 8–24 transients for one spectrum during the presence of a peak in the NMR cell. This low number of transients limits the detectable amount of sample to 5–10 μg per compound.

The flowing eluent is not an ideal matrix for the acquisition of the NMR spectra. Turbulences will cause inhomogenities of the magnetic field, thus leading to deterioration in the spectral resolution. Solvent gradients are typically used in LC separations. As the chemical shift of solvent and sample resonances depend on the solvent composition, the steadily changing composition

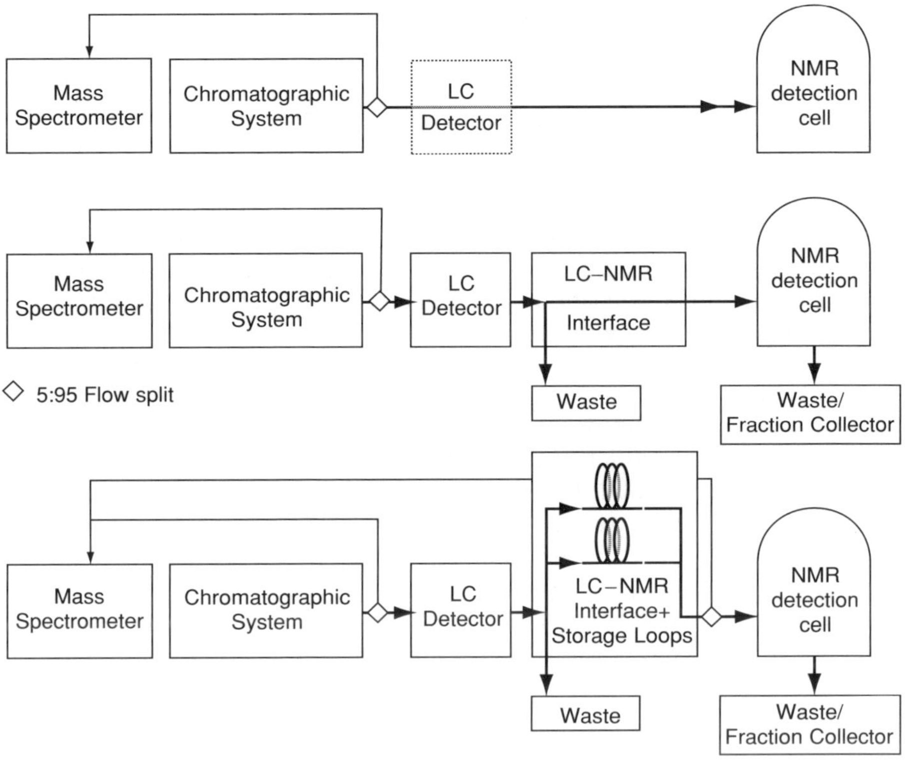

Figure 2.1 Schematic set-ups of the different working modes in LC–NMR and LC–NMR/MS

of the eluent will lead to changes of the chemical shifts during the acquisition process.

In a mixture of acetonitrile and D_2O the signal for acetonitrile has a chemical shift of 1150 Hz with 50 % acetonitrile, and of 1300 Hz with 60 % acetonitrile. This corresponds to a shift of ca. 15 Hz per %. With weak solvent gradients of only 1–2 % per minute, this will already cause a considerable shift of the samples signals, even during the elution of a peak. This will not only affect the spectrum quality of the sample signals, but also the performance and stability of the solvent suppression will suffer from this phenomenon.

2.2.2 DIRECT STOP-FLOW

In this mode, the eluent is directly flowing from the chromatographic system into the NMR probe. As only selected peaks are measured in the NMR spectrometer the separation is monitored in parallel with an LC detector (typically a UV detector). A peak is selected from the chromatogram recorded by the LC detector. The separation is interrupted after a certain time delay which is necessary to allow the peak to 'move' from the LC detector to the NMR detection cell. Now, all kinds of 1D and 2D NMR measurements can be carried out. In order to measure further peaks, the separation is continued until the next peak is positioned in the NMR detection cell. The result is a set of NMR spectra for certain selected peaks of the chromatogram.

The samples remain static in the flow cell and the conditions will remain stable during the whole NMR experiment. The parameters can be precisely adapted for the measurement of each individual sample. These include, in particular, the homogenization of the magnetic field and the adjustment of the solvent suppression parameters.

While the NMR experiment for a certain peak is being performed, further peaks remain in the chromatographic system. Diffusion may occur and will broaden these peaks and therefore decrease the concentration or even destroy the separation of two closely eluting peaks. This effect is dramatically reduced if solvent gradients are used. During the measurement of early peaks, the whole system is filled with a solvent mixture where later (analyte) peaks are not 'dissolved' but 'adhere' to the stationary phase of the column. Diffusion is minimized so that gradient separations can be often extended to run times of several hours. This is illustrated in Figure 2.2. In trace (b), a test chromatogram is shown. In the second separation (a) a higher sample amount was injected and

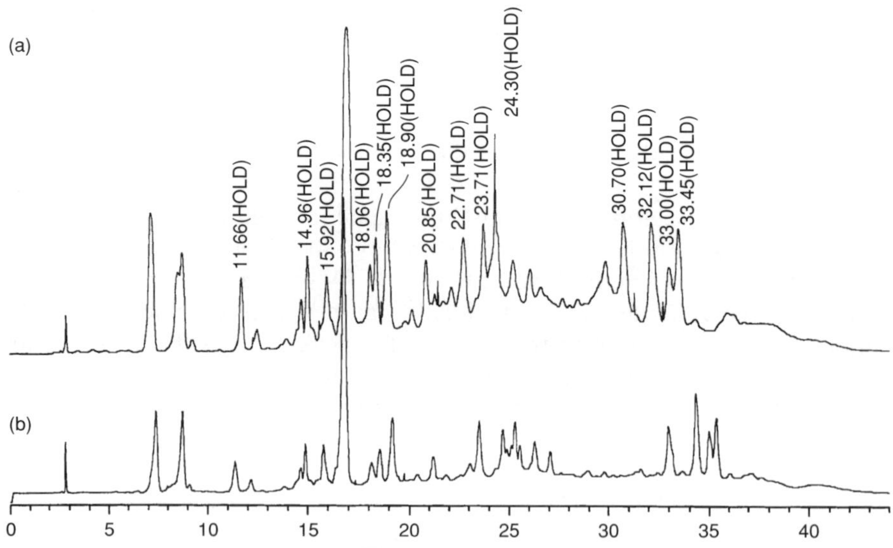

Figure 2.2 Effect of the stop-flow procedure on a chromatographic separation: (a) chromatogram with 14 stops for NMR measurements of 0.5–1 h each, with a total run time of 9.5 h; (a) uninterrupted chromatogram

the chromatographic experiment was stopped and continued for a total of 14 times. Through the duration of the NMR experiments of 0.5–1 h each, the total run time of the chromatogram was extended to ca. 9.5 h. Observe that the separations for the peaks at 33.0 and 33.48 min, which had the longest residual times on the column, are comparable to those of the uninterrupted chromatogram.

The volume of the NMR detection cell is relatively large (30–240 μl) when compared with the peak volumes and other void volumes in the chromatographic system. This leads to a considerable broadening of the peaks when they pass the NMR detection cell. It takes a long time until a peak is completely washed out of the flow cell, i.e. a tailing is observed. This is especially critical when traces of a high-concentration first peak interferes with the spectrum of a minor compound.

2.2.3 LOOP STORAGE/LOOP TRANSFER

In this case, the eluent is directly flowing from the chromatographic system into a storage device. As only selected peaks are measured in the NMR system the separation is monitored, in parallel, with an LC detector (typically a UV detector). A peak is selected from the chromatogram recorded by the LC detector, and the storage loop is isolated after a certain time delay which is necessary to allow the peak to 'move' from the detector into the loop. Without interrupting the separation, further peaks can be 'trapped' in the subsequent storage loops.

At a later stage, after the separation is finished, the loop contents are transferred in arbitrary order into the NMR spectrometer. Now, all kinds of 1D and 2D NMR measurements can be carried out. The result is a set of NMR spectra for certain selected peaks of the chromatogram. As the peaks are 'collected' in the sample loops, the separation is not influenced by the overall process, no start/stop disturbances, nor diffusion due to waiting times can occur.

The transfer process is completely independent from the NMR measurements. This means that the samples can be prepared for the NMR measurements while the NMR spectrometer is used for other purposes. It is even possible to perform the chromatographic experiments in a separate laboratory.

Once the peaks are 'stored' and 'isolated' in the loops, the measurement times for the individual NMR experiments are not limited by diffusion effects.

The separation conditions are typically tested and planned with aqueous systems, and a relatively low on column loading. For the best quality spectra, the NMR stage requires the use of D_2O, instead of H_2O, which will lead to a different pH value. This different solvent system, and the fact that a different chromatographic system is used for the LC–NMR experiments and the development of the separation, will lead to differences in the observed chromatograms. In addition, the total on-column loading will be in the mg range if the minor compounds are the target of the NMR analysis. This is the case in all kinds of LC–NMR experiments and will make the evaluation of the chromatograms and the peak selection difficult. In loop storage, all possibly interesting peaks can now be 'stored', and due to the arbitrary order of the measurement of the peaks the NMR spectra for those peaks already measured can be used to identify the compound of interest in the chromatogram.

The contamination of samples with previously eluted peaks, as observed in the direct stop-flow mode, can be avoided here. The volume of the storage loops is similar to that of the NMR detection cell, but as the volume is 'formed' by a capillary rather than by a cavity with a large inner diameter, the broadening effects are dramatically reduced. In addition, the peaks are first 'stored' in separate, previously washed loops so that they contain the clearly separated peaks. Between measurements of the samples, the NMR detection cell can be washed with an arbitrary amount of solvent. This procedure is shown in Figure 2.3. The technique is very useful for the measurement of closely eluting peaks and systems with large concentration differences. The effects of the different transfer methods of the peaks into the NMR detection cell is demonstrated in Figure 2.4. In this figure, peaks 1 and 2 have a retention time difference of 0.4 min, corresponding to 400 μl of sample with a flow rate of 1 ml/min. In the direct stop-flow spectrum of peak 2, a carry-over of ca. 20 % of the original concentration of peak 1 is visible, whereas in the spectrum of the second peak acquired after loop transfer the cross-contamination is not detectable.

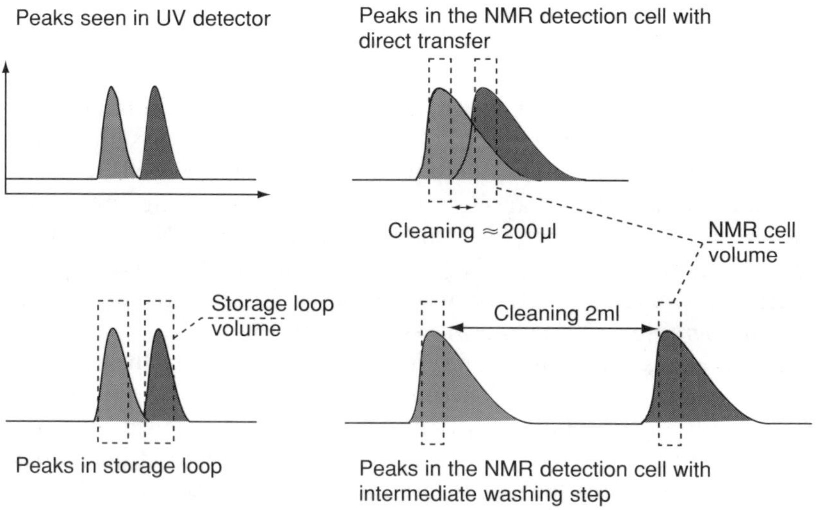

Figure 2.3 Peak broadening effects of the (a) direct stop-flow and (b) loop-storage/loop transfer procedures. The amounts of washing solvent required is defined by the chromatographic separation of the peaks in the direct stop-flow mode, while being user-defined in the loop transfer mode

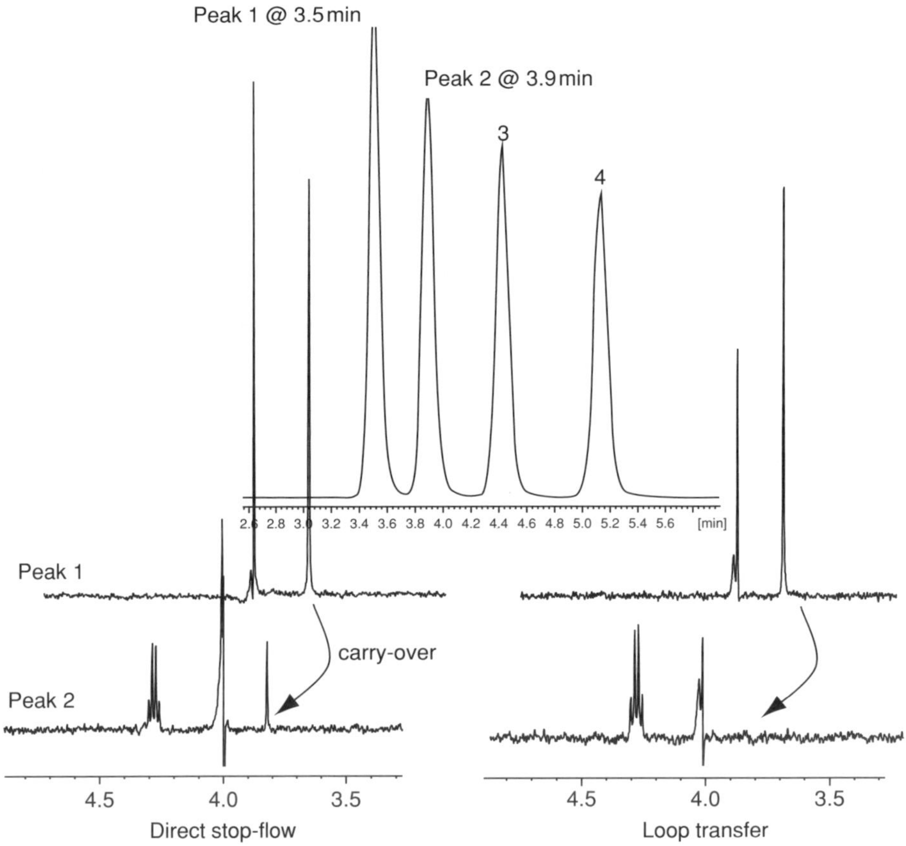

Figure 2.4 NMR spectra of closely eluting peaks ($\Delta t = 0.4$ min) in different working modes

2.2.4 CONCLUSIONS

On-flow

Due to the limited time available for NMR data acquisition and the additionally reduced stability under flowing conditions, the on-flow mode is limited to the acquisition of 1D spectra of the major peaks from a chromatographic separation. Minor compounds are normally not accessible. As no interruption or control of the chromatographic stage is necessary, the experiments can be carried out with standard chromatography equipment without the necessity of special equipment or software. The whole chromatogram is 'covered' by the NMR spectra and all NMR-active compounds are detected.

Direct Stop-flow

The diffusion of peaks is of critical concern with isocratic separations and very long-lasting NMR measurements. Highly concentrated compounds may pollute the NMR detection cell and make measurement of the following lower-concentration peaks difficult. A reproducible chromatographic stage is therefore necessary to allow the identification of the peaks in the chromatogram. Direct submission from the chromatograph to the NMR spectrometer provides the best possibilities for measuring unstable samples. The acquisition of NMR data can start only seconds after the separation is done. The static conditions provide stability and the best NMR conditions for the acquisition of all kinds of high-resolution 1D and 2D NMR spectra.

Loop Storage/Loop Transfer

Peaks are intermediately stored before the measurement, and thus very labile compounds may undergo decomposition or isomerization during this procedure.

The chromatographic stage is not interrupted and therefore no stop–start effects will create disturbances. The peaks are 'separated' in the storage loops, and therefore the NMR measurement time is not limited and will not decrease the performance of other peaks. In complex chromatograms the chance of finding the peak(s) of interest is dramatically increased. As in the direct stop-flow mode, the static conditions provide stability and the best NMR conditions for the acquisition of all kinds of high-resolution 1D and 2D NMR spectra.

2.3 USE OF MASS SPECTROMETRY IN THE SET-UP

The first attempts to include a mass spectrometer in the LC–NMR set-up were made several years ago. Nowadays, LC–NMR/MS systems are commercially available, and the usage of this kind of set-up will be briefly evaluated below [1–7].

The first idea for including an MS system in such a set-up was to yield all information about a sample with just one sample preparation, i.e. to provide one hyphenated experiment. A closer look at the practical situation shows that due to the large sensitivity difference between the NMR and MS instruments, of at least 10^3 to 10^6, and even higher, the performance of the mass spectrometer would be hindered by the NMR spectrometer. The speed and sensitivity of the MS unit would provide detailed data concerning all compounds, even in the on-flow mode, whereas only the major compounds would be visible in the NMR spectra.

In the stop-flow mode, the additional time could be used to acquire MS^n spectra of higher order, although these experiments would still be finished long before the NMR experiments have been completed. In practice, the mass spectrometer would have to wait most of the time for the NMR instrument to move on to the next peak.

Finally, use of the loop-storage/loop-transfer technique provides an escape from this dilemma. The chromatogram is acquired without any interruption, the mass spectrometer acquires MS and MS^2 spectra online, and provides spectroscopic information which allows reliable peak selection for the storage procedure. The MS unit is used above all as a very selective and sensitive detector. After the chromatographic separation, it is available for other purposes and the collected samples are then independently analysed in the NMR spectrometer.

If desired, the mass spectrometer can be again connected to the LC–NMR system during the loop transfer procedure. During loop storage, the linewidth of the chromatographic peak determines the duration of the MS experiments – typically much less than 1 min. Now the peak is 'available' as an isolated sample and can be injected with much lower speed into the MS unit. The peak profile, which is now broadened to several minutes, allows the acquisition of mass spectra of higher order. With these additional spectra, one can not only check for any decomposition during the loop storage process, but can also gain detailed MS information about the transferred sample.

The application of the mass spectrometer in the loop storage/loop transfer mode makes the best use of both systems. The data provided during the chromatographic stage can be used to identify the peaks and to ensure that the NMR spectrometer time is not wasted. The mass spectrometer is not blocked during longer-running NMR experiments.

2.4 MEASUREMENT PROCEDURES

Following our discussions of the principle differences between the various working modes of LC–NMR, we will now describe the practical aspects of the measurements, and in particular the possibilities of using automation routines.

Due to the intrinsically low sensitivity of NMR spectroscopy when compared to UV or MS, LC–NMR measurements typically can take an hour or more to complete. Therefore, it makes no sense for an operator to sit and wait for completion of the NMR experiment to activate another chromatographic separation – this has to be done automatically.

The analysis of a sample in a complex system such as LC–NMR or even LC–NMR/MS can be split into several steps, as follows:

- Sample preparation and introduction ('injection') into the chromatographic system
- Chromatographic separation

- Peak detection and selection
- MS measurements
- NMR measurements
- Sample recovery.

Here we can find several stages where automation procedures should be used. These are already partially well known from standard LC analysis. However, as the number of analysed samples, the types of analysis and various other requirements are completely different to these standard applications, the usage of such procedures therefore needs to be discussed in some detail.

2.4.1 SAMPLE PREPARATION AND INTRODUCTION ('INJECTION') INTO THE CHROMATOGRAPHIC SYSTEM

The time-consuming factor in LC–NMR is the NMR spectroscopy stage. In comparison to MS or UV, the sensitivity of the NMR spectrometer is low and therefore relatively long experimental times are required, which will often exceed 30 min for one single measurement. In addition, the NMR spectrometer requires careful adjustment of the various measurement parameters in order to obtain good results. If we also consider that the injection of one sample normally produces several peaks/fractions which are measured individually in the NMR spectrometer, the total experimental time for one original sample will in most cases exceed one hour or more. From the above mentioned facts it is obvious that automation of the sample introduction procedure does not play the most important role in LC–NMR or LC–NMR/MS. Sophisticated procedures for high throughput are not currently needed; however, with a loop collection system, e.g. with 36 loops, it might already be feasable to collect peaks from several separations overnight and then start the NMR measurements the next day in loop transfer mode. In this case, an autosampler is needed for injection.

The injection of the sample into the LC–NMR system can be carried out by an autosampler or a manual injection valve. The only difference which has to be considered is the fact that the amount of sample required for the NMR system is larger than that required for UV detection. Therefore, it is often necessary to inject sample volumes which exceed 100 μl. In such cases, it is mandatory to dissolve the sample in the starting solvent phase in order to avoid additional LC gradients created by the sample solvent being injected.

2.4.2 CHROMATOGRAPHIC SEPARATION

The basic conditions of a chromatographic separation conducted for an LC–NMR experiment do not differ from those required for (routine) analytical chromatography. However, in this case, additional important aspects need to be considered, as follows:

- If the peaks/fractions are directly ‘submitted’ to the NMR spectrometer for measurement under static conditions, the chromatographic data system must be capable of ‘pausing’ the separation. This includes not only an interruption of the flow but also the gradient generation and acquisition of the chromatogram. Otherwise, later peaks will be lost and subsequent measurements of multiple peaks from ***one*** chromatogram will not be possible.

- The on-line acquired chromatogram is used to select a peak for the NMR measurement. The software must export the chromatograms for the necessary peak search.

- If LC–NMR/MS is carried out, the data obtained from the mass spectrometer is often used to identify peaks of interest. For this purpose, it is preferable that the chromatography software itself must be able to create and display chromatograms from the mass spectra.

- A solvent system has to be used that is suitable for NMR measurements, i.e. one which will generate as few NMR signals as possible.

The chromatographic equipment which is responsible for the separation includes the pump, and in many systems a column oven. The parameters which affect the separation are the flow rate, the solvent composition and the LC gradients. Many different software packages are available which allow completely unattended automatic chromatography. Such systems also include control of the sample injection process.

What is lacking in most of the software programs which are only designed for (standard) chromatography is the ability to interrupt the separation as described above. While ‘trigger functions’ to start and stop external instruments are relatively easy to implement, sophisticated communication routines for peak recognition, the import of mass spectra or feedback about the measured peaks are, however, not standard features. From this, it becomes obvious that for coupled systems such as LC–NMR, and especially LC–NMR/MS, the use of dedicated chromatography software is necessary. These software packages should preferably already contain the functionalities of automatic peak detection and selection, rather than the need to export the data to a separate program.

2.4.3 PEAK DETECTION AND SELECTION

The acquisition of UV spectra using a DAD system normally covers the whole chromatogram. The result is a 2D plot with spectra for all peaks visible in the chromatogram. During most coupled experiments, only certain selected peaks or positions in the chromatogram are measured under static conditions. The rest of the chromatogram is discarded.

In order to allow selection of these peaks, the separation is monitored by an LC detector, in most cases a conventional UV detector, which displays a chromatogram of the separation. The chromatography software must allow selection of certain positions in the chromatogram for further measurement.

The void volumes in a coupled system are relatively high. The NMR probe, or the storage loop which is located further downstream, is reached ca. 10–40 s after the peak first appears in the LC detector. The software must calculate such delays depending on the system parameters and the actual flow rate of the separation. When the peak has 'reached' the desired position, the necessary actions for storage or measurement must be started.

In order to allow this process to take place in a convenient, reproducible and precise manner, the software must have the following features:

- *Automatic peak detection*. If the chromatography system runs unattended, the software must automatically find the desired peaks for the measurement. The predefinition of certain positions in the chromatogram, based on retention times from a reference chromatogram, is not sufficient. It should also be noted that:
  - Reference chromatograms are not available in all cases.
  - A deviation of 5 s from the actual peak maximum can immediately reduce the signal by nearly 100 %.
  - The retention times vary from separation to separation. This is especially the case when working under stop-flow conditions and/or with high column loadings.
  - Even if the user is present to select the peaks manually, an automation routine is very useful. If the delay time for the peak to 'pass' between the LC detector and the NMR probe or storage loop is short, it is difficult to define the peak maximum manually.
- *Calculation of the timings for the 'movements' of the peak between the different positions in the hyphenated system*. The time taken for the peak to 'reach' the NMR unit or the storage loop depends on the void volume between the LC detector and the NMR spectrometer loop. This is also a function of the flow rate. For precise and reproducible positioning of the peaks, the software must allow interactive selection of the peaks from the chromatogram and the automatic calculation of these time delays based on the actual parameters.
- *Control of an interface for the selection of the flowpath and the positioning of the peaks in the system.*

– For the loop-storage working mode, an additional instrument containing the storage loops needs to be under automatic control.

– For stop-flow measurements, an interface is necessary. Turning off the LC pump will not allow the precise 'positioning' of a peak in the NMR probe, as the whole system is still pressurized and the expanding liquid 'washes' the peak out of the probe.

2.4.4 MASS SPECTROMETRIC MEASUREMENTS

The use of a mass spectrometer in an LC–MS set-up is now a standard technique. However, if a mass spectrometer is brought into an LC–NMR/MS system, some further problems will occur which are mainly due to the relatively high sample loadings used in NMR spectroscopy. In the case of an ion-trap mass spectrometer, some of the already established routines are as follows:

- Automatic adjustment of the accumulation time. This prevents overloading of the trap and therefore maintains good mass resolution over a large range of concentrations.

- Due to the high sensitivity of the mass spectrometer, even MS^n data can be acquired online during the chromatography experiments. Routines are available which isolate the highest peaks in predefined mass ranges and submit them for automatic MS^2 measurements.

- For the analysis of minor compounds by LC–NMR, the column is often loaded with sample amounts of 1 mg or larger. In such cases, the ion source can be disconnected from the flow path during the elution of the main compound. This prevents contamination of the source and trap, which would make the analysis of later-eluting minor compounds difficult.

- For efficient use of the mass spectrometer, data exchange between the MS software and the LC–NMR software must be possible. Preferably, peak selections based on the MS chromatograms is carried out by the LC–NMR software.

If the mass spectrometer acquires spectra in parallel with the NMR spectrometer, the main flow which has transported the sample is stopped. The NMR spectra are then acquired under static conditions. To deliver fractions for the mass spectrometer, an additional syringe pump is necessary which must be triggered by the (computer) software.

The example presented in Figure 2.5 shows how the mass spectrometer can be used, instead of the unselective UV detector, to identify the components of

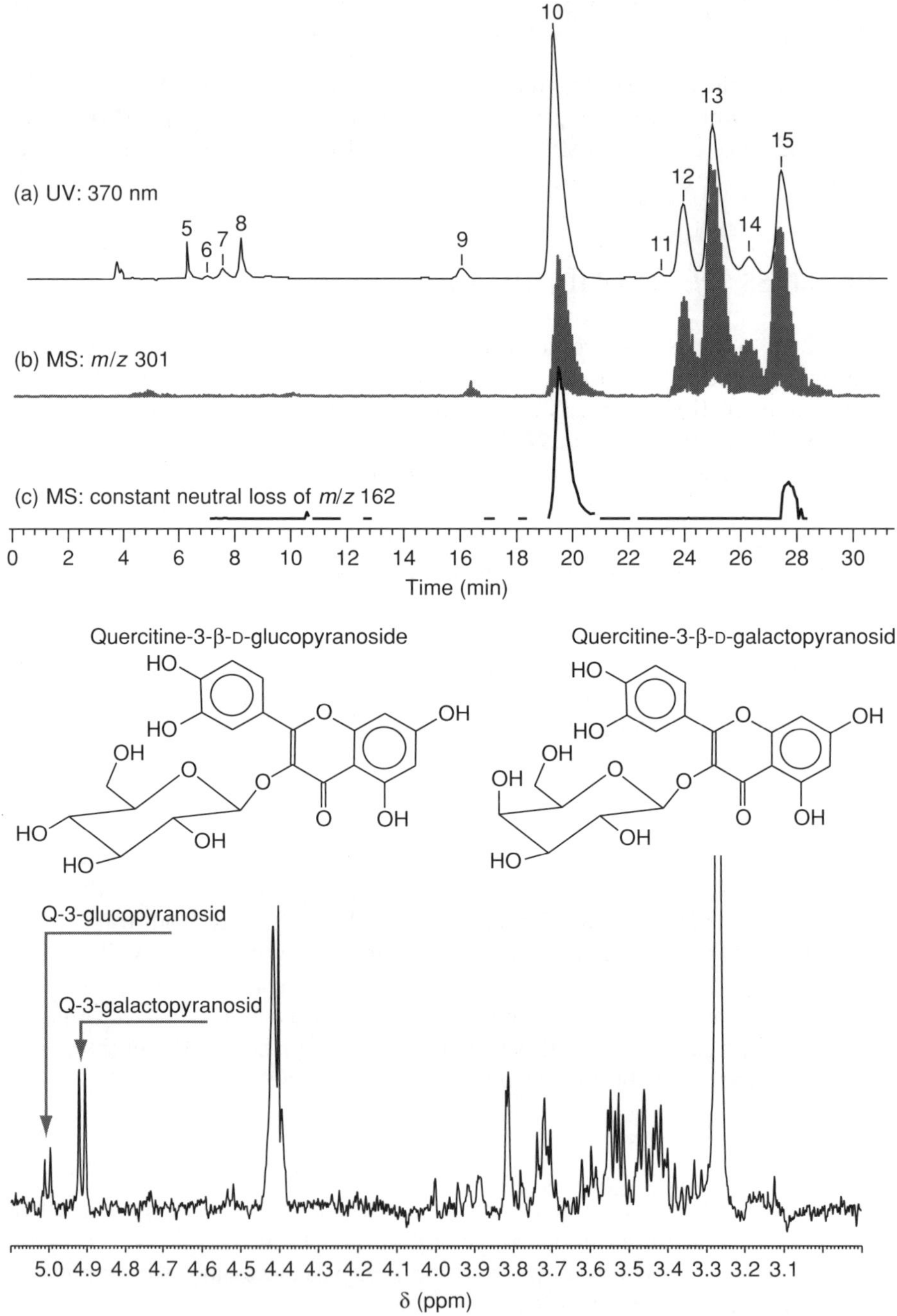

Figure 2.5 LC–NMR/MS analysis of glycosides of quercetine

interest. Under these circumstances, not only the (parent) mass but also fragments or fragmentation patterns can be employed in the analysis. In the example given, various glycosides of quercetine were of interest [8]. The UV chromatogram (Figure 2.5(a)) shows multiple peaks unselectively. The chromatogram shown in Figure 2.5(b), which is created from the mass spectrum at an *m/z* of 301, only shows, selectively, those compounds which contain the quercetine fragments. In a further step, the chromatogram shown in Figure 2.5(c) was created from the constant neutral loss of ion *m/z* 162. Therefore, only peaks that contain a C6 sugar give a response. Based on this information, the peak which occurs at 20 min was selected for the NMR measurement. This example demonstrates clearly that the results obtained from both techniques complement each other. Despite the high sensitivity and the information content of the mass spectrometric analysis, it becomes obvious that only with the resulting NMR spectrum was it possible to discover that, in the peak actually two compounds – the glucopyranosid and the galactopyranosid – were co-eluting.

2.4.5 NUCLEAR MAGNETIC RESONANCE MEASUREMENTS

NMR spectrometers are equipped with automatic sample changers. Such devices have a capacity of 6–120 samples, and the software packages controlling these sample changers allow unattended operation. In addition, the acquisition of complex 2D and 3D NMR data is possible. The NMR software which normally controls the sample changer can be easily modified to allow it to recognize the software of the chromatographic equipment as being the 'sample changer'. This means requesting/accepting samples from the chromatograph and sending back information when the NMR experiments are finished. An additional benefit is that the user is already familiar with the major part of the automation software from standard sample tube operation. The major differences for NMR experiments under LC–NMR conditions concern the sample matrix and the sample amount. These special experimental conditions are handled by the (computer) software sub-routines which control the acquisition of the individual experimental data.

The sample matrix in LC–NMR/MS is the eluent. This – from the viewpoint of NMR – complex matrix differs dramatically from the solvents which are commonly used in (routine) NMR analysis:

- Special solvent suppression techniques are necessary to handle the large background signals arising from such solvents.

  In conventional NMR measurements, the sample is prepared manually, and therefore the amount of sample is known and the resulting experiment time can be estimated:

- In LC–NMR, only the total amount of the sample injected into the chromatography system is known. The amount actually represented by the measured peak can only be estimated by using the UV response from the chromatogram, and cannot be used quantitatively. The same, of course, is true for the MS response, and here ionization properties often define the LC peak intensity.

- The typically very low sample concentrations employed in LC–NMR analyses make it absolutely essential to optimize the NMR conditions to give the best sensitivity.

In general, we have to consider that even under optimized conditions the signal-to-noise ratios (S/N) of the NMR spectra are already close to the detection limit. Any mistake in the adjustment of the experimental conditions can lead to a total loss of results. It is also obvious that, for longer runs, LC gradients must be used, in order to keep the eluting LC peak volume reasonable. Such gradients mean constantly changing the composition of the eluent system and therefore this also always requires changing the conditions for NMR solvent signal suppression.

The necessary steps during the acquisition of an NMR spectrum under LC–NMR conditions are discussed below.

Tuning and Matching

The capacity and impedance of the NMR probe has to be tuned to the specific sample being analysed. If a sample in a different solvent is used, this normally requires a retuning of the NMR probe. There are commonly three possibilities for solving or at least overcoming this problem, as follows:

- Use a probe design with minimized effects from different solvents, which normally involves an overall loss of sensitivity.
- Use a probe design which allows automatic tuning matching, via the use of motors mounted on the probe body.
- Tune the probe to one sample and then use an adapted 90° pulse for other solvents where the tuning is not optimal. This, of course, also leads to sensitivity losses.

In LC–NMR this problem does not arise in the same way as it does in standard sample changer automation systems, where a change between such different solvents as water and chloroform is possible. In LC–NMR, the samples are eluted from a column in a mixture of solvents, in which only the relative composition changes:

- The difference in the length of the 90° pulse of acetonitrile/D_2O mixtures containing between 0 and 100% of D_2O is only 5%.
- Most spectra acquired in LC–NMR are only 1D proton spectra, where a deviation of the pulse length of 5% is negligible.

Shimming

Changing solvents also has an effect on the homogeneity of the magnetic field. Again, the observed changes in LC–NMR systems are not as large as those found in normal sample changer automated devices:

- The LC–NMR probe is a fixed set-up, where the sample container is the flow cell and is fixed in the probe. Different filling heights or qualities of sample tubes will not affect the homogeneity as in normal sample changer operation.
- The solvent mixture is always the same and varies only in its relative composition.
- Standard LC–NMR probes of 4 or 3 mm diameter have only a length of 3 or 2mm in the *x–y* direction, and therefore the adjustment of off-axis shims is not necessary in most cases. Where required, this is only normally done once.

However, there is one major drawback implemented by the use of solvent mixtures. The commonly used solvent system of acetonitrile/water displays a strange behaviour. Typically, the line shape of the water and the acetonitrile is different. As normally the water is used in its deuterated form, the spectrometer is also locked on to the D_2O component. Trying to shim such a sample by using the lock level, and therefore the line shape of the D_2O, provides a good line shape for the water. However, for the acetonitrile component, severe distortions can be observed, and the sample signals show a similar behaviour like the signals of the acetonitrile.

This makes the straightforward shim procedure on the lock-level impossible. As most NMR probes are nowadays equipped with *z*-gradients, shimming is usually accomplished by gradient-shimming on the acetonitrile signal.

Solvent Suppression

Samples which are measured in sample tubes are normally dissolved in deuterated solvents. In LC–NMR, the sample is measured in the solvents which are used to carry out the chromatographic separation. The prices of some deuterated solvents, such as water or chloroform, has dramatically dropped in recent years, so that these solvents can be replaced by their deuterated forms. However, this is not the case for the other commonly used solvents, i.e. acetonitrile and methanol. Therefore, in LC–NMR analysis the NMR spectrometer still has to handle non-deuterated solvents, which requires solvent suppression.

The most commonly used solvent suppression technique is presaturation. Modern NMR spectrometers are normally equipped with at least two independent frequency channels. Using these, there are usually no problems in conducting experiments where two solvents are presatureated at the same time. However, the number of solvents which can be suppressed is limited by the available hardware.

It would be impossible to handle ternary solvent systems such as acetonitrile/water/methanol or mixtures which contain further additives such as triethylamine. Therefore most techniques nowadays use modulated shaped pulses for multiple presaturation or pulse frequency generation (PFG) techniques. With basic hardware, it is then possible to suppress an arbitrary number of solvents.

Suppression is mainly used for water in aqueous samples. Such a technique does not require any complicated adjustments of the suppression parameters. In this case, the spectrometer is locked on to the D_2O. The signal of the HDO/H_2O is found at nearly exactly 4.7 ppm. Adjustment of the receiver gain (RG) is therefore all that is needed before the experiment can be started.

The situation in LC–NMR is different. A solvent mixture is used and in most cases two or more solvents are suppressed. The solvents which are delivered from the LC pump are not completely isolated from the humidity of the ambient air. Additional water is brought into the system via the acetonitrile, the sample itself and any additives such as ammonium acetate which are used in the undeuterated form. Therefore, the HDO signal is relatively strong and has to be suppressed as well.

The position of the solvent resonances depend on pH, solvent compositions and temperature. Especially in gradient separations the exact solvent composition in which the sample is dissolved is unknown. In addition, variation in the solvent position (which also depends on variations in other parameters) is strong. Depending on the absolute composition and temperature, a solvent change of 1 % can already shift the acetonitrile signal by more than 15 Hz. The complexity of the dependency of the solvent signals on the conditions makes it absolutely necessary to adjust the parameters for each sample individually. The adjustment procedure consists of the following steps:

(i) acquisition of a solvent search spectrum without solvent suppression;

(ii) a solvent search based on user-defined parameters which exclude potential sample signals;

(iii) special solvent search procedures, so that the minor solvents (e.g. HDO) are reliably found in the presence of the major non-deuterated solvents;

(iv) manipulation of the shaped pulses with the found frequencies;

(v) adjustment of the receiver gain to the amount of non-deuterated solvent and/or efficiency of the solvent suppression.

An example of a spectrum with automated solvent suppression is given in Figure 2.6. Normally, one has to deal with a mixture of two solvents, where often one of the solvents is water and is replaced by D_2O. This leads to one or two intense solvent lines which have to be suppressed. In this figure, a worst-case scenario is presented. The mixture consists of two non-deuterated solvents (acetonitrile and methanol) which show, in total, three intense solvent lines. The solvent suppression parameters were adjusted automatically, and an additional digital filtering to remove the residual solvent signals was also carried out.

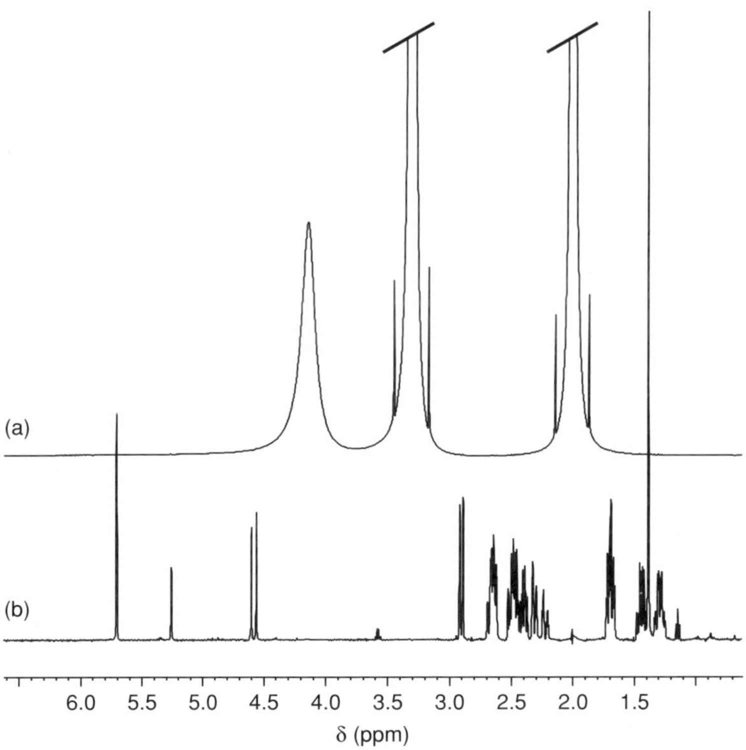

Figure 2.6 NMR spectra obtained under LC–NMR conditions: (a) an unsuppressed spectrum showing acetonitrile (@ 2.0 ppm), methanol-CH_3 (@ 3.3 ppm) and –OH signals (@ 4.2 ppm); (b) the corresponding 1D proton spectrum acquired with (WET) solvent suppression (on the three solvent lines a digital solvent filtering procedure was also applied). The sample used was 140 μg of cortisone in a acetonitrile/methanol/D_2O (5:5:1) solvent mixture

Signal-to-Noise Control

In conventional NMR spectroscopy, a defined amount of sample is brought into a sample tube and measured at a predefined experimental time. Most of the resulting NMR spectra are acquired with only low numbers of transitions and the signal-to-noise values of these spectra are of the order of several 100s or 1000 to 1.

In LC–NMR/MS analysis, the original sample is a mixture, and the actually measured peak represents only a fraction of this. The sample amount in this fraction is unknown, and only a rough estimation based on the response of the UV detector is possible.

In addition, the sample concentrations employed in LC–NMR are in the very low μg and even upper ng regions. This will yield spectra where the signal-to-noise ratios are already close to the detection limit, even after an acquisition time of one hour and more.

Setting up a fixed number of transitions either wastes instrument time if peaks are more concentrated than expected or leads to a complete loss of results for lower-concentration peaks. A sensible procedure which measures the spectrum up to a predefined signal-to-noise level, or even stops the acquisition if the sample concentration is too low, helps dramatically to save instrument time.

2.4.6 SAMPLE RECOVERY

After separation of the sample and acquisition of spectroscopic data, the sample can be collected in a fraction collector. This allows further investigation, especially by mass spectrometry, if the MS spectra were not already collected on-line.

Another important issue is that a mass spectrum of the collected sample confirms that no decomposition, either during the storage procedure or during measurement in the NMR spectrometer, has occurred. Most fraction collectors have built-in software, which can detect and collect peaks. Due to the special conditions of stop-flow measurements, it is preferable that the fraction collector should also be controlled by the central software. This also means that the measured NMR spectra and the collected fractions are easier to correlate.

2.5 CONCLUSIONS

As preferably ***one*** user should be able to operate the whole system, it becomes clear that a highly integrated system is needed. Even if the peak selection is carried out by the user, the positioning of the sample, followed by the start of the NMR and MS experiments, must take place automatically. An integrated LC–NMR/MS system therefore has to contain the following components:

- A chromatographic system with autosampler and fraction collector under the control of software specially adapted for hyphenated experiments.
- An interface controlling the flow path using switching valves controlled by software. The latter also carries out peak detection and is preferably a part of the main chromatography software.
- An NMR spectrometer under the control of software which allows the automatic acquisition of spectra with solvent suppression. Communication with the chromatography software is also necessary.
- A mass spectrometer with software which allows close interactions with the chromatography programs.

REFERENCES

1. Wilson, I. D., Morgan, E. D., Lafont, R., Shockcor, J. P., Lindon, J. C., Nicholson, J. K. and Wright, B., *Chromatographia*, 1999, **49**, 374.
2. Scarfe, G. B., Wright, B., Clayton, E., Taylor, S., Wilson, I. D., Lindon, J. C. and Nicholson, J. K., *Xenobiotica*, 1999, **29**, 77.
3. Clayton, E., Taylor, S., Wright, B. and Wilson, I. D., *Chromatographia*, 1998, **47**, 264.
4. Taylor, S., Wright, B., Clayton, E. and Wilson, I. D., *Rapid Commun. Mass Spectrom.*, 1998, **12**, 1732.
5. Scarfe, G. B., Wright, B., Clayton, E., Taylor, S., Wilson, I. D., Lindon, J. C. and Nicholson, J. K., *Xenobiotica*, 1998, **28**, 373.
6. Scarfe, G. B., Wilson, I. D., Spraul, M., Hofmann, M., Braumann, U., Lindon, J. C. and Nicholson, J. K., *Anal. Commun.* 1997, **34**, 37.
7. Shockcor, J. P., Unger, S. E., Wilson, I. D., Foxall, P. J. D., Nicholson, J. K. and Lindon, J. C., *Anal. Chem.*, 1996, **68**, 4431.
8. Lommen, A., Godejohann, M., Venema, D. P., Hollman, P. C. H. and Spraul, M., *Anal. Chem.*, 2000, **72**, 1793.

3 Biomedical and Pharmaceutical Applications of HPLC–NMR and HPLC–NMR–MS

JOHN C. LINDON and JEREMY K. NICHOLSON
Department of Biological Chemistry, Faculty of Medicine, Imperial College of Science, Technology and Medicine, University of London, London, UK

and

IAN D. WILSON
AstraZeneca Pharmaceuticals, Alderley Park, Macclesfield, Cheshire, UK

3.1 INTRODUCTION

High performance liquid chromatography (HPLC) is one of the principal methods for the analysis of chemical mixtures but conventional detectors used to monitor the separation, based on refractive index, UV, fluorescence and electrochemical properties, are unable to provide more than a limited amount of information on molecular structure. Real advances in on-line minor component structure determination have only resulted from the relatively recent advent of the reliable hyphenation of HPLC and mass spectrometry (MS). This new technology is now widely exploited and there has been a huge growth of applications of HPLC–MS, particularly in the pharmaceutical industry, and especially in the identification and quantification of drugs and their metabolites. These advances notwithstanding, mass spectrometry by itself does not always provide unambiguous structural identification, and NMR spectroscopic data are also often required. However, conventional NMR spectroscopic analysis has until recently required time-consuming isolation and purification steps, and sometimes this can conflict with the efficient HPLC–MS approach.

High-resolution NMR spectroscopy is demonstrably one of the most important methods of structure elucidation with an extensive range of biochemical and chemical applications. The direct linking of HPLC with NMR spectroscopy has been a remarkable success story as discussed elsewhere in this book and has transformed the technique from a research tool to the stage where routine analytical applications are possible [1,2]. As the number of HPLC–NMR systems have increased in user laboratories, there has been a logical

On-Line LC–NMR and Related Techniques
Edited by K. Albert.

extension of the hyphenation of HPLC–NMR and HPLC–MS into a single combined system for structure elucidation. It should be noted that even HPLC–NMR can be considerably more time-consuming than HPLC–MS. The operation of HPLC–NMR and HPLC–NMR–MS are also explained in detail elsewhere in this volume (see Chapters 1 and 2), as are the operational difficulties and practical solutions for this double hyphenation (see Chapter 2).

This present chapter will summarise the application of HPLC–NMR and HPLC–NMR–MS to studies of biomedical and pharmaceutical interest and include some limited examples of further hyphenation (hypernation) of UV, NMR, MS and IR spectroscopy that illustrate the potential of these combinations in this field [3,4]. This chapter excludes examples in the drug metabolism field involving the separation, identification, and in some cases, quantification of metabolites in biofluids and their extracts as these are considered elsewhere in this volume. However, because of their general relevance, HPLC–NMR–MS studies of the chemical reactivity of such molecules will be discussed.

3.2 TECHNICAL AND OPERATIONAL OVERVIEW

Before going on to describe some of the applications of these techniques, it is worth very briefly recapping some of the operational aspects as they are applied to biomedical and pharmaceutical applications.

Direct on-line coupling of an NMR spectrometer as a detector for chromatographic separation, analogous to the use of MS for such applications, has required the development of technical features such as flow-probe hardware, efficient NMR solvent suppression pulse sequences and new software.

It is necessary to detect signals from low concentrations of analytes in the presence of large ^{1}H NMR signals from the HPLC solvents. The solution to this problem has been the development of new techniques for suppressing the solvent NMR resonances. These are able to cope with mixed solvents, such as methanol/water, acetonitrile/water and even more complex solvent combinations, including the problems which are associated with eluent proportions changing during a gradient run [5,6]. A related factor which delayed the implementation of practical HPLC–NMR spectroscopy was the earlier need to use deuterated solvents for chromatography to overcome the problem of high dynamic range detection and these solvents are relatively expensive, except for microbore separations. The solution to this difficulty for reversed-phase HPLC–NMR has come about because modern NMR spectrometers can perform solvent NMR resonance suppression very efficiently, thus negating the need for deuterated solvents. In practice, D_2O is still often used to prepare eluents rather than H_2O, simply because this makes multiple-solvent suppression easier (D_2O currently costs about \$200/1).[†] In addition, acetonitrile-d_3 is being used increasingly in pharmaceutical laboratories because the cost of this solvent is negligible in relation to the other operating costs of such a laboratory and it allows a substantial gain in quality of the results.

The commonly used solvents for reversed-phase HPLC separations are methanol and acetonitrile. Both of these solvents give rise to a singlet resonance in the ^1H NMR spectrum which can be suppressed easily. However, the ^{13}C satellite peaks, caused by the one-bond ^1H–^{13}C spin couplings from the 1.1% of molecules with the naturally abundant ^{13}C isotope at the methyl carbon, remain following suppression of the main peak and are often much larger than the signals for the analytes, and thus these must also be suppressed.

A recent practical development, which may have useful applications in HPLC–NMR, is the finding that when chromatography is performed at elevated temperatures (greater than 100 °C) the organic modifier content can be greatly reduced or even eliminated altogether. Providing that the analyte (and column) is stable under this condition, this can permit the use of 100% D_2O, thus simplifying the solvent suppression in the NMR spectra. To date, there have been a limited number of applications of HPLC–NMR and HPLC–NMR–MS using superheated chromatography to pharmaceutical studies and these are described later in this chapter [7–9]. The use of superheated water causes surprisingly few problems when coupled to NMR spectroscopy, compared with, for example, supercritical fluid chromatography (SFC). In practice, the only modification that is required is the insertion of the HPLC column in an oven capable of operating at the required temperature and in general such an oven can easily be adapted from an old gas chromatograph, such as a Pye 104 model. The use of a gas chromatograph in this respect has a number of advantages as the oven is equipped to perform temperature gradients, which can be used to modify the eluotropic strength of the D_2O (in the same way as increasing the organic modifier content in conventional systems). The operation of the column at temperatures above the boiling point of the eluent might seem at first sight to require significant modification of the system in order to ensure that it remains as a liquid rather than vaporising. However, surprisingly little back-pressure is required in order to maintain the D_2O as a liquid (again in contrast to SFC). Indeed, it has been found that the approximately 2 m of PEEK tubing required to connect the system to the flow probe is quite sufficient to generate the required pressure, thereby eliminating the need for a flow probe adapted to withstand pressure. This length of tubing also acts as a heat exchanger, serving to ensure that the solvent is at ambient temperature when it reaches the spectrometer. Thus, the temperature gradients used to obtain the required chromatographic separation have no effect on subsequent spectroscopy.

Because of the necessary layout of an NMR laboratory, HPLC–NMR requires the use of relatively long column-to-detector transfer times and the

[†] Current price as of December 2001.

use of relatively high volume flow cells (by chromatographic standards). For HPLC–NMR probes, a compromise has to be made between the needs of chromatography and NMR spectroscopy for the detection volume of the flow cell and the whole issue of flow rates and flow cell sizes for coupled chromatography–NMR spectroscopy has been addressed elsewhere in this volume. This, of course, makes it impossible to spin the sample, as is performed in conventional probes, to improve magnetic field inhomogeneities. Practically, however, this turns out not to be a problem as field homogeneity with these small volumes is good. Furthermore, modern computer-controlled methods for optimising field homogeneity reduce the requirement for spinning. A detailed analysis of the flow and NMR requirements for optimum operation of HPLC–NMR has been given recently [10,11].

The low sensitivity of NMR spectroscopy in relation to the quantities of metabolites often separated using HPLC with UV detection must be recognised. However, as will be seen later, impressive sensitivities can be achieved by using HPLC–NMR in the stop-flow mode. The detection limits of HPLC–NMR are continually being revised downwards as new technical advances are made [12]. Recently, these have included the use of high magnetic field strengths (operating at 750 and 800 MHz for ^{1}H NMR spectroscopy), the incorporation of digital filtering and oversampling into NMR data acquisition and the introduction of microbore HPLC methods. By combining the use of digital electronics with microbore HPLC [13,14], it appears that the detection limits for structural characterisation are in the region of 5 ng, even at the lower ^{1}H NMR observation frequencies of 500 and 600 MHz.

In principle, it is possible to effect NMR detection for any of the magnetically active nuclei, but those of most importance in pharmaceutical studies are ^{1}H, ^{2}H, ^{19}F, ^{13}C, ^{15}N and ^{31}P. In addition, because of the generally low levels of minor components, the most sensitive nuclei, i.e. ^{1}H, ^{19}F and ^{31}P, have been used most extensively. The use of ^{13}C NMR in HPLC–NMR can be facilitated through indirect detection of ^{13}C resonances via the much more sensitive ^{1}H NMR signals of attached protons. A major benefit of using ^{19}F NMR spectroscopy for the detection of fluorine-containing molecules is that the background, unlike that for ^{1}H NMR spectroscopy, is usually negligible.

HPLC–MS has been employed for many years but only since the advent of electrospray ionisation has it become a truly robust and routine method for the analysis of mixtures. Now that NMR spectroscopy has been coupled to HPLC, it has, for the first time, become possible to acquire both NMR and MS data simultaneously from a single chromatographic analysis.

Solvent selection for HPLC–NMR–MS has to be a compromise between the ideal requirements of each instrument. Thus, for HPLC–NMR the use of inorganic buffers, e.g. sodium phosphate, for pH modification is preferred because no additional signals are introduced into the NMR spectrum although this type of buffer system is currently incompatible with most HPLC–MS systems using an electrospray interface. An alternative acidic modifier is trifluoroacetic acid (TFA), which has no protons to cause interferences in the NMR spectrum and it has been shown that 0.1 % TFA can be used with HPLC–MS for a limited range of analytes present at high concentration (> 1 μg on column) in the positive-ion mode. However, with acidic analytes ion suppression can be total and, even at high sample concentrations, MS data cannot be obtained. Formic acid has been found to provide a suitable compromise between the needs of MS on the one hand and NMR on the other. The interference of the single proton of formic acid, which has a sharp, easily suppressed NMR singlet near δ 8.5, gives minimal interference in the resulting NMR spectra and enables MS data to be acquired for acidic analytes.

There are two ways to configure the NMR and mass spectrometers, i.e. either in parallel or in series. As NMR spectroscopy is a relatively insensitive technique, large volumes and high concentrations of analytes are used wherever possible to compensate and to reduce analysis time. As a result, generally 4.6 mm HPLC columns are used to avoid problems of overloading. This means that flow rates of the order of 0.5–1.0 ml/min must be employed to meet the requirements of NMR spectroscopy without compromising the chromatography. Such high flow rates can easily be accommodated by modern mass spectrometers. However, as electrospray-MS is concentration-dependent, as opposed to mass-sensitive, operating the NMR and mass spectrometers in parallel, and thus splitting the flow such that a minor fraction goes to the latter, has little effect on sensitivity but greatly enhances the source lifetime, thus allowing the mass spectrometer to be operated at optimum sensitivity for longer. If the flow is split prior to the NMR unit, with the length of the capillary to the mass spectrometer adjusted so that the analyte peak has just passed completely through the latter as it fills the NMR flow cell, the mass spectrometer can be used to supplement the UV data to direct the NMR experiments. Furthermore, splitting in this manner enables the use of stop-flow NMR spectroscopy with minimum degradation of the integrity of the chromatography [15].

Running in series, i.e. with the sample-destructive MS stage after the non-destructive NMR stage, allows for the completion of all NMR experiments, whether on-flow or stop-flow, before mass spectrometric analysis begins, but introduces the possibility of peak dispersion before such analysis for any peaks trapped between the NMR and MS units when the flow is stopped. Series operation causes the NMR flow cell and its connections to be operated at higher pressures than they were designed for, with the consequent possibility that leaks are more likely. Series operation also fails to take advantage of the mass spectrometrist's ability to flag up peaks of interest quickly. Nevertheless, some applications have used serial coupling of NMR spectroscopy and MS to HPLC because of the ease of disassembling the components for separate mass spectrometry studies, and have obtained useful results.

Given the strength of the magnetic field surrounding an NMR magnet, there is obviously the potential for interference with the operation of the mass spectrometer. There is also the potential for the presence of the mass spectrometer to interfere with the operation of the NMR spectrometer. However, the development of actively shielded or so-called 'supershielded' superconducting NMR magnets has alleviated this potential problem considerably. At present, the minimum distances which are possible between the NMR magnet and the HPLC components and between the NMR magnet and a typical ion trap or quadrupole mass spectrometer, are typically 0.5 and 1.0 m, respectively, for a 400 NMR MHz magnet, and 1.5 and 2.0 m, respectively, for a 600 MHz NMR magnet.

It is also worth briefly reiterating here the main operational modes for HPLC–NMR and HPLC–NMR–MS. There are currently five main options which can be employed for HPLC–NMR using either isocratic or gradient elution. These are continuous-flow, stop-flow, 'time-sliced' stop-flow, peak collection into capillary loops for post-chromatographic analysis, and automatic peak detection with UV-detected triggered NMR acquisition.

The simplest of these is continuous-flow detection, but this is usually only practical when using ^{1}H or ^{19}F NMR for detection unless isotopically enriched compounds are available. However, there are examples of HPLC–NMR studies using ^{2}H and ^{31}P NMR detection in the drug metabolism field. Where continuous-flow NMR detection is used for gradient elution, the NMR resonance positions of the solvent peaks shift with the changing solvent composition. For effective solvent suppression, these solvent resonance frequencies must be determined as the chromatographic run proceeds.

If the retention times of the analytes are known, or there is an efficient method for their detection on-line, such as UV, MS or radioactivity, stop-flow HPLC–NMR becomes a viable option. In the stop-flow technique, all the usual techniques available for high-resolution NMR spectroscopy can be used. In particular, these include valuable techniques for structure determination such as 2-dimensional NMR experiments which provide correlation between NMR resonances based on mutual spin–spin coupling such as the well-known COSY or TOCSY techniques. In practice, it is possible to acquire NMR data on a number of peaks in a chromatogram by using a series of stops during elution without on-column diffusion causing an unacceptable loss of chromatographic resolution.

There are two further special categories of stop-flow experiment. First, fractions eluting from the column can be stored in capillary loops for later off-line NMR and MS studies ('peak picking'). Secondly, the flow can be halted at short intervals during the passage of the eluting peak through the NMR flow cell ('time-slicing') in a manner analogous to the use of a diode-array UV detector to obtain spectra from various portions of the peak. This allows chromatographic peak purity to be estimated. Time-slicing is most useful where the separation is poor, or where the compounds under study have weak/no UV chromophores which make it difficult to determine the retention times.

Fully automated analysis is also an option wherein the samples are placed in an autosampler and predefined HPLC–NMR experiments are performed. This is covered in detail elsewhere in this volume, but in summary, the software allows automatic detection of UV peaks in the chromatogram based on predetermined 'time-windows' or peak intensities. The successful detection of each UV peak triggers the system to stop the flow at an appropriate time to isolate the peak in the NMR flow probe. Then data relating to the peak (intensity, retention time, etc.) are transferred to the NMR host computer and used to define the parameters for the automatically acquired NMR spectrum. This automatic NMR operation includes field homogeneity optimisation, setting and optimisation of all NMR acquisition parameters, and the predefinition of the resultant signal-to-noise ratio required in the spectrum. The measurement of 2-dimensional (2D) NMR spectra can also be performed.

In the case of HPLC–NMR–MS experiments, there are some additional considerations. So far, the principal MS ionisation method has been electrospray in either positive- or negative-ion mode (using either single quadrupole, triple quadrupole or ion-trap mass spectrometers) and this puts further constraints on the chromatographic solvent systems as outlined earlier. However, other types of mass spectrometers are increasingly being employed in conjunction with HPLC–NMR and these include time-of-flight (TOF) instruments and Fourier-transform (FT) mass spectrometers. Future prospects include the application of inductively coupled plasma mass spectrometer (ICPMS) systems to produce HPLC chromatograms with elemental specificity [16].

When using HPLC–NMR, the chromatography is often developed off-line from the NMR unit by using non-deuterated solvents. It is not always simply a matter of replacing non-deuterated solvents with deuterated solvents to reproduce the chromatography for HPLC–NMR or HPLC–NMR–MS as this can give rise to changes in retention times. For this reason it is standard practice to perform an initial chromatographic run with a small injection volume (e.g. 10 μl) and then scale-up (e.g. to 50 μl) for stop-flow NMR experiments, when optimum conditions have been established. In addition, data can be acquired while mixing the eluent just prior to the mass spectrometer stage with a non-deuterated solvent to back-exchange the deuterium atoms in exchangeable situations (e.g. NH and OH groups) for hydrogens. In this way, if these initial data cannot be readily understood, the number of exchangeable hydrogens in any compound can be counted as it elutes [15].

Finally, the double hyphenation of NMR and MS to HPLC brings some additional benefits of MS [17–20]. These include the often superior sensitivity of MS, although this is not a universal factor, and the ability of MS to be used to search for particular diagnostic groups or fragments.

While the coupling of NMR spectroscopy and MS in the same system is undoubtedly the most productive of the possible multiple hyphenations, others have been performed which deserve consideration. Probably the easiest to implement is the inclusion of a UV diode-array detector (DAD) to enable UV spectra to be acquired and all that this requires is the replacement of the conventional variable-wavelength detector with the DAD. These detectors are in fact now becoming standard fixtures in HPLC–NMR systems. UV spectroscopy provides limited chemical structural information, but spectra can be obtained relatively easily and probably the only real difficulty with working with such detectors is the relative ease with which they can be overloaded when working with samples containing compounds with strong UV chromophores that have been concentrated to enable NMR spectra to be obtained as efficiently as possible. As well as UV spectroscopy, it is now possible to obtain IR spectra for components present in column eluates by two different methods, either directly on-flow or via on-line deposition of the eluent with concomitant removal of the solvent, followed by subsequent acquisition of the spectra. Both methods have been used in combination with ^{1}H NMR spectroscopy.

The first reported example of the combination of HPLC–IR–NMR [21] was to a mixture of polymer additives separated on a size-exclusion column with a deuterated chloroform solvent system at a flow rate of 1 ml min^{-1}. Following NMR spectroscopy, the eluent was deposited on to a rotating germanium disc, driven at a constant rate by a stepping motor. When chromatography was complete the disc was removed and placed in a scanning module enabling FT-IR spectra to be obtained in an FT-IR spectrometer. Excellent on-flow NMR and IR spectra were obtained for the test mixture (2,6-di-*tert*-butyl-4-methyl phenol (BHT), octadecyl-3-(di-*tert*-butyl-4-methylphenyl)propionate (Irganox 1076) and di-iso-octylphthalate).

Shortly afterwards, this work was extended by the incorporation of a mass spectrometer into the system, thus enabling on-line NMR and MS data to be obtained with on-line collection of the eluent for off-line FT-IR spectroscopy [22]. The incorporation of the mass spectrometer required the addition of a small proportion of ammonium acetate, dissolved in methanol, to the deuterated chloroform used as the eluent in order to promote the ionisation of the analytes. The inclusion of methanol and ammonium acetate to the solvent obviously introduced new signals into the NMR spectra, and in addition resulted in the loss of exchangeable protons from the analytes which had been observable when chloroform alone was used as the solvent. This work demonstrated the feasibility of multiple hyphenation ('hypernation') but the off-line nature of the FT-IR data acquisition, with the inevitable delay inherent in off-line analysis, represents a slight disadvantage. In addition, volatile components may well be lost as the solvent is evaporated. This can be a problem that, together with analyte instability, is exacerbated with such interfaces when reversed-phase eluents are used since these require heating in order to ensure removal of the solvent.

These studies led on to the investigation of on-flow FT-IR using attenuated total reflectance flow cells [23]. Although it was initially believed that the use of reversed-phase eluents would not be feasible with IR spectroscopy because of the strong interference resulting from the presence of water, the use of D_2O

makes this much less problematic, and in fact the use of D_2O may represent a useful innovation for on-line coupling of IR spectroscopy to reversed-phase HPLC. In this type of system, UV-DAD and FT-IR detection were performed prior to splitting the flow in the ratio 95:5. The major portion of the flow was then directed to the flow probe (120 μl volume) of a 500 MHz NMR spectrometer, while the minor portion was sent to a time-of-flight (TOF) MS. The advantage of using TOF was that accurate mass data could be obtained, thus enabling atomic composition data to be determined for a range of analytes. In this prototype system, the various flow rates had not been optimised and the quantities of sample used to ensure NMR and IR detection meant that the UV and MS detectors were somewhat overloaded. However, complete spectral data were obtained for a range of compounds, and the relatively similar sensitivities of the IR and NMR detectors suggested that this combination alone would be quite useful.

3.3 APPLICATIONS IN COMBINATORIAL CHEMISTRY

Characterisation of the structure and conformation of small biologically active molecules is part of the standard approach to lead generation in drug design studies. In particular, it is now possible to automatically synthesise many thousands of small molecules and then rapidly measure their effects in a given pharmacological test system. The power of such techniques comes from the immense number of compounds which can be generated and screened for activity. Two studies have evaluated HPLC–NMR in the field, one based on a mixture of 27 closely related tripeptides [24] and the other on two separate mixtures of four aromatic compounds and three pentapeptides [25].

The tripeptide application [24] comprised a synthetic mixture of the 27 combinations of the tripeptide formed from alanine (A), methionine (M) and tyrosine (Y), as the C-terminal amide, $H_2N.CH(R_1).CONH.CH(R_2).CONH.CH(R_3).CONH_2$, where R_1, R_2 and R_3 take all combinations of A, M and Y.

HPLC analysis was carried out using a 250 × 4.6 mm id Spherisorb ODS-2 column at 35°C with elution using isocratic D_2O/phosphate buffer at pH 2.5 for 5 min, followed by a linear gradient of acetonitrile to 50 % after 50 min with UV monitoring of the eluent peaks at 225 nm. The NMR measurements were carried out with solvent suppression at a ^1H NMR frequency of 600 MHz. The assignment of the resonances were based on those of standard compounds such as A–A–A–OH and Y–Y–Y–OH using standard 1-dimensional (1D) spectroscopy and 2-dimensional (2D) heteronuclear correlation methods such as ^1H–^{13}C and ^1H–^{15}N HMQC spectra.

The ^1H NMR on-flow detected separation is shown in Figure 3.1. This is a contour plot with ^1H NMR chemical shift on the horizontal axis and retention

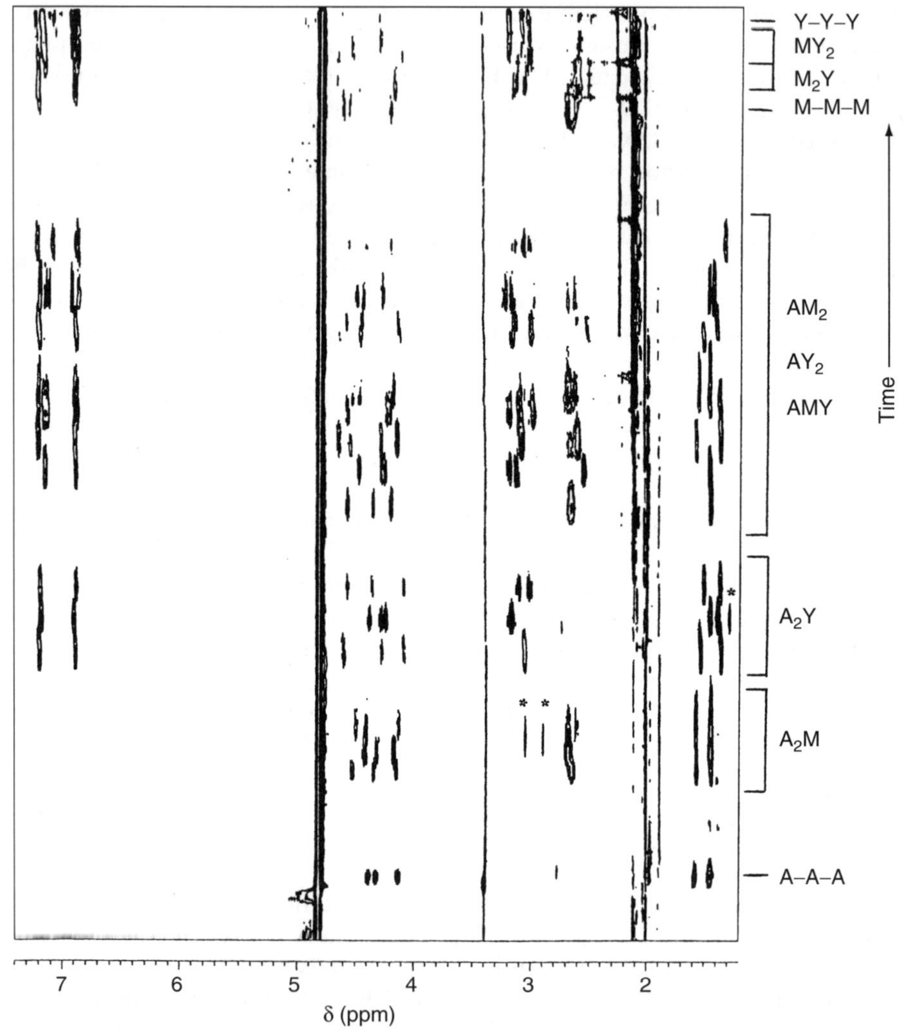

Figure 3.1 On-flow 600 MHz ^{1}H NMR spectral detection of the HPLC separation of a tripeptide mixture [24]. The horizontal axis corresponds to the ^{1}H NMR spectrum and the vertical axis represents time, with the total acquisition period being 50 min. The asterisks denote non-peptide impurity peaks, and the labels at the right-hand side denote the classes of tripeptide, e.g. A_2M refers to the three compounds, A–A–M–NH_2, A–M–A–NH_2 and M–A–A–NH_2

time on the vertical axis. This figure includes all of the resonances, including those of the residual peaks from suppressed HDO (δ 4.8) and acetonitrile (δ 2.0). The chemical shifts of the individual amino acids fall into characteristic ranges. Thus, all alanyl methyl signals appear between δ 1.6

and 1.35. The methionyl β-CH_2 and γ-CH_2 signals appear in a narrow range between δ 2.6–2.7 and the methionyl *S*-methyl signal gives rise to a singlet near δ 2.1, but this is obscured by the suppressed acetonitrile resonance. The tyrosyl β-CH_2 signals appear at around δ 3.0–3.2 and the aromatic signals give the resonances around δ 6.9 and 7.3 for the protons *ortho* and *meta* to the hydroxyl substituent, respectively. All of the α-CH resonances for all of the peptides appear between δ 4.0 and 4.6. On expansion, those due to alanyl residues appear as quartets and those arising from methionyl or tyrosyl residues appear as triplets.

Based on chemical shifts and peak multiplicities, the on-flow HPLC–NMR characterisation of the majority of the components in the mixture of 27 tripeptides was achieved and demonstrated that this approach is likely to be an effective method for compound mixtures. The elution positions of all of the alanyl-containing peptides were determined, with the exception of A–M–M–NH_2, which may have co-eluted with another peptide or may have been synthesised in a much smaller quantity. The only other tripeptides for which assignments have not been obtained are the MY_2–NH_2 isomers and two of the three M_2Y–NH_2 isomers. These eluted towards the end of the gradient run and are not as well resolved under these HPLC conditions. Additionally, with changes in the relative chemical shifts of the solvent signals, the intensities of the non-*N*-terminal α-CH protons and the methionyl β-methylene signals from these peptides may have been reduced by the effects of the solvent suppression irradiation of the water and acetonitrile resonances, respectively. With further optimisation of the elution conditions, it is possible that all 27 analytes could have been resolved and characterised.

A similar application has been the investigation of a model mixture of four (dimethoxybenzoyl)glycines with the same molecular weight [25]. Both continuous-flow and stop-flow ^1H NMR spectroscopies at 500 MHz were employed to characterise the four molecules using a gradient elution of acetonitrile/D_2O containing phosphoric acid. In addition, three model mixtures of pentapeptides were also investigated (Glu–Asn–X–Glu–Phe–OH, Phe–Asn–X–Glu–Phe–OH and Val–Asn–X–Glu–Phe–OH), where in each case the X refers to Asp, Gln, Ile, Lys or Thr. These three mixtures each of five pentapeptides were characterised by using stop-flow ^1H NMR spectrocopy at 500 MHz and it was estimated that there was less than 50 μg of each component in each mixture. In these cases, the HPLC elution again used a gradient of acetonitrile/D_2O but with trifluoroacetic acid in the aqueous phase.

More recently, significant effort has been put into using direct-flow NMR probes such as HPLC–NMR probes for analysing the components of combinatorial chemistry libraries but without the need for an HPLC separation first [26]. In this case, 88 samples from a combinatorial chemistry synthesis were placed in a 96-well plate and after the NMR spectra were measured the samples could be returned to the sample plate for further analysis or use. Although the exact nature of the compounds used in this study were not revealed, some of the cells contained the results of a reaction between a cyclic anhydride and a range of eleven different primary amines to form the urea structure shown below and it was possible to evaluate how well the reaction had proceeded.

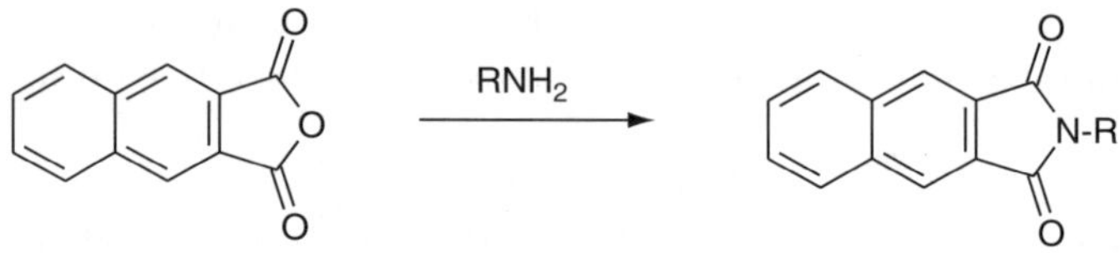

The aliquots of the solution-state chemical synthesis samples were directly injected into a standard HPLC–NMR probe by using a robotic liquid handler. The NMR software was used to automatically find and suppress the intense NMR signals from any non-deuterated solvents used, typically using the WET sequence [5]. Unlike the characterisation of impurities in organic compounds (see the next section) or drug metabolites (see the appropriate chapter in this volume) where the proportions of the analytes can be very different, combinatorial chemistry samples tend to be all of similar quantity and this simplifies the analysis in that it is not usually necessary to worry overly about carry-over of material from sample to sample, nor it is necessary to readjust the NMR spectrometer receiver gain after every sample, thus saving considerable machine time.

3.4 APPLICATION TO CHEMICAL IMPURITIES

The manufacture and quality control of a drug are controlled by a variety of national regulatory authorities. As well as drug efficacy, there is also a strong emphasis on the purity of final drug substances and it is necessary to obtain full characterisation and identification of any impurities at the level of $\geq 0.1\,\%$ of the UV peak area using HPLC analysis [27]. In order to characterise such impurities, currently it is necessary to isolate individual components by preparative HPLC. This work is often time-consuming and expensive, and yet still may not give conclusive identification. Furthermore, it is possible for the impurities to be degraded during sample extraction and purification. There is, therefore, a considerable need to develop and validate new methods for determining product purity.

Following the development of the technique and with the availability of commercial equipment, one of the first published real applications of HPLC–NMR was concerned with the identification of an impurity in a synthetic drug precursor [28]. In this case, a 400 MHz HPLC–NMR system was tested with a small molecule typical of pharmaceuticals:

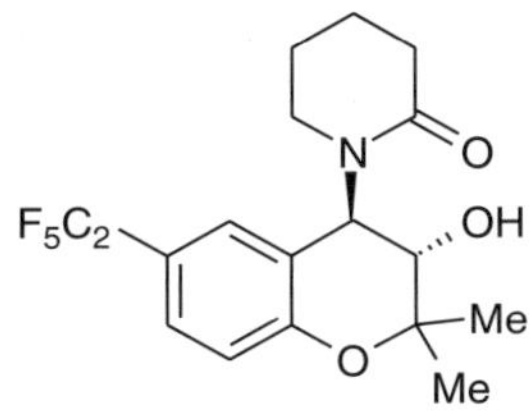

spiked with 9 % and 4 %, respectively, of the following compounds:

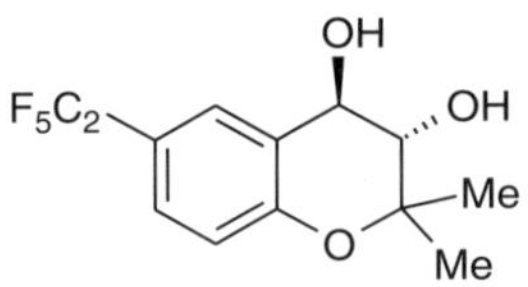

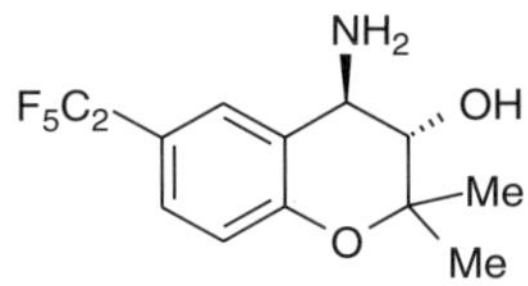

Having shown that the HPLC–NMR system could separate and characterise these three substances by using both continuous-flow and stop-flow approaches, a more realistic test was applied. This successfully characterised the two components shown below with the impurity at a level of 3 % using stop-flow HPLC–NMR:

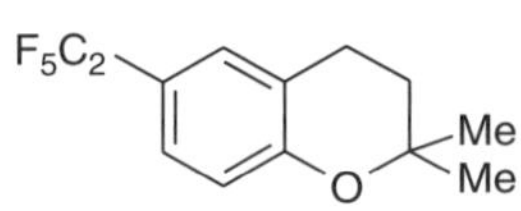

97%

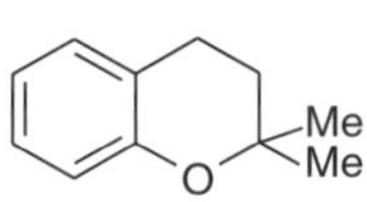

3%

The identification of degradation products of drugs in dosage formulations is an important part of the drug development process. As a result of such studies, it has been known for drug candidate structures to be modified or dosage formulations changed. Degradation products can have differing biological activities to the parent compound and, indeed, different toxicity profiles. HPLC–NMR and HPLC–electrospray ionisation mass spectrometry have been used to identify six degradation products in a dosage formulation of a matrix metalloprotease inhibitor [29]. This enzyme has been implicated in cancer, arthritis and other connective tissue disorders. The compound under study, PGE4410186, belonged to a class of hydroxamic acid derivatives, as shown below:

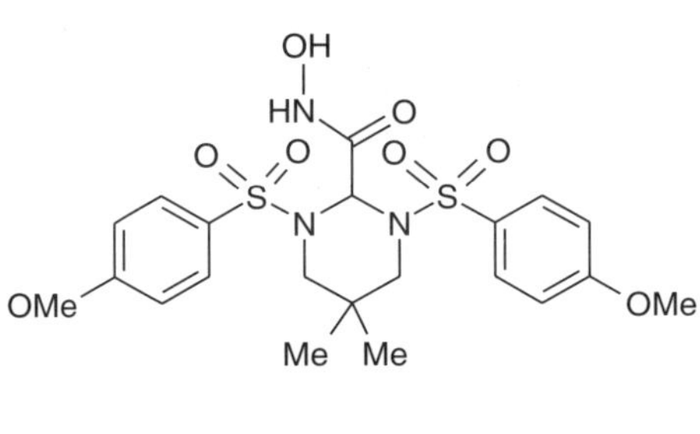

The dosage formulation used was 10 mg ml^{-1} of the compound in a solution of 10% ethanol, 65% PEG-400 and 25% aqueous phosphate pH 7.4 buffer. The HPLC experiment with UV detection, employing a gradient elution based on water/acetonitrile/formic acid, showed the parent compound eluting first, followed by six degradation products. The mass spectral information was sufficient to suggest structures for three of the degradation products but there were isomeric possibilities for three others. Stop-flow ^{1}H HPLC–NMR at 600 MHz was then used to confirm the structures of all of the six degradation substances and it was possible to distinguish between the isomeric structures suggested by the MS results. The partial structures, which all differed in the central part of the molecule, are shown below:

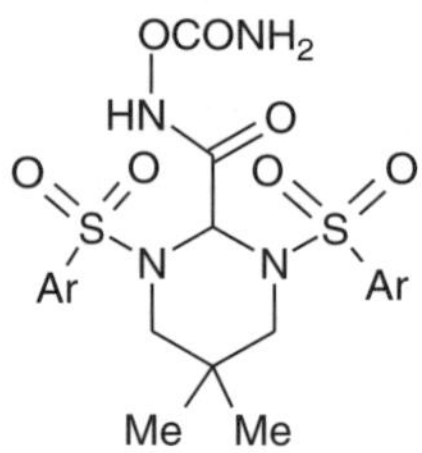

Deg-1

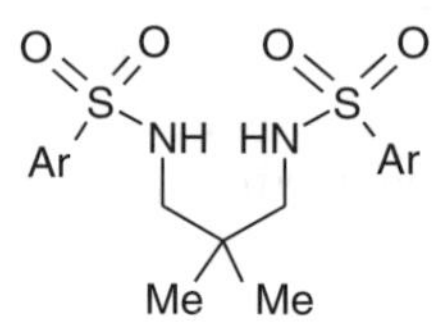

Deg-2

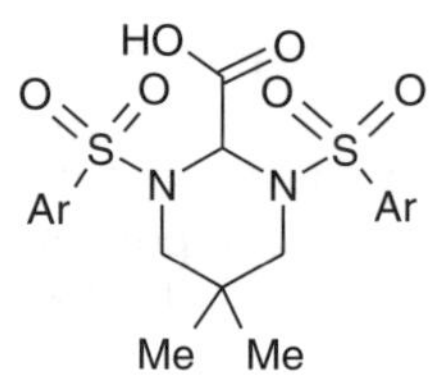

Deg-3

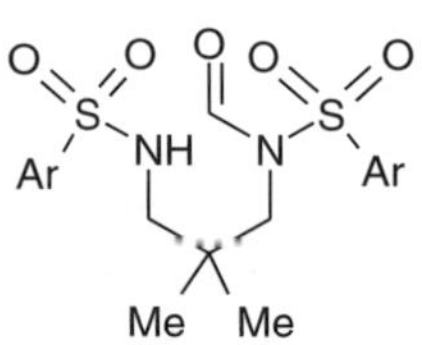

Deg-4

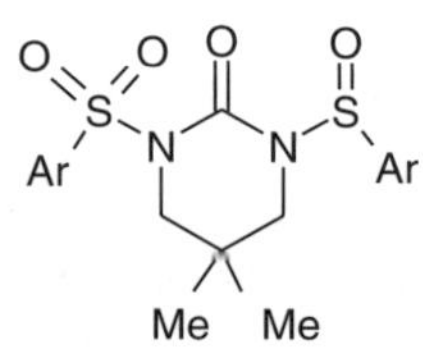

Deg-5

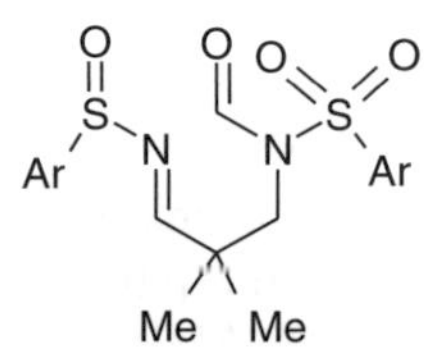

Deg-6

In situations where impurities have low or no UV chromophores, HPLC–NMR provides a unique window on the separation process. This has been exemplified using a drug substance known as SKF-99085, where HPLC–MS, HPLC–NMR, refractive index, light scattering and gas chromatography methods were all used to quantify and identify impurities [30]:

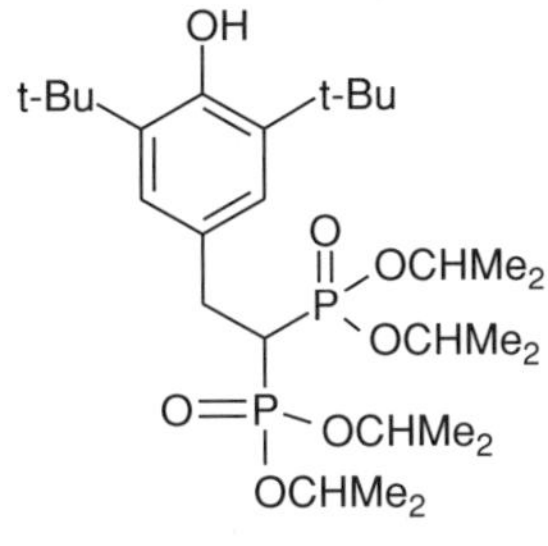

SKF-99085

HPLC with UV detection at 280 nm showed essentially only the required drug compound but a quantitative assay against a reference material showed a level of only 96.6 % of that expected. A number of impurities were detected by using HPLC with MS and refractive index detection and the structures of the major components were elucidated by using 500 MHz ^{1}H HPLC–NMR spectroscopy.

The approach to characterisation of impurities in bulk drug materials from a production batch is also exemplified by a study on fluticasone propionate [31] which has the structure and numbering system shown below:

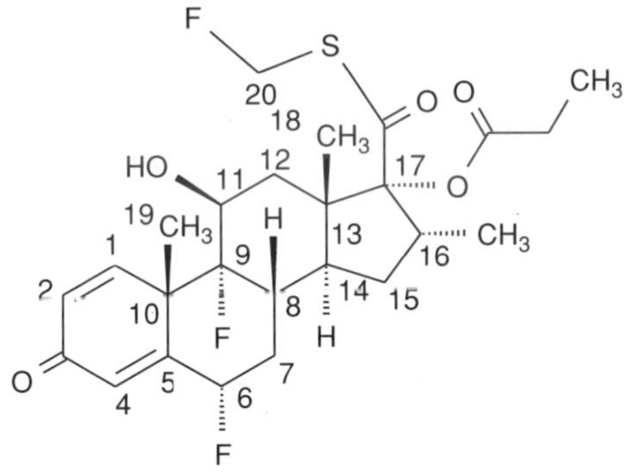

An HPLC method was developed using a 250 × 4.6 mm column packed with 5 μm ODS-2. Gradient elution was performed using D_2O (containing 0.05 % trifluoroacetic acid) and acetonitrile. The HPLC–^{1}H NMR spectra were acquired at 600 MHz in the stop-flow mode and also by time-slicing. In the time-slice mode, the HPLC elution was halted at 15 s intervals and ^{1}H NMR spectra were acquired over a single HPLC peak. Separate HPLC–MS data were also acquired on a single quadrupole mass spectrometer equipped with a standard thermospray source operated in the positive-ion mode. The chromatographic details were the same as used for HPLC–NMR spectroscopy except for substitution of H_2O by D_2O in the mobile phase. The common fragment ion of 435, with the structure shown below, was used to identify fluticasone-related material:

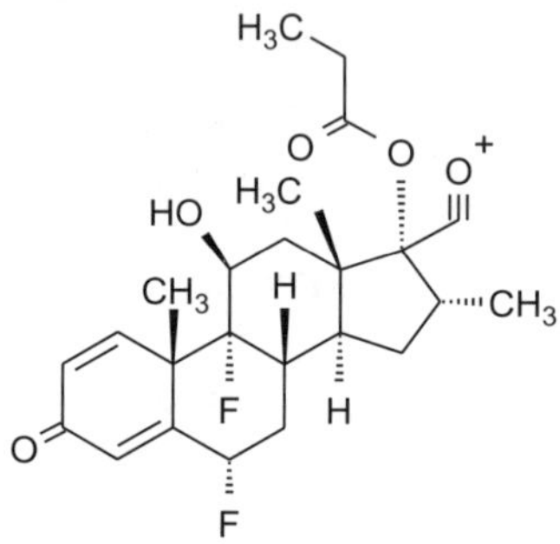

Based on that of authentic material, the UV-detected chromatogram showed that the peak from fluticasone propionate eluted first and four impurity peaks (see below for structures) had later retention times of 20, 26, 30 and 36 min, respectively:

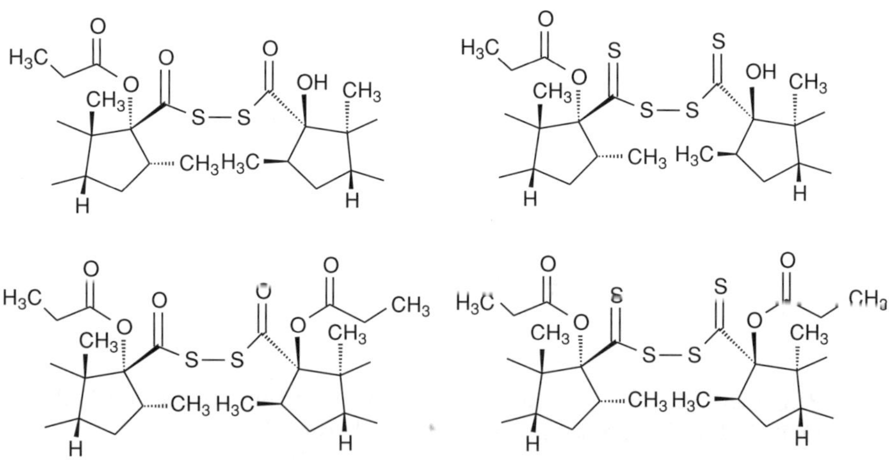

Identification of the different peaks in the HPLC chromatogram was achieved by 1D ^{1}H HPLC–NMR spectroscopy in the stop-flow mode and by HPLC–MS using positive-ion thermospray ionisation. All of the proposed structures for the four impurity peaks were consistent with the known chemical synthesis of fluticasone. The purity of the HPLC UV peak of fluticasone propionate itself was investigated by HPLC–NMR using the time-slice mode of operation. The elution was halted every 15 s over the peak and a ^{1}H NMR spectrum acquired. Near the leading edge of the UV peak where the signal-to-noise ratio is poor, through the main part of the peak where the signal-to-noise increases, and finally to the trailing edge of the peak where the signal-to-noise is again lower, the spectra showed no evidence for components other than fluticasone propionate.

This shows that directly coupled HPLC–NMR and HPLC–NMR–MS can enhance the ability to characterise impurities in a pharmaceutical product. As such, this approach is potentially significant as a general tool for purity analyses and would be expected to be important in speeding up production chemistry processes and for regulatory affairs. In the fluticasone study, it was shown that HPLC–NMR spectroscopy could detect and characterise impurities below the 0.1 % peak level, which is the relevant limit for submissions to regulatory authorities [27].

However, NMR spectra on the peaks which were at a level of 0.2 % or less of the parent drug required considerable and time-consuming data acquisition in order to achieve acceptable signal-to-noise ratios. Although this appears to involve a high cost in NMR analysis time it would be justified in cases such as that found with fluticasone propionate where there are few alternative analytical approaches. It may, therefore, be beneficial to concentrate the impurities, if stable, before the HPLC–NMR analysis by the application of solid-phase extraction chromatography, column switching or concentration/enrichment of the impurities by preparative HPLC. Such techniques can be feasible because of the availability of large amounts of sample from drug production batches.

The fluticasone study focused on the evaluation of HPLC–NMR spectroscopy for characterising impurities in raw bulk drug substances. However, the technique could also be applicable to the investigation of formulated drug substances where impurities often appear as a result of the drug substance reacting with the formulation compound or on samples from degradation studies. These types of adduct can sometimes be unstable during sample extraction and purification. This area is of considerable practical importance in the pharmaceutical industry and is worthy of further study.

Directly coupled HPLC–NMR spectroscopy has been used in a number of other studies of chemical impurities. An impurity in a bulk drug sample of the glycinamide ribonucleotide transformylase inhibitor AG2034, shown below:

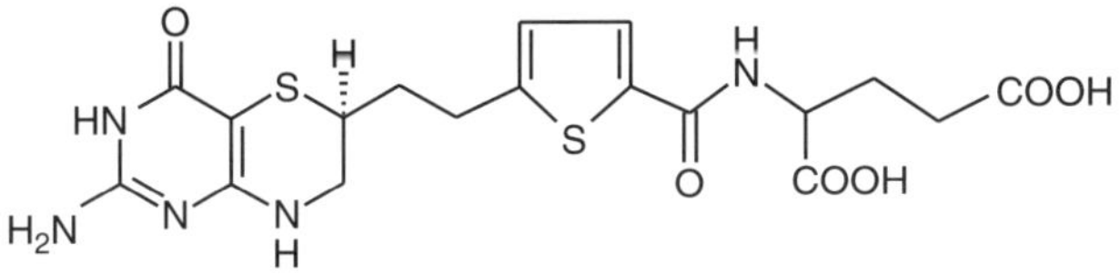

has been characterised by using both 1D and 2D TOCSY analysis [32]. The studies were actually performed on an impurity-enriched sample of the final drug synthetic precursor and this demonstrated that the impurity was a dimeric molecule of the desired parent structure.

The technique has also been used to characterise the synthetic route for fluticasone in the preparation of [*S*-fluoromethyl-^{18}F] fluticasone propionate for use as a radio-tracer substance for lung deposition studies employing positron emission tomography [33].

HPLC–^{1}H NMR spectroscopy using an isocratic separation with acetonitrile and D_2O in the ratio 7:13 has also been used to study photo-isomerisation of the natural material azadirachtin extracted from the seeds of the neem tree and which is a powerful insect anti-feedant [34]. This has a complex structure with an (*E*)-2-methylbutyl-2-enoate fragment but which after exposure to UV iradiation is converted to the *Z*-isomer:

E *Z*

Another different area of application where there is a need to characterise low-level impurities occurs in environmental studies. A recent study has used HPLC–NMR spectroscopy and HPLC–MS in the MS^n mode to identify organic components in the industrial waste-water effluent of a textile company. Generally in the past, GC–MS and HPLC–MS have been used for such analyses where the identity of many of the pollutants are not known beforehand. However, it has recently been shown that HPLC–NMR can provide useful information and if this is used with complementary information from HPLC–MS then this is the most efficient approach.

This study evaluated the impurity profile of untreated water from a textile plant in Portugal [35]. The organic material was concentrated by extraction from 1 l of water into dichloromethane and HPLC–NMR and HPLC–MS experiments were carried out using a reverse-phase separation with an acetonitrile/D_2O gradient elution with ^{1}H NMR spectroscopic observation at 600 MHz. For the HPLC–NMR studies, the samples were further fractionated into two pools according to their HPLC retention times. The HPLC–NMR studies were carried out in the stop-flow mode and the combination of NMR and MS results yielded the identification or tentative identification of 14 compounds, comprising mainly surfactants, anthraquinone dyes and nonylphenol-related molecules.

3.5 APPLICATION TO CHIRAL SEPARATIONS OF PHARMACEUTICAL MIXTURES

Many pharmaceutical products are chiral molecules, either as single isomers or more commonly as racemic mixtures. In addition, many formulated products are mixtures of active compounds together with a number of additives such as excipients. For chiral molecules, the pressure to develop single isomer forms as therapeutics in preference to racemic mixtures arises from the fact that one enantiomer is usually more biologically active than the other and also that enantiomers can have very different toxicity profiles.

Chiral HPLC has been coupled directly to NMR spectroscopy to demonstrate its application to the separation and characterisation of different isomers present in a drug substance using, as an example, atracurium besylate, a neuromuscular blocking agent used widely in surgery [36]. Atracurium besylate, (2, 2′–(3,11-dioxo-4,10-dioxatrideca-methylene)-bis-(2-methyl-1,2,3,4-tetrahydropapaverinium benzenesulfonate), is prepared from racemic 1,2,3,4-tetrahydropapaverine and has four chiral centres. However, because of the symmetry of the molecule, atracurium has 10 distinct species with the general structure shown below where the configuration at C1 can be *R* or *S*. Since the final stage of synthesis is quaternisation at N2, the isomers have been distinguished, simply for convenience, by the configuration of the substituents at the C1–N2 bond, such that when a tetrahydroisoquinoline residue has the benzyl group at C1 and the 3,11-dioxo-4,10-dioxatridecamethylene chain in a *cis*-configuration, this is called a '*cis*'-residue. Thus, the structure below shows atracurium in the *R-cis/R-cis* isomer form:

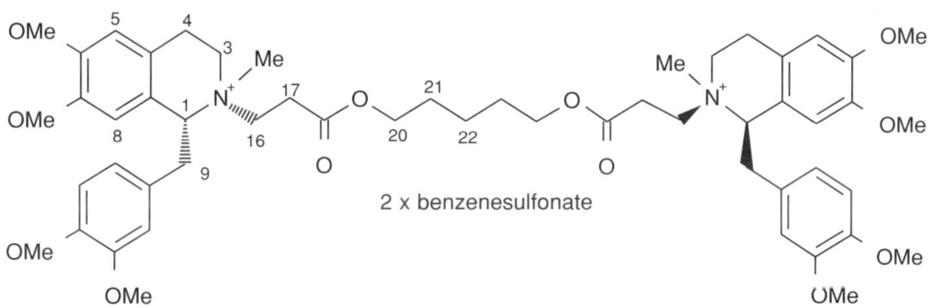

In achiral media, different NMR spectra are expected for each of the four types of enantiomeric pairs and for the two *meso* compounds. In general, because of the synthetic approach, the ratio of *cis*- to *trans*-residues is about three, which, assuming that quaternisation at one tetrahydroisoquinoline residue does not affect quaternisation at the other, leads to proportions of *cis–cis*, *cis–trans* and *trans–trans* isomers in the approximate ratio of 11:6:1. The preference for *cis*-residues has been proved by using nuclear Overhauser effect NMR measurements and by X-ray crystallography on related substances [37,38]. The ^{1}H NMR chemical shifts, principally of the H8 proton, can be affected not only by whether the C1–N2 configuration in the residue is *cis* or *trans* but also by the configuration of the remote tetrahydroisoquinoline unit.

In summary, therefore, after on-line chiral HPLC separation, NMR spectroscopy has been used to characterise compounds in terms of the *cis*- and *trans*-isomers and to identify the racemic pairs on the basis of their identical NMR spectra. In addition, HPLC–circular dichroism (HPLC–CD) was used to identify the absolute configuration of the enantiomers based on the known CD spectrum of *R*-laudanosine hydrochloride, a closely related molecule. The application to pharmaceutical analysis of HPLC separation with on-line CD detection alone has been the subject of one publication [39].

Chiral HPLC was performed isocratically at a flow rate of 3 ml min^{-1} using two 250 × 4.6 mm, 5 μm Chiracel OD-H columns connected to a Chiracel OD guard column of 50 × 4.6 mm id and an eluent comprising 60:40 v/v 0.5 M sodium perchlorate buffer (pH 2.0): acetonitrile.

A full assignment of the 750 MHz ^{1}H NMR spectrum of the atracurium besylate mixture sample was achieved using standard 1D and 2D NMR methods. From the ^{1}H NMR spectrum of the mixture it was possible to resolve and assign different chemical shifts for the *cis–cis*, *cis–trans*, *trans–cis* and *trans–trans* isomers for various protons in the molecule. The HPLC–NMR spectroscopic data were also acquired at 750 MHz in the stop-flow mode. For HPLC–CD spectroscopy, the separation was performed as for the HPLC–NMR experiments using a CD spectrometer with a specially constructed HPLC cell (5 mm path length, 2 mm aperture, 16 μl volume), including quartz doublet focusing and defocusing optics, connected to the HPLC column via a PEEK capillary. Spectroscopic detection was performed in stop-flow mode at 236 nm (at which the atracurium isomers have a CD peak). The instrument was continually flushed with dry nitrogen to ensure optimal spectroscopic performance.

Figure 3.2(a) shows the resulting UV chromatogram used for HPLC–NMR giving good resolution for nine out of the ten isomers. These eluted with retention times of between 60–100 min. This chromatogram also shows some

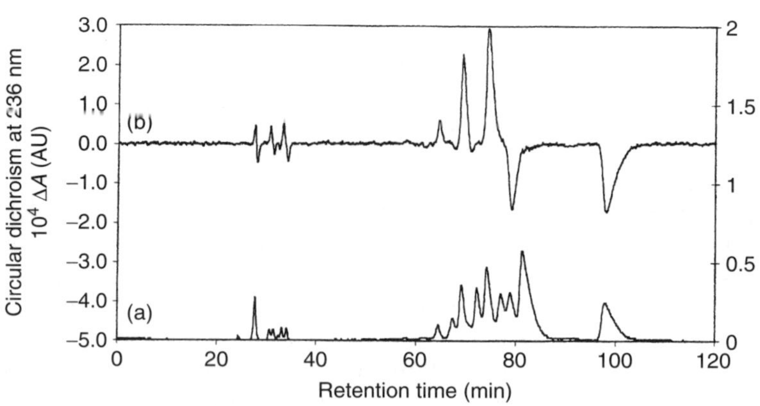

Figure 3.2 (a) HPLC separation of the atracurium mixture, using UV detection at 280 nm. (b) HPLC separation of the atracurium mixture, using CD detection at 236 nm, and with the same chromatographic conditions as for UV detection. The left-hand vertical scale refers to the CD detection, while the right-hand vertical scale refers to the UV detection. The peaks with retention times of between 60 and 100 min arise from atracurium isomers and correspond in order to the NMR spectra A to K shown in Figure 3.5 below

early eluting substances in the atracurium besylate mixture which were seen at retention times of between 20–35 min. These peaks are minor impurities or degradation products related to atracurium.

Expansions of key reporter resonances in the 750 MHz ^{1}H NMR spectra obtained in stop-flow mode are given in Figure 3.3. It is clear from the chiral HPLC chromatogram that although some peak resolution for nine out of the ten species has been achieved, most of the peaks show considerable overlap. It might therefore have been expected that the NMR spectra produced from the individual chiral HPLC peaks would not show single isomers. However, it has been found that these NMR spectra were, in general, remarkably pure. Although the HPLC peaks were in some cases as much as 8 min wide, the NMR flow probe used in this work only had a 65 μl volume. Hence, only a small slice from the centre of each HPLC peak was collected in the NMR flow cell for spectral data acquisition, thus giving maximum probability of a spectrum uncontaminated by other components. The peaks were labelled A–K in order of retention time, with peak D being assigned to a small unresolved shoulder immediately after peak C.

Peaks A and B were identified as the *trans–trans* isomers by NMR spectroscopy. This is because not only are they the smallest peaks in the chromatogram,

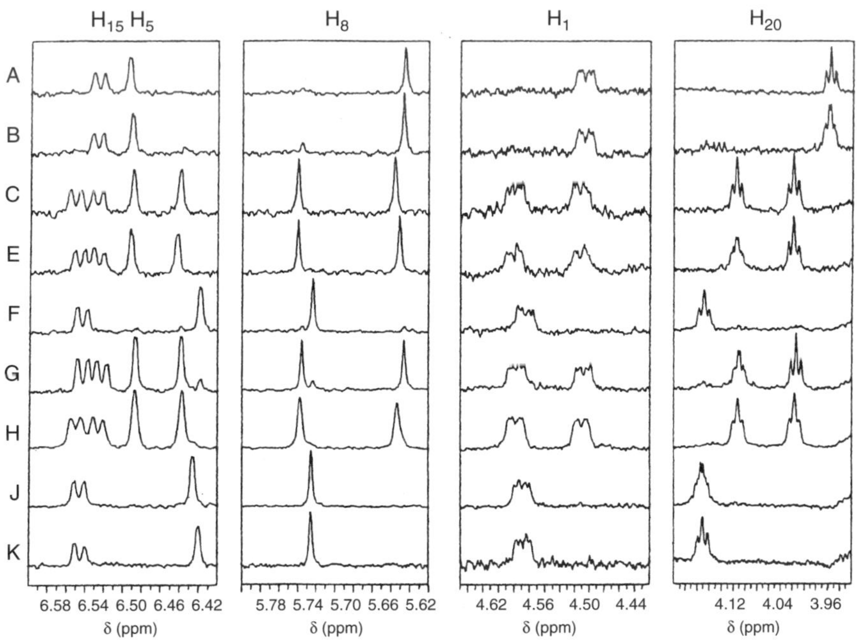

Figure 3.3 Expansions of the 750 MHz ^{1}H NMR spectra for key peaks in the chiral HPLC–NMR spectra for the different isomers of atracurium besylate

but also give a high field shift for a single H8 proton at $\delta \sim 5.6$ which is consistent with the assignments of the *trans*-isomers from the whole mixture. It was not possible to obtain an NMR spectrum for peak D. Peaks F, J and K are identified as the *cis*-isomers, based on the fact that these peaks are the largest in the HPLC chromatogram and give a single lower-field ^{1}H chemical shift for the H8 proton at $\delta \sim 5.74$ which is also consistent with the assignments of the *cis*-isomers from the whole mixture. This leaves peaks, C, E, G and H which have been identified as the *cis–trans* isomers because they each give two H8 resonances in their spectra for the different *cis*- and *trans*-parts of the isomer. Other key resonances in the ^{1}H NMR spectrum, including H5, H1, H20, H21 and H22, also show differences which are consistent with the assignments made on the whole mixture to confirm the stereochemistry. The identification of the enantiomeric pair from the three *cis–cis* isomers was possible from the NMR spectra in that the spectra from components F and K were essentially identical and that from component J showed distinct chemical shift differences (e.g. as indicated by resonances for H20 in particular, see Figure 3.3). From this, it can be deduced that peaks F and K were the *R/cis–R/cis, S/cis–S/cis* enantiomeric pair and that peak J was the *R/cis–S/cis meso* compound. Similarly, it could be shown that peaks C and H formed a *cis–trans* enantiomeric pair, as did peaks E and G as indicated by resonances for H5 and H20 in particular. Finally, although good NMR spectra of only two out of the three *trans–trans* isomers were obtained (peaks A and B), these were clearly different and thus one is from the *meso* compound and the other is one of the enantiomeric-pair isomers.

The results from the HPLC–CD spectroscopy are given in Figure 3.2(b). This shows an HPLC–CD trace on the same time-scale as the UV chromatogram. The result was a series of CD peaks, some of which were positive, others negative and some which gave virtually no CD spectrum. An additional CD spectrum was also collected on a reference compound *R*-laudanosine hydrochloride which gave a negative CD spectrum. From this, it was therefore possible to confirm that for the previously identified *cis–cis* isomers, peak F had the *S/S* configuration, peak K was the *R/R* compound and peak J was the *R/S-meso* compound. Similar arguments could be applied to the peaks from the *trans–trans* isomers in the chromatogram in that peak A was *S/S*, Peak D was *R/R* and Peak B was the *R/S-meso* compound. Finally, for the *cis–trans* isomers, again in agreement with the NMR results, peak C was *S/S*, peak E was *R S*, peak G was *R/S* and peak H was *R/R*.

HPLC–NMR spectroscopy was useful for identifying the isomeric configuration at the C1–N2 bond, for identifying the enantiomeric pairs of compounds and for distinguishing them from the *meso* forms. The HPLC–CD experiments were complementary in that, while unable to distinguish the C1–N2 isomers (*cis* or *trans*), it was possible to determine the absolute stereochemistry at C1 at each tetrahydroisoquinoline residue as either *R/R, S/S* or *R/S*, based on the sign of the CD response at a chosen wavelength. A consistent finding was that the *S*-isomers eluted before the *R*-isomers and the *trans*-forms eluted before the *cis*-

forms. By these means, a full characterisation of all of the ten isomers of atracurium was achieved.

3.6 APPLICATION TO NATURAL PRODUCTS

Natural products have been, and remain, a rich source of leads for the pharmaceutical industry and many marketed drugs are either natural products or are modifications of such substances. Hence, considerable effort is spent in isolating and characterising chemicals from natural sources which can be tested in a variety of biological screens. Often, it is necessary to carry out laborious extraction and purification steps and the advent of directly coupled HPLC–NMR has been explored as an alternative technique for natural product identification. The use of HPLC–NMR, and other hyphenated techniques such as HPLC–MS–MS, for identification of natural products from plant sources has been reviewed by Wolfender and co-workers [40,41].

The application of directly coupled HPLC–NMR and HPLC–CD, used separately, has been employed extensively for characteristion of natural products, as is described in more detail in another chapter in this volume [42].

One example which relates to a pharmaceutically relevant mixture concerns HPLC–NMR–MS studies undertaken on extracts of the plant *Hypericum perforatum L* (St. John's Wort) [43]. In this study, a total of 15 components were identified by using a combination of retention time, stop-flow ^{1}H NMR spectroscopy and MS data. These included quercetin-galacturonide, quercetin-rutinoside, quercetin-galactoside, quercetin-arabinoside, quercetin-rhamnoside, quercetin and 13-II8-biapigenin. Some of the later-eluting minor constituents (13′-II-biapigenin, hypericin, protohypericin, pseudohypericin, protopseudohypericin, hyperforin and adhyperforin) were present in insufficient quantities for NMR spectroscopic detection and were identified by using MS data alone. The separation was carried out using a gradient of D_2O (containing 20 mM ammonium acetate) and acetonitrile–0.1% acetic acid from 95% aqueous to 95% organic over 40 min. Two of these compounds, tentatively identified as quercetin-arabinoside and quercetin-galacturonide, had not previously been described in extracts of this species.

Microbial production of secondary metabolites is also an important source of novel therapeutic agents. However, the physiological and biochemical factors that determine the onset of production of a specific secondary metabolite in a particular species are incompletely understood. Generally, a range of analytical techniques, often elaborate, time-consuming and involving extensive sample pre-treatment, have to be developed in order to monitor the details of the metabolic changes and substrate consumption that accompany secondary metabolite production. In order to provide rapid multi-parametric information about the microbial fermentation process, ^{1}H HPLC–NMR has been applied to characterise microbial metabolites directly in the broth supernatants from a wild-type strain of *S. citricolor* [44]. This species produces aristeromycin, the carbocyclic analogue of adenosine, a secondary metabolite with antibiotic properties. The 600 MHz ^{1}H NMR spectrum of the broth was particularly complex in the chemical shift region between δ 4.2 and δ 3.4, since it contains major sugar signals obscuring minor metabolites. Hence, an HPLC method with on-line ^{1}H NMR detection was employed to physically separate components which gave signals in this region, specifically trehalose and fructose, but citrate, succinate, acetate and uridine were also identified.

Other studies of natural products, which will be described in detail elsewhere in this book, involve the characterisation of isomeric structures, an application for which NMR spectroscopy is particularly powerful. For example, Albert *et al.* have used HPLC–NMR spectroscopy to determine the structures of isomers of vitamin A acetate [45]. The parent molecule has all-*trans* double bonds as shown below and by brief heating of this substance a number of isomers with a single *cis* double bond and one isomer with two *cis* double bonds can be formed:

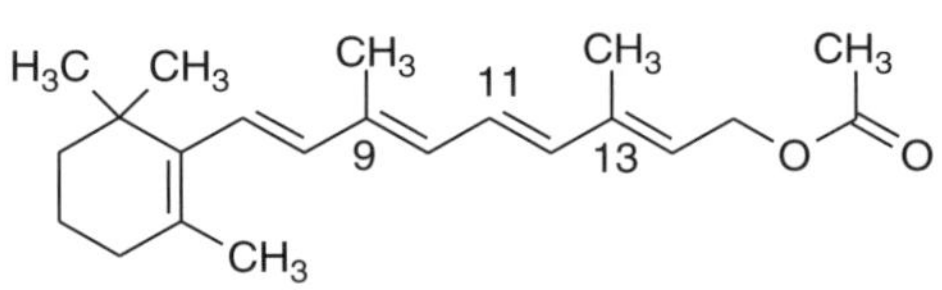

By using continuous-flow 400 MHz HPLC–^{1}H NMR spectra obtained using a cyanopropyl–silica gel column with heptane as the eluent, and by examination of the spectra extracted from the pseudo-2-dimensional representation, it was possible to identify the all-*trans* molecule, as well as the 9-*cis*, 11-*cis*, 13-*cis* and 11, 13 di-*cis* isomers.

More recently, this same group have used a combination of HPLC–NMR and HPLC–MS to identify analogues of vitamin E found in a palm oil extract rich in such tocopherol derivatives [46]. For the MS, a new technique called coordination ion spray was used in which the addition of silver ions to the HPLC eluent caused the enhanced formation of ions of such non-polar compounds. In addition, the HPLC utilised a C_{30} column in order to gain suitable chromatographic resolution. The compounds in the mixture comprised vitamin E itself, tocopherol, and five related compounds, some of which were isomeric. Tocopherol has the structure shown below:

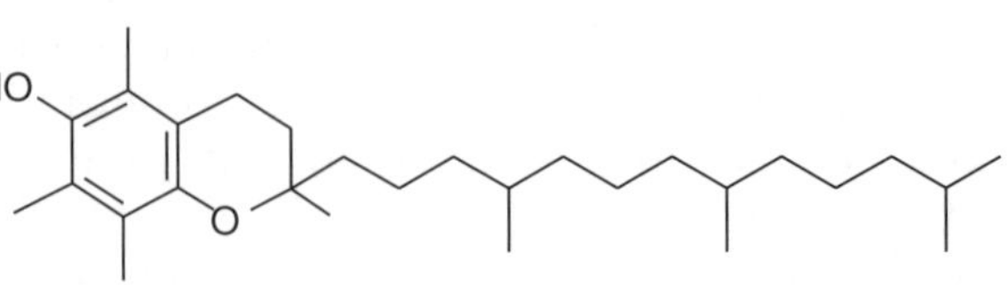

The other five molecules which were identified using continuous-flow HPLC–^{1}H NMR spectroscopy at an observation frequency of 400 MHz were α-tocoenol with a mass of 428 (a double bond at C11′), α-tocotrienol with a mass of 424 (double bonds at C3′, C7′ and C11′), β-tocotrienol with a mass of 410 (double bonds at C3′, C7′ and C11′, but with loss of a methyl on the phenyl ring), γ-tocotrienol, also with a mass of 410 (like β-tocotrienol but with loss of a different methyl group on the phenyl ring) and δ-tocotrienol with a mass of 396 (like β-tocotrienol but with the loss of two methyl groups from the phenyl ring).

3.7 APPLICATION TO CHEMICAL REACTIVITY OF DRUG GLUCURONIDES

By far, the largest body of work to date using HPLC–NMR and HPLC–NMR–MS is in the field of drug metabolism where the methodology has been used extensively for the identification of metabolites in studies from clinical trials involving human subjects, the investigation of model drugs in animals *in vivo* and also through the use of *in vitro* systems such as liver microsome incubations. The results in the literature are summarised elsewhere in this volume.

However, many drug metabolites have a significant degree of chemical reactivity. For example, many drugs containing carboxylate groups form β-1-*O*-acyl glucuronides as major metabolites. Such ester glucuronides are potentially reactive due to the susceptibility of the acyl group to nucleophilic reactions and they can undergo hydrolysis, acyl migration and covalent adduct formation [47,48]. The acyl migration reactions result in positional isomers and anomers as shown in Figure 3.4 and these may be reactive towards serum proteins with toxicological consequences. The acyl group migrates successively to the 2-, 3- and 4-hydroxyl groups of the glucuronic acid moiety, thereby allowing the formation of both α- and β-anomers of the positional isomers (see Figure 3.4).

Synthetic fluorobenzoic acid and trifluoromethylbenzoic acid glucuronide conjugates were initially chosen as model compounds of carboxylate-group-containing drugs and an HPLC method was developed for the simultaneous determination of the 1-, 2-, 3- and 4-positional isomers of the acyl glucuronides, and their α- and β-anomers for 2-, 3- and 4-fluorobenzoic acids, together with the aglycones formed via hydrolysis. A typical result is shown in Figure 3.5, which depicts the continuous-flow 750 MHz ^{1}H HPLC–NMR characterisation of the glucuronides from an equilibrium mixture of transacylated glucuronides of 4-fluorobenzoic acid, measured in the continuous flow-mode [49]. In this figure, the ^{1}H NMR frequency is on the horizontal axis, while the chromatographic retention time is on the vertical axis. Each of the glucuronide isomers elutes separately and can be identified from its NMR spectrum. It has been noted that in general the elution order of transacylated glucuronides is β-4-*O*-acyl-, α-4-*O*-acyl-, α-3-*O*-acyl-, β-3-*O*-acyl-, β-2-*O*-acyl- and α-2-*O*-acyl-, irrespective of the nature of the carboxylic-acid-containing moiety.

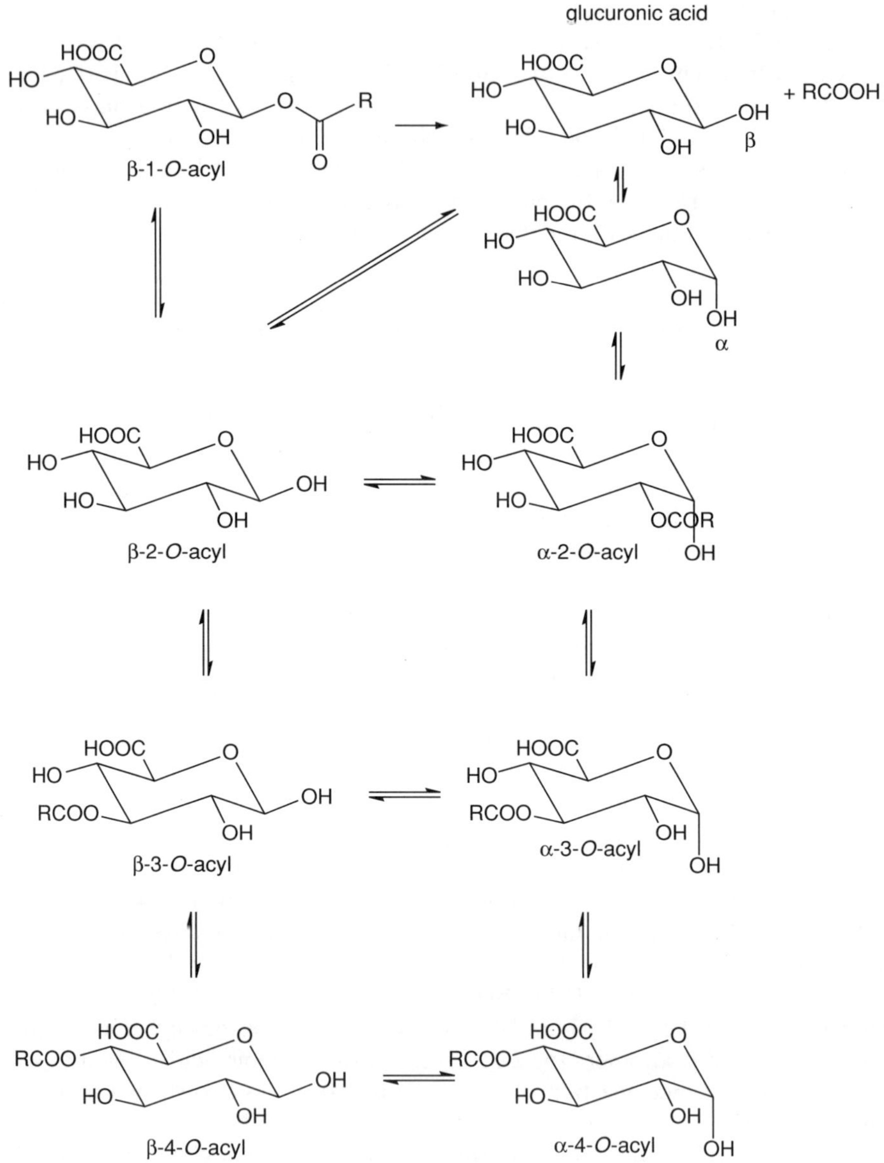

Figure 3.4 The reaction scheme depicting acyl migration in glucuronide conjugates. Recently, the back reaction from the α-2-*O*-acyl isomer to the α-1-*O*-acyl isomer has been identified

This directly coupled HPLC–NMR method has been used to investigate the acyl migration kinetics of individual isomers of 2-, 3- and 4-fluoro-, and 2- and 3-trifluoromethylbenzoyl-D-glucopyranuronic acid separated from an equilibrium mixture of the β-1-*O*-acyl isomer, the α- and β-2-*O*-acyl isomers, the

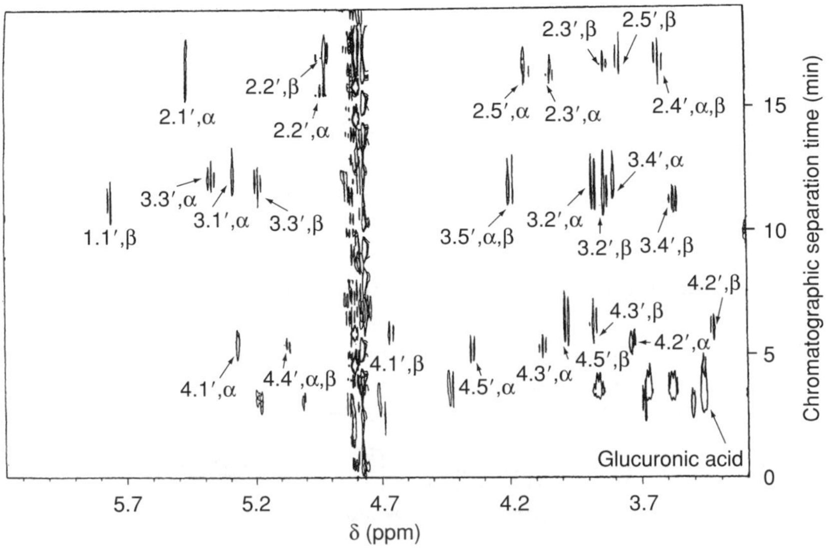

Figure 3.5 750 MHz ^{1}H continuous-flow HPLC–NMR spectrum of an equilibrium mixture of transacylated glucuronides of 4-fluorobenzoic acid. The nomenclature 2,1′-α denotes the 1′ proton in the α-anomer of the 2-*O*-acyl glucuronide

α- and β-3-*O*-acyl isomers, and the α- and β-4-*O*-acyl isomers at pH 7.4 and 25 °C [49–52]. Both continuous-flow HPLC–NMR and stop-flow methods have been used. For detailed kinetic studies, each isomer was separated by using reversed-phase HPLC and then led into an NMR flow probe in a 600 MHz NMR spectrometer. The flow was stopped and sequential ^{1}H NMR spectra collected, thus allowing the direct observation of the appearance of the glucuronide positional isomers of that particular glucuronide isomer which had been isolated. This is illustrated in Figure 3.6, which shows the build-up of other products following the introduction of the β-4-*O*-acyl-glucuronide of 2-fluorobenzoic acid into the NMR probe after HPLC separation. The rate constants for the decomposition of the various isomers were determined and the acyl migration reactions were simulated by using a mathematical model of the kinetics of the glucuronide rearrangement (incorporating nine first-order rate constants determining acyl migration reactions and six first-order rate constants describing the mutarotation of the 2-, 3- and 4-positional isomers.

The acyl migration of the glucuronide metabolite of the model drug 6,11-dihydro-11-oxodibenz (*b*,*e*) oxepin-2-acetic acid has also been investigated in urine and pH 7.4 buffer using directly coupled 600 and 750 MHz stop-flow HPLC–^{1}H NMR spectroscopy [53,54].

More recently, the acyl migration reaction has been studied for a range of non-steroidal anti-inflammatory compounds. For example, a detailed study has

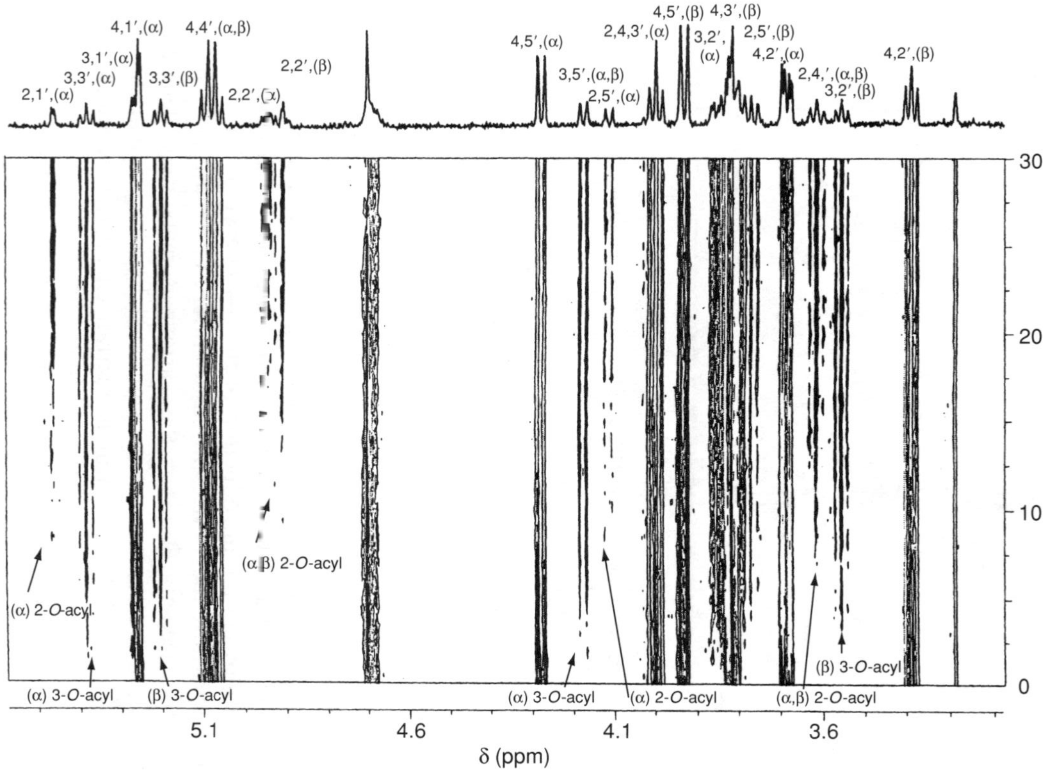

Figure 3.6 600 MHz ^{1}H NMR spectrum obtained in stop-flow mode after the β-4-*O*-acyl-glucuronide of 2-fluorobenzoic acid had been isolated in the NMR flow cell. This shows the successive formation of 3-*O*-acyl- and 2-*O*-acyl isomers as a function of time

been made of the reactions of *S*-naproxen-β-1-*O*-acyl glucuronide in phosphate buffer at pH 7.4, purified from human urine after oral administration of a therapeutic dose of approximately 400 mg. In this case, new signals were observed in the 600 MHz ^{1}H NMR spectra and these were assigned to the α-1-*O*-acyl glucuronide isomer and the structure was confirmed by using 2D TOCSY NMR spectra and parallel HPLC–MS studies to provide the molecular weight [55]. By stopping the chromatographic peaks of individual isomers in the NMR detector, it was possible to show that the α-1-*O*-acyl isomer also underwent transacylation and the reverse reaction from the mixture of 2-*O*-acyl isomers was possible. This was unexpected as it had been postulated previously that there was no back-reaction to the 1-*O*-acyl isomers from the 2-*O*-acyl isomers [48]. The use of HPLC–NMR at 600 MHz observation has enabled the determination of the elution order of the *S*-naproxen glucuronide isomers and these again follow the general pattern shown above [56]. Finally, the kinetics of the various transacylation reactions have been elucidated by using both HPLC–NMR and HPLC–UV methods [57]. In total, nine kinetic rate constants for six reactions could be determined using UV spectroscopic quantitation of the isomers in the HPLC mobile phase containing acetonitrile. Replacing H_2O by D_2O and increasing the relative proportion of acetonitrile both caused a decrease in the reaction rates. The α-1-*O*-acyl glucuronide was the most reactive isomer. When the same mobile phase was used for parallel HPLC–NMR and HPLC–UV analyses, the reaction rate constants were identical (with in the limits of experimental error).

It is clear that directly coupled HPLC–NMR spectroscopy offers a unique analytical approach for obtaining structural information of interconverting compounds in a complex mixture of isomers. This method will be of value in the elucidation of the reactivity of drug glucuronides in terms of acyl migration and enables an investigation of the potential for protein binding. Furthermore, this HPLC–NMR approach to the study of glucuronide acyl migration reactions allows unique kinetic information to be obtained relating to glucuronide reactivity and such an approach will be useful in future structure–activity studies on the toxicity of drug ester glucuronides.

3.8 APPLICATION TO FUTILE DEACETYLATION REACTIONS

It has been shown by using NMR spectroscopy in conjunction with isotope-labelling studies that there is a significant degree of deacetylation followed by reacetylation (futile deacetylation) of paracetamol metabolites *in vivo* in the rat [58]. If this also occurs in humans, then it may help to explain the observed incidence of nephrotoxicity of paracetamol in that the process would result in levels of the potent nephrotoxin 4-aminophenol *in vivo*. Confirmation of the levels of futile deacetylation in individual metabolites of isotopically labelled paracetamol in man has been achieved by using directly coupled HPLC–NMR spectroscopy at 600 MHz. In this study, a solid-phase extract of a 0–4 h urine sample after dosing with paracetamol-d_3 was separated by using HPLC with a methanol–water gradient elution. Methanol was used instead of the more usual acetonitrile to avoid the large methyl NMR resonance of the latter which would obscure any observation of transacetylated products. Good ^{1}H NMR spectra were obtained from the sulfate and glucuronide conjugates of paracetamol-d_3 and quantitation of the level of transacetylated products for both of these metabolites was obtained by integration of the observed acetyl peak relative to the peaks from the aromatic protons. In man, it was shown that paracetamol glucuronide underwent transacetylation to an extent of 1 %, while for the sulfate conjugate the level was 2 % [59]. This compares with the results obtained for the sulfate in the rat of about 10 % [58].

Further work has been carried out on the now withdrawn drug, phenacetin or 4-ethoxyparacetamol. Using the same HPLC–NMR approach, the level of futile deacetylation in phenacetin in the rat was much higher than for paracetamol. This has been quantified in the major metabolites, paracetamol sulfate and paracetamol glucuronide, as 30 and 36 %, respectively, using HPLC–NMR spectroscopy. The level of futile deacetylation for paracetamol and a further, tentatively assigned, metabolite, *N*-hydroxyparacetamol sulfate, was quantified at 32 % [60].

A final example of the use of this strategy to examine futile deacetylation comes from work on the beta-blocking drug practolol where a radio-labelled $^{13}C/^{14}C$ drug was dosed to rats and the urine profiled by HPLC with UV, radioactivity, NMR and MS–MS detection [61]. As shown by the structure below, practolol can be considered to be an analogue of paracetamol with a beta-blocker side-chain:

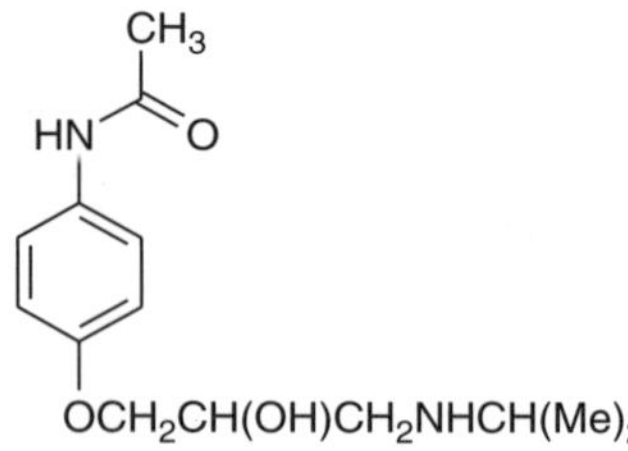

The amount of futile acetylation observed for this compound in the rat, at ca. 7–10 % for parent compound and metabolites, was less than that seen for phenacetin and similar to that found for paracetamol itself. The bulk of the radio-label was rapidly excreted in urine as practolol itself (albeit with 7–10 % reacetylation) and the remainder as either the ring-hydroxylated metabolite or its glucuronide conjugate. As in previous examples, deuterated methanol was used in the mobile phase rather than methanol in order to be able to more easily observe the acetyl resonances of practolol and related compounds.

3.9 APPLICATION TO TRAPPING OF REACTIVE INTERMEDIATES

The metabolism of paracetamol proceeds through a reactive intermediate *N*-acetyl-*p*-benzoquinone imine (NAPQI) which is formed by cytochrome P450 oxidation of the drug. NAPQI is very reactive and is believed to be responsible for the generation of the toxic species which are responsible for the liver damage seen after high doses of paracetamol. It can form covalent adducts with glutathione and is also a powerful oxidant and thus covalent binding to thiols and oxidation of protein thiols have both been implicated in the hepatotoxicity of paracetamol. Cellular glutathione can protect against liver toxicity and it was shown that this was due to the reaction of NAPQI at the 3-position. Although the 1-, 2- and 4-positions of paracetamol could also, in principle, react, it was assumed that only the 3-substituted molecule was formed. More recently, it has been postulated that the *ipso* (1)-position and also the 2-position of paracetamol can be implicated in NAPQI reactivity. The reaction scheme for this is shown in Figure 3.7. This reaction has now been investigated by using directly coupled HPLC–NMR at 500 MHz [62].

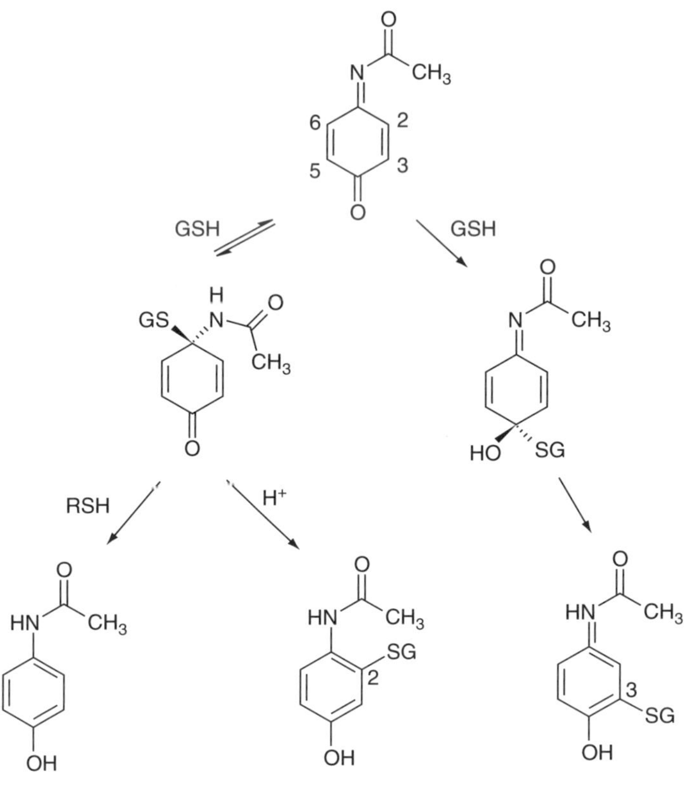

Figure 3.7 Products from the reaction of glutathione (GSH) with NAPQI

The two reactants, (NAPQI and glutathione (GSH)) were mixed in equimolar proportions and the product mixture separated by using directly coupled HPLC–NMR in the stop-flow mode with an eluent of 5% acetonitrile in D_2O-based ammonium acetate buffer. Three chromatographic peaks were observed, all with the same mass-to-charge ratio in the mass spectrum, characteristic of glutathione adducts to paracetamol. The third peak showed a ^1H NMR spectrum identical to authentic 3-glutathionyl paracetamol. The second peak gave an NMR spectrum consistent with the 2-position adduct, while the first peak showed a different ^1H NMR spectral pattern (AA′XX′) for the aromatic protons, consistent with a symmetrical structure and hence 1-substitution. Thus, all three isomers were shown to be produced with the *ipso* derivative being most abundant and the 2′-isomer the least abundant. By holding the *ipso*-NAPQI-GSH adduct in the flow probe of the NMR spectrometer for 1 h, the rate at which it decomposed to the other isomers and other species could be monitored.

3.10 APPLICATION TO UPTAKE AND TRANSFORMATION OF XENOBIOTICS BY PLANTS

The identification of the metabolites of drugs in humans and animal species forms the basis of a separate chapter in this volume. However, some studies have been carried out on metabolism of model herbicide compounds in plants and these are reviewed briefly here.

Directly coupled HPLC–NMR–MS has been used to characterise the major metabolites of 5-trifluoromethylpyridone, which can also exist as the tautomer 2-hydroxy-5-trifluoromethylpyridine, in hydroponically grown maize plants [63]. Such a compound is a model for many herbicides. This molecule is metabolised to the glucoside and this can be further conjugated to yield a malonylglucoside. Both tautomers can potentially yield conjugation products, as shown in the scheme given in Figure 3.8.

In this study, the retention times of the parent compound, the *N*-glucoside and the *O*-malonylglucoside were identified by using ^{19}F NMR spectroscopy and on-flow detected HPLC–NMR, with this being followed by stop-flow HPLC–NMR–MS.

Later, a study was made using 5-nitropyridone where the absence of a ^{19}F NMR ‘handle’ meant that it was necessary to use the information from the on-line MS results to identify the chromatographic peaks of interest [64]. By monitoring the fragment with $m/z = 143$ in the mass spectrometer, relevant peaks for study using NMR spectroscopy could be identified. In this study, three metabolites were characterised, namely the *N*-glucoside, *N*-malonylglucoside and *O*-malonylglucoside conjugates of the parent compound.

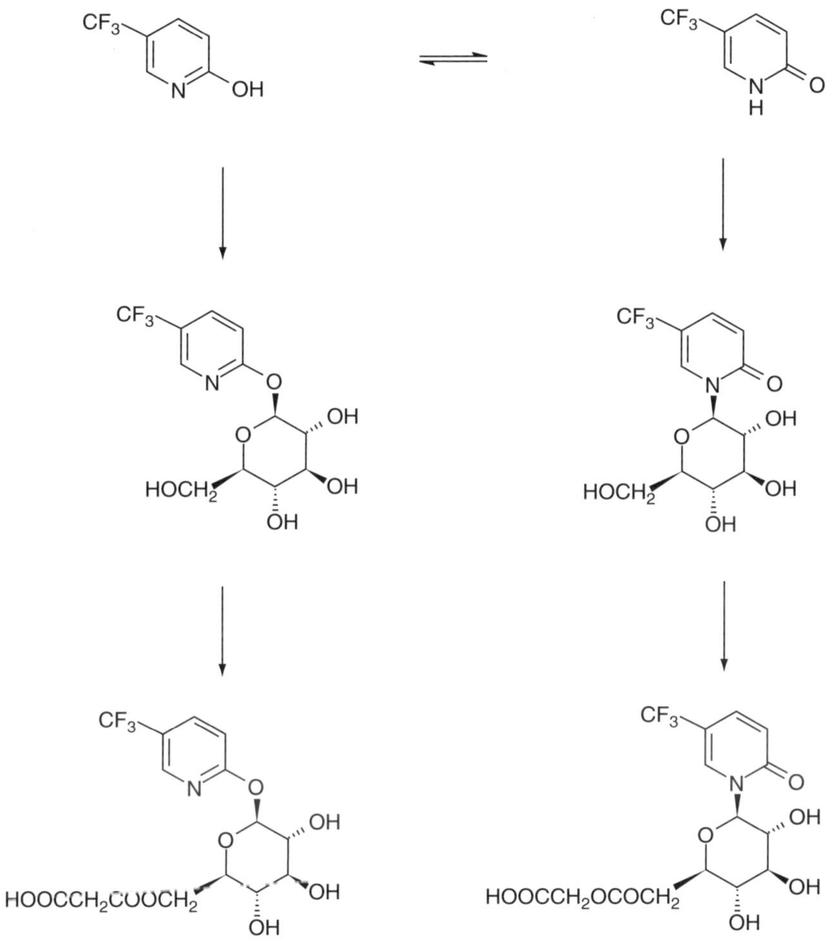

Figure 3.8 Possible metabolites for 2-hydroxy-5-trifluoromthylpyridine in maize plants

3.11 SEPARATION OF LIPOPROTEINS AND THEIR CHARACTERISATION USING HPLC–NMR

Disorders in lipoprotein metabolism are critical in the etiology of several disease states, such as coronary heart disease and atherosclerosis. Thus, there is considerable interest in the development of novel methods for the analysis of lipoprotein complexes. A simple chromatographic method for the separation of high-density lipoprotein (HDL), low-density lipoprotein (LDL) and very low-density lipoprotein (VLDL) from intact serum or plasma has been reported recently [65]. The separation was achieved by using an hydroxyapatite column and elution with pH 7.4 phosphate buffer with 100 μl injections of whole

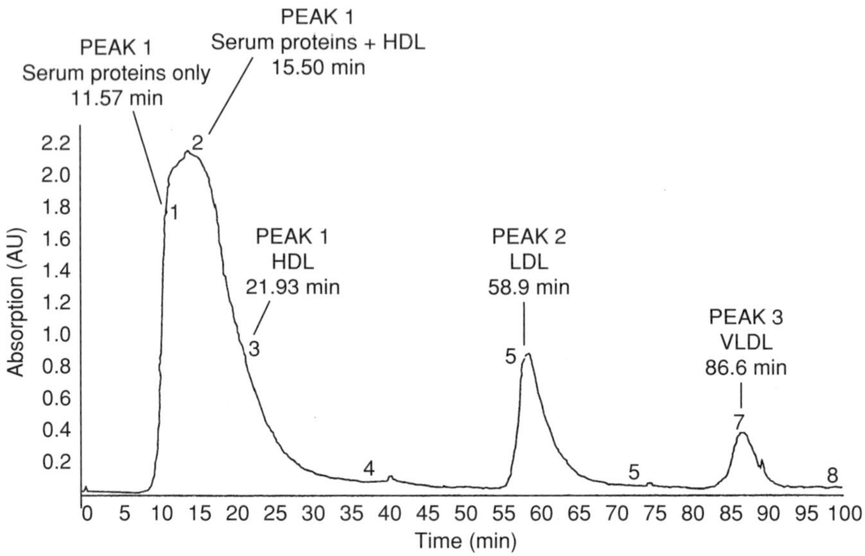

Figure 3.9 UV-detected (280 nm) chromatogram of lipoproteins separated from whole blood serum. The points indicated as '1' to '8', show where the flow was stopped so that ^{1}H NMR spectra could be acquired

plasma. A typical UV-detected chromatogram is shown in Figure 3.9. Coelution of HDL with plasma proteins such as albumin occurred and this clearly limits quantitation of that species by HPLC peak integration. It was also shown, for the first time, that directly coupled HPLC–^{1}H NMR spectroscopy could be used to confirm the identification of the three major lipoproteins. The full chromatographic run-time was 90 min, with stop-flow 600 MHz NMR spectra of each lipoprotein being collected using 128 scans, in 7 min. The ^{1}H NMR chemical shifts of the lipid signals were identical to those obtained from conventional ^{1}H NMR spectra of freshly prepared lipoprotein standards, as shown in Figure 3.10, thus confirming that the lipoproteins were not degraded by the HPLC separation and that their gross supramolecular organisation was intact.

3.12 SUPERHEATED-WATER HPLC–NMR AND HPLC–NMR-MS STUDIES ON PHARMACEUTICALS

As alluded to earlier, even though the major problems associated with the need for suppression of the signals for the solvents used in HPLC–NMR have been addressed to the extent that the techniques can now be routinely employed, avoidance of the problem still remains an area of interest. In recent years, there has been a resurgence of interest in the use of hot and even superheated water

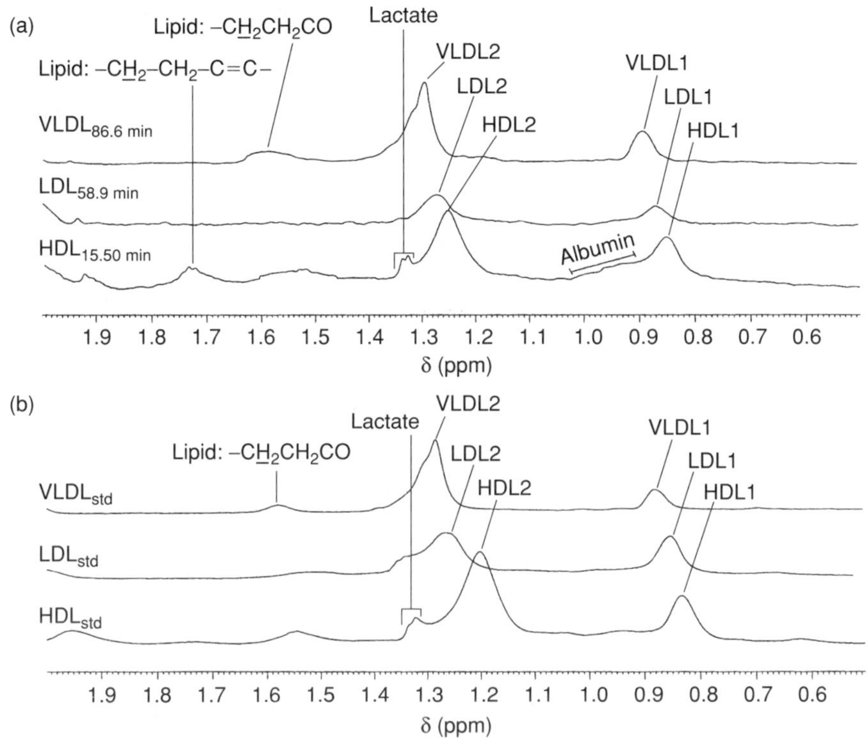

Figure 3.10 Comparison of lipoprotein standards and HPLC-resolved lipoproteins. (a) Partial 600 MHz HPLC–^{1}H NMR spectra (δ 0.5–2.0) of the HDL, LDL and VLDL fractions separated by using directly coupled HPLC–NMR spectroscopy at 15.5, 58.9 and 86.6 min, respectively. (b) Partial 600 MHz ^{1}H NMR spectra (δ 0.5–2.0) of standard (std) HDL, LDL and VLDL. Abbreviations: VLDL1, LDL1 and HDL1 indicate the resonances from the terminal methyl groups of the mobile fatty acid chains bound in the various lipoproteins; VLDL2, LDL2 and HDL2 indicate resonances from the methylene groups of the mobile fatty acid chains

(i.e. water used as the solvent at temperatures greater than 100 °C). This has certain advantages as it means that reversed-phase chromatography can be performed on moderately polar molecules with the elimination or, at least a significant reduction in, the level of any organic modifier. This not only reduces the cost, if expensive deuterated solvents would otherwise have to be used, but also simplifies the solvent suppression that has to be performed and increases the amount of spectral information that can be obtained. Such chromatography has, however, some potential drawbacks. In particular, the range of column types that can be used is significantly reduced due to the generally poor high-temperature stability of conventional silica-based phases. Thus, the lifetime of a typical C18-bonded silica column is of the order of a few hours or days when water (or D_2O) is used as an eluent at temperatures in excess of 100 °C. This means that, until recently, column choice was limited to phases such as organic polymers, graphitised carbon or zirconia-based materials. Recent experience with such materials is that their overall chromatographic performance is often not as good as their silica-based alternatives. However, recently a range of more stable silica-based materials has become available which seem to provide acceptable separations coupled with a reasonable column lifetime at elevated temperatures. In addition, the analyte under study has to be stable under the conditions employed for the separation and some are not suitable. For example, acetylsalicylic acid (aspirin) is rapidly hydrolysed to salicylic acid [66].

The first example of HPLC–NMR using superheated D_2O as the sole eluent was reported in 1998 [67]. The separation was of a simple mixture of barbiturates on a 5 μm polystyrene–divinylbenzene column at an oven temperature of 200 °C and a flow rate of 1 ml min^{-1}. A mixture of the three analytes, barbitone, amylobarbitone and heptabarbitone, each at a level of 100 μg, were injected on to the column and stop-flow ^{1}H NMR spectra were obtained, as shown in Figure 3.11. These spectra illustrate well the lack of interference from organic solvents. This preliminary study was followed by an HPLC–NMR–MS investigation of the use of superheated water for the chromatographic separation of a mixture of caffeine, phenacetin and paracetamol on a C18 bonded column, and salicylamide on a polystyrene–divinylbenzene column [66]. In the case of salicylamide, the column was operated at 180 °C with ca. 35 bar backpressure. Under these conditions, the compound eluted at 4.7 min and spectroscopy was performed under stop-flow conditions. This enables both 1D and 2D ^{1}H NMR spectra to be obtained without difficulty. In the same study, positive electrospray mass spectra were also obtained by diverting a portion of the flow (ca. 5 %), just prior to entry into the NMR flow probe, to a triple quadrupole mass spectrometer. The mixture of paracetamol, caffeine and phenacetin was separated using a temperature gradient from 80 to 130 °C at 8°C min^{-1}, resulting in retention times of 0.94, 3.06 and 4.58 min, respectively with stop-flow spectra obtained for each.

Most recently, a further study has been performed using superheated-water HPLC with NMR and MS to analyse a mixture of sulphonamides [68]. The chromatography was performed as before with D_2O-phosphate buffer (pD 3.0) as eluent. A temperature gradient from 160 to 200 °C at 2°C min^{-1} was employed. A mixture of four sulphonamides, i.e. sulacetamide, sulphadiazine, sulfamerazine and sulfamethazine, was separated in this system with UV, NMR and MS detection. It rapidly became clear from a study of the spectroscopic data that while sulfacetamide and sulfadiazine gave the expected NMR and mass spectra, those for sulfamerazine and sulfamethazine did not. These compounds gave spectra that were 3 and 6 mass units higher than expected,

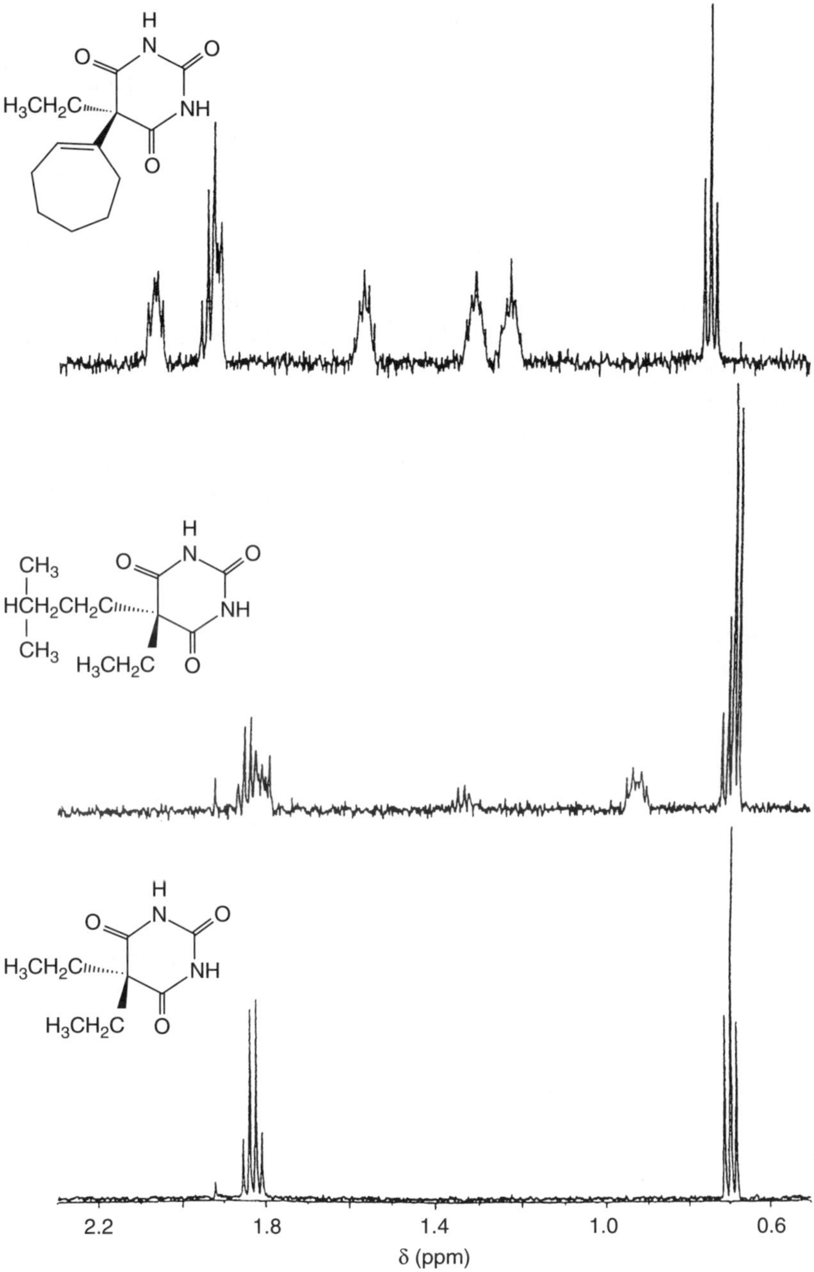

Figure 3.11 Separation of a mixture of three barbiturates using HPLC–NMR with superheated D_2O as eluent, showing the ^1H NMR spectra of the three separated substances

respectively, while the NMR spectra showed a complete absence of the expected 4- and 4,6-methyl groups on the pyrimidine ring. This strongly suggested that, during chromatography, the protons on these methyl groups had been exchanged for deuterium atoms as a result of the high temperature. Similarly, when sulfisomidine (the 2,4-dimethyl analogue) was chromatographed under these conditions there was almost complete conversion to the deuterated material. It is interesting to speculate how long this selective and efficient deuteration would have gone unnoticed without the use of NMR and MS for detection.

These investigations of the use of superheated water with NMR spectroscopy are still at an early stage and have not yet been applied to real problems. Many questions remain to be answered concerning the suitability of this chromatography for thermally labile compounds, and with the current stationary phases available the technique is probably limited to moderately polar compounds. However, the technique is readily implemented and may in time prove to be a useful addition to the armoury of HPLC–NMR methods.

3.13 APPLICATION OF HYPERNATION TO A MIXTURE OF NON-STEROIDAL ANTI-INFLAMMATORY DRUGS

As discussed above, as well as the combination of HPLC–NMR with MS there have been a number of alternative/additional combinations of spectroscopic techniques described. In one of these, isocratic reversed-phase HPLC with a D_2O–acetonitrile (1:1) solvent system at acidic pH (1 % deuterated formic acid) was coupled to UV–DAD, FT-IR, NMR and TOF-MS, thus enabling a very complete set of spectroscopic data to be obtained for the four components of a mixture of non-steroidal anti-inflammatory drugs [23]. This mixture was comprised of naproxen, flurbiprofen, indomethacin and ibuprofen and a mixture comprising about 2 mg of each component dissolved in D_2O was injected on to the column. Despite the high column loading, excellent chromatographic peak shape was observed for all of the analytes. The stop-flow mode was used to obtain ^1H NMR spectra. In addition to the spectral data obtained (UV, IR, ^1H NMR and MS), the use of TOF-MS enabled accurate mass data to be obtained, so allowing the atomic compositions to be calculated. A example of the type of data afforded by this instrumental arrangement is shown in Figure 3.12, which shows the results obtained for naproxen.

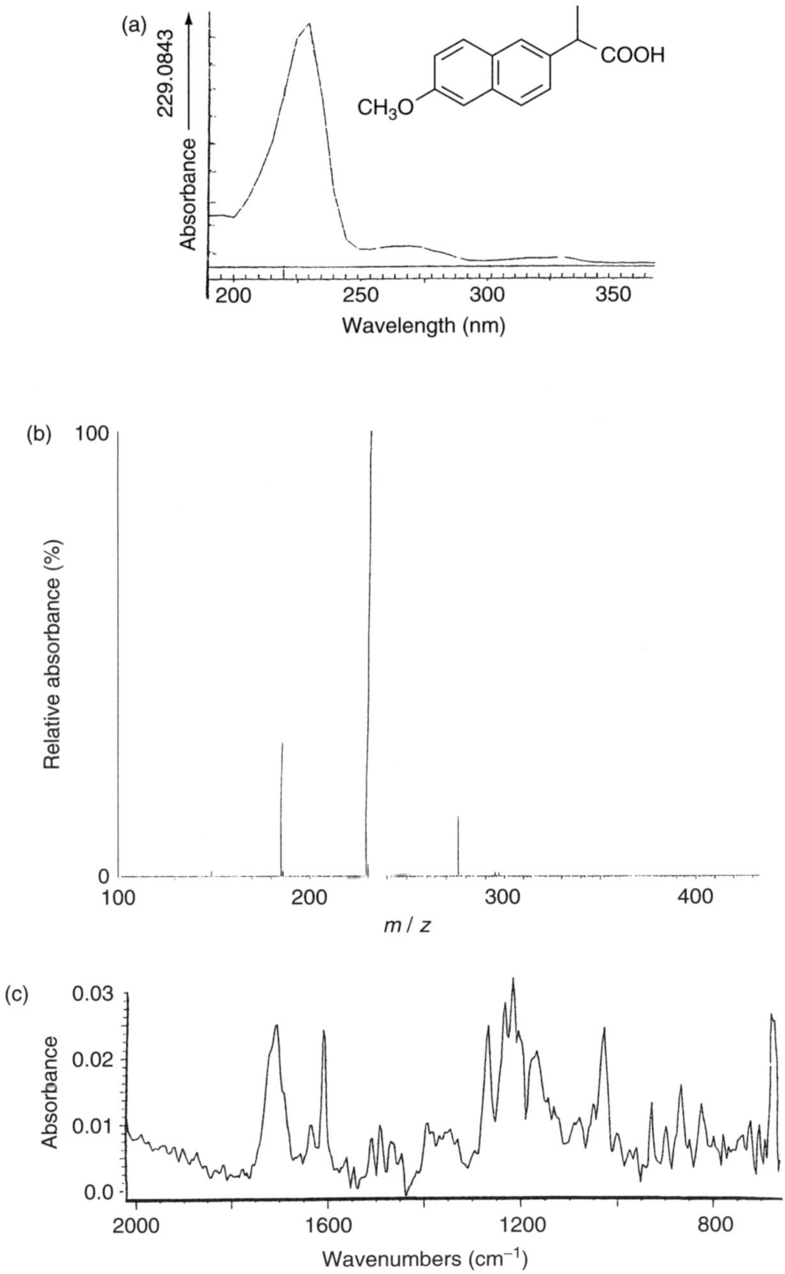

[**Figure 3.12** *cont.*]

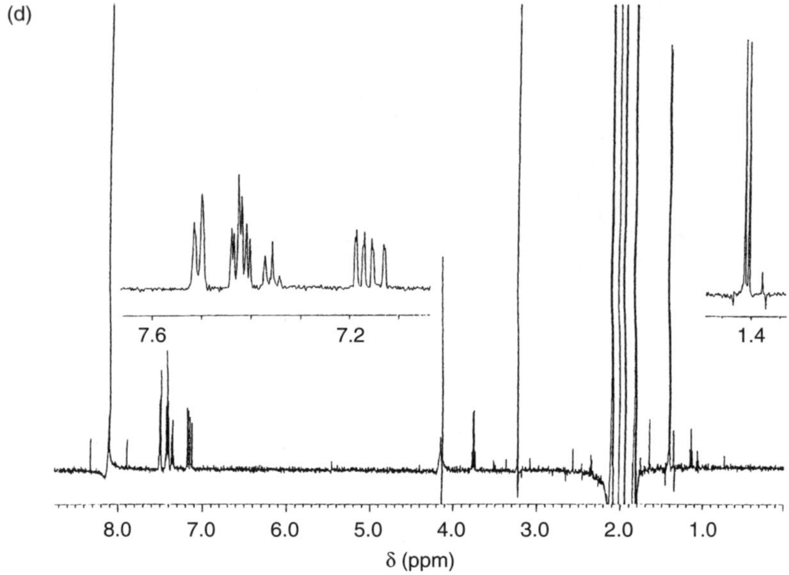

Figure 3.12 A summary of the data obtained for naproxen using a combined HPLC–UVDAD–FTIR–NMR–TOFMS approach: (a) – UV spectrum; (b) mass spectrum; (c) IR spectrum; (d) ^{1}H NMR spectrum

3.14 CONCLUDING REMARKS

HPLC–UV–NMR can now be considered to be a routine analytical technique for pharmaceutical mixture analysis and for many studies in the biomedical field. HPLC–UV–NMR–MS is becoming more routine with a considerable number of systems now installed worldwide, but the chromatographic solvent systems are limited to those compatible with both NMR spectroscopy and mass spectrometry. The increased use of HPLC–UV–IR–NMR–MS is possible, but it is unlikely to become widespread, and the solvent problems are more complex. The future holds the promise of new technical advances to improve efficiency, and to enhance routine operation. These approaches include the use of small-scale separations, such as capillary electrochromatography, greater automation, and higher sensitivity and lower NMR detection limits through the use of NMR detectors cooled to cryogenic temperatures.

REFERENCES

1. Lindon, J. C., Nicholson, J. K. and Wilson, I. D., *Prog. NMR Spectrosc.*, 1996, **29**, 1.
2. Albert, K., *J. Chromatogr., A*, 1995, **703**, 123.

3. Logan, T. M., Murali, M., Wang, G. S. and Jolivet, C., *Magn. Reson. Chem.*, 1999, **37**, 512.
4. Ludlow, M., Louden, D., Handley, A., Taylor, S., Wright, B. and Wilson, I. D., *J. Chromatogr., A*, 1999, **857**, 89.
5. Smallcombe, S. H., Patt, S. L. and Keifer, P. A., *J. Magn. Reson.*, 1995, **117**, 295.
6. Liu, M., Mao, X., Ye, C., Huang, H., Nicholson, J. K. and Lindon, J. C., *J. Magn. Reson.*, 1998, **132**, 125.
7. Smith, R. M., Chienthavorn, O., Wilson, I. D. and Wright, B., *Anal. Commun.*, 1998, **35**, 261.
8. Smith, R. M., Chienthavorn, O., Wilson, I. D., Wright, B. and Taylor, S. D., *Anal. Chem.*, 1999, **71**, 4493.
9. Smith, R. M., Chienthavorn, O., Saha, S., Wilson, I. D., Wright, B. and Taylor, S. D., *J. Chromatogr., A*, 2000, **886**, 289.
10. Griffiths, L., *Anal. Chem.*, 1995, **67**, 4091.
11. Griffiths, L., *Magn. Reson. Chem.*, 1997, **35**, 257.
12. Sweatman, B. C., Farrant, R. D., Sanderson, P. N., Philippe, I., Salman, S. R., Nicholson, J. K. and Lindon, J. C., *J. Magn. Reson. Anal.*, 1995, **1**, 9.
13. Wu, N., Webb, L., Peck, T. L. and Sweedler, J. V., *Anal. Chem.*, 1995, **67**, 3101.
14. Behnke, B., Schlotterbeck, G., Tallarek, U., Strohschein, S., Tseng, L.-H., Keller, T., Albert, K. and Bayer, E., *Anal. Chem.*, 1996, **68**, 1110.
15. Taylor, S., Wright, B., Clayton, E. and Wilson, I. D., *Rapid Commun. Mass Spectrom.*, 1998, **12**, 1732.
16. Nicholson, J. K., Lindon, J. C., Scarfe, G., Wilson, I. D. Abou-Shakra, F., Castro-Perez, J., Eaton, A. and Preece, S., *Analyst*, 2000, **125**, 235.
17. Pullen, F. S., Swanson, A. G., Newman, M. J., and Richards, D. S., *Rapid Commun. Mass Spectrom.*, 1995, **9**, 1003.
18. Shockcor, J. P., Unger, S. E., Wilson, I. D., Foxall, P. J. D., Nicholson, J. K. and Lindon, J. C., *Anal. Chem.*, 1996, **68**, 4431.
19. Burton, K. I., Everett, J. R., Newman, M. J., Pullen, F. S., Richards, D. S. and Swanson, A. G., *J. Pharm. Biomed. Anal.*, 1997, **15**, 1903.
20. Holt, R. M., Newman, M. J., Pullen, F. S., Richards, D. S. and Swanson, A. G. *J. Mass Spectrom.*, 1997, **32**, 64.
21. Ludlow, M., Louden, D., Handley, A., Taylor, S. and Wright, B., *Anal. Commun.*, 1999, **36**, 85.
22. Ludlow, M., Louden, D., Handley, A., Taylor, S., Wright, B. and Wilson, I. D., *J. Chromatogr., A.*, 1999, **857**, 99.
23. Louden, D., Handley, A., Taylor, S., Lenz, E., Miller, S., Wilson, I. D. and Sage, A., *Anal. Chem.*, 2000, **72**, 3992.
24. Lindon, J. C., Farrant, R. D., Sanderson, P. N., Doyle, P. M., Gough, S. L., Spraul, M. and Hofmann, M., *Magn. Reson. Chem.*, 1995, **33**, 857.
25. Chin, J., Fell, J. B., Jarosinski, M., Shapiro, M. J. and Wareing, J. R., *J. Org. Chem.*, 1998, **63**, 386.
26. Keifer, P. A., Smallcombe, S. H., Williams, E. H., Salomon, K. E., Mendez, G., Belletire, J. L. and Moore, C. D., *J. Comb. Chem.*, 2000, **2**, 151.
27. Impurities in New Drug Substances, International Conference on Harmonisation of Technical Requirements for Registration of Pharmaceuticals for Human Use. Recommended for Adoption. 7 Feb. 2002. ICH Steering Committee, Geneva, Switzerland.
28. Roberts, J. K. and Smith, R. J. *J. Chromatogr., A.*, 1994, **677**, 385.
29. Peng, S. X., Borah, B., Dobson, R. L. M., Liu, Y. D. and Pikul, S., *J. Pharm. Biomed. Anal.*, 1999, **20**, 75.
30. McCrossen, S. D., Bryant, D. K., Cook, B. R. and Richards, J. J., *J. Pharm. Biomed. Anal.*, 1998, **17**, 455.

31. Mistry, N., Ismail, I. M., Smith, M. S., Nicholson, J. K. and Lindon, J. C., *J. Pharm. Biomed. Anal.*, 1997, **16**, 697.
32. Potts, B. C. M., Albizati, K. F., Johnson, M. O. and James, J. P., *Magn. Reson. Chem.*, 1999, **37**, 393.
33. Aigbirhio, F. I., Carr, R. M., Pike, V. W., Steel, C. J. and Sutherland, D. R., *J. Labelled Compd. Radiopharm.*, 1997, **39**, 567.
34. Johnson, S., Morgan, E. D., Wilson, I. D., Spraul, M. and Hofmann, M., *J. Chem. Soc., Perkin Trans.*, 1, 1499 (1994).
35. Preiss, A., Sänger, U., Karfich, N., Levsen, K. and Mügge, C., *Anal. Chem.*, 2000, **72**, 992.
36. Mistry, N., Roberts, A. D., Tranter, G. E., Francis, P., Barylski, I., Ismail, I. M., Nicholson, J. K. and Lindon, J. C., *Anal. Chem.*, 1999, **71**, 2838.
37. Lindon, J. C. and Ferrige, A. G. *Tetrahedron*, 1980, **36**, 2157.
38. El-Sayad, H. A., Swaringen, R. A., Yeowell, D. A., Crouch, R. C., Hurlbert, S., Miller, R. W. and McPhail, A. T., *J. Chem. Soc., Perkin* Trans., 1, 1982, 2067.
39. Blessington, B., Beiraghi, A., Lo, T. W., Drake, A. and Jonas, G., *Chirality*, 1992, **4**, 227.
40. Wolfender, J.-L., Ndjoko, K. Hostettmann, K., *Curr. Org. Chem.*, 1998, **2**, 575.
41. Hostettmann, K., Potterat, O. and Wolfender, J.-L., *Chimia*, 1998, **52**, 10
42. Bringmann, G., Messe, K., Wohlfarth, M., Kraus, J., Dumbuya, K. and Rückert, M., *Anal. Chem.*, 1999, **71**, 2678.
43. Hansen, S. H., Jensen, A. G., Cornett, C., Bjornsdottir, I., Taylor, S., Wright, B. and Wilson, I. D., *Anal. Chem.*, 1999, **71**, 5235.
44. Abel, C. B. L., Lindon, J. C., Noble, D., Rudd, B. A. M., Sidebottom, P. J. and Nicholson, J. K., *Anal. Biochem.*, 1999, **270**, 220.
45. Albert, K., Schlotterbeck, G., Braumann, U., Händel, H., Spraul M. and Krack, G., *Angew. Chem. Int. Ed. Engl.*, 1995, **34**, 1014.
46. Stroschein, S., Rentel, C., Lacker, T., Bayer, E. Albert, K., *Anal. Chem.*, 1999, **71**, 1780.
47. Smith, P. C., Mcdonagh, A. F. and Benet, L. Z., *J. Clin. Invest.*, 1986, **77**, 934.
48. Williams, A. M., Worall, S., Jersey, J. and Dickinson, R. G., *Biochem. Pharmacol.*, 1992, **43**, 745.
49. Sidelmann, U. G., Gavaghan, C., Carless, H. A. J., Farrant, R. D., Lindon, J. C., Wilson, I. D. and Nicholson, J. K., *Anal. Chem.*, 1995, **67**, 3401.
50. Sidelmann, U. G., Hansen, S. H., Gavaghan, C., Carless, H. A. J., Lindon, J. C., Farrant, R. D., Wilson, I. D. and Nicholson, J. K., *Anal. Chem.*, 1996, **68**, 2564.
51. Sidelmann, U. G., Gavaghan, C., Carless, H. A. J., Spraul, M., Hofmann, M., Lindon, J. C., Wilson, I. D. and Nicholson, J. K., *Anal. Chem.*, 1995, **67**, 4441.
52. Sidelmann, U. G., Nicholls, A. W., Meadows, P., Gilbert, J., Lindon, J. C., Wilson, I. D. and Nicholson, J. K., *J. Chromatogr., A.*, 1996, **728**, 377.
53. Sidelmann, U. G., Lenz, E. M., Sanderson, P. N., Hofmann, M., Spraul, M., Lindon, J. C., Wilson, I. D. and Nicholson, J. K., *Anal. Chem.*, 1996, **68**, 106.
54. Lenz, E. M., Greatbanks, D., Wilson, I. D., Spraul, M., Hofmann, M., Lindon, J. C., Troke, J. and Nicholson, J. K., *Anal. Chem.*, 1996, **68**, 2832.
55. Corcoran, O., Mortensen, R. W., Hansen, S. H., Troke, J. and Nicholson, J. K., *Chem. Res. Toxicol.*, 2001, **14**, 1363.
56. Mortensen, R. W., Corcoran, O., Cornett, C., Sidelmann, U. G., Troke, J., Lindon, J. C., Nicholson, J. K. and Hansen, S. H., *J. Pharm. Biomed. Anal.*, 2001, **24**, 477.
57. Mortensen, R. W., Corcoran, O., Cornett, C., Sidelmann, U. G., Lindon, J. C., Nicholson, J. K. and Hansen, S. H., *Drug Metab. Disp.*, 2001, **29**, 375.
58. Nicholls, A. W., Caddick, S. T., Wilson, I. D., Farrant, R. D., Lindon, J. C. and Nicholson, J. K., *Biochem. Pharmacol.*, 1995, **49**, 1155.

59. Nicholls, A. W., Farrant, R. D., Shockcor, J. P., Unger, S. E., Wilson, I. D., Lindon, J. C. and Nicholson, J. K., *J. Pharm. Biomed. Anal.*, 15 (1997) 901.
60. Nicholls, A. W., Lindon, J. C., Farrant, R. D., Shockcor, J. P., Wilson, I. D. and Nicholson, J. K., *J. Pharm. Biomed. Anal.*, 1999, **20**, 865.
61. Scarfe, G. B., Lindon, J. C., Nicholson, J. K., Martin, P., Wright, B., Taylor, S., Lenz, E. and Wilson, I. D., *Xenobiotica*, 2000, **30**, 171.
62. Chen, W., Shockcor, J. P., Tonge, R., Hunter, A., Gartner, C. and Nelson, S. D., *Biochemistry*, 1999, **38**, 8159.
63. Bailey, N. J. C., Stanley, P. D., Hadfield, S. T., Lindon, J. C. and Nicholson, J. K., *Rapid Commun. Mass Spectrom.*, 2000, **14**, 679.
64. Bailey, N. J. C., Cooper, P., Hadfield, S. T., Lenz, E. M., Lindon, J. C., Nicholson, J. K., Stanley, P. D., Wilson, I. D., Wright, B. and Taylor, S. D., *J. Agric. Food Chem.*, 2000, **48**, 42.
65. Daykin, C. A., Corcoran, O., Hansen, S. H., Bjornsdottir, I., Cornett, C., Connor, S. C., Lindon, J. C. and Nicholson, J. K., *Anal. Chem.*, 2001, **73**, 1084.
66. Smith, R. M., Chienthavorn, O., Wilson, I. D., Wright, B. and Taylor, S. D., *Anal. Chem.*, 1999, **71**, 4493.
67. Smith, R. M., Chienthavorn, O., Wilson, I. D. and Wright, B., *Anal. Commun.*, 1998, **35**, 261.
68. Smith, R. M., Chienthavorn, O., Saha, S., Wilson, I. D., Wright, B. and Taylor, S. D., *J. Chromatogr., A*, 2000, **886**, 2890.

4 Application of On-Line LC–NMR and Related Techniques to Drug Metabolism Studies

JOHN P. SHOCKCOR

Department of Biological Chemistry, Faculty of Medicine, Imperial College of Science, Technology and Medicine, University of London, London, UK

4.1 INTRODUCTION

NMR spectroscopy and mass spectrometry are the principal techniques for the profiling and characterization of xenobiotic metabolites. Typically, the study of xenobiotic metabolites began with profiling a biological fluid or an *in vitro* incubation broth using LC–MS, followed, if necessary, by isolation of the metabolites using HPLC for further characterization by NMR spectroscopy. Major technical improvements in console electronics and probe design, higher magnetic field strengths, multidimensional NMR pulse sequences and improved solvent suppression methods have allowed NMR to have an impact in this process at a much earlier stage by vastly improving sensitivity and lowering the absolute amount of sample needed for analysis. However, isolation of even the relatively small quantities (1–10 μg) of a metabolite needed for NMR analysis often represents the rate-determining step in the process of metabolite structure elucidation. The application of the hyphenated techniques LC–NMR and LC–NMR–MS has made it possible to eliminate the isolation step and provided a new set of powerful tools for the study of drug metabolism [1–33]. In the following sections, the basic techniques of LC–NMR and LC–NMR–MS will be discussed, along with a practical example of how these techniques can be applied to the study of drug metabolism.

4.2 LC–NMR TECHNIQUES

4.2.1 CONTINUOUS-FLOW LC–NMR

The simplest method of operation is continuous-flow detection. This mode of operation is generally only practical when using ^{1}H or ^{19}F NMR for detection

On-Line LC–NMR and Related Techniques
Edited by K. Albert.

unless enriched compounds are used. Continuous-flow LC–NMR is a series of one-dimensional (1D) spectra acquired for 16 to 32 transients into 2K to 8K data points. Total acquisition time for each transient is typically around 1 s. The data are multiplied by a line-broadening function of 1 to 3 Hz to improve the signal-to-noise ratio and zero-filled by a factor of two before Fourier transformation in the F2 domain only. This results in a contour plot of intensity with ^{1}H or ^{19}F NMR chemical shift on the horizontal axis and chromatographic retention time on the vertical axis. If continuous-flow detection is required during a solvent gradient elution, the nuclear magnetic resonance positions of the solvent peaks will shift as the solvent proportions change. For effective solvent suppression, it is therefore necessary to determine these solvent resonance frequencies as the chromatographic run proceeds. This is accomplished by measuring a single exploratory scan as soon as a chromatographic peak is detected in real time during the chromatographic run and then applying solvent suppression irradiation at these frequencies as the peak elutes. The data from a continuous-flow experiment performed on a sample of concentrated human urine after dosing with paracetamol (**1**) is shown in Figure 4.1. The data are plotted in a pseudo-two-dimensional (2D) format with the axes representing chemical shift in ppm and the retention time of the chromatographic run. Slices extracted from the 2D plot show the 1D spectra of the glucuronide (**2**) and sulfate (**3**) conjugates of paracetamol (see Scheme 4.1).

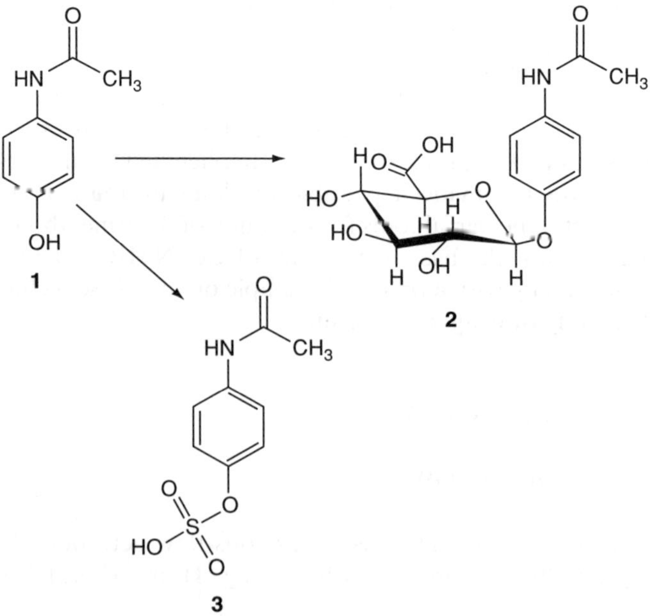

Scheme 4.1

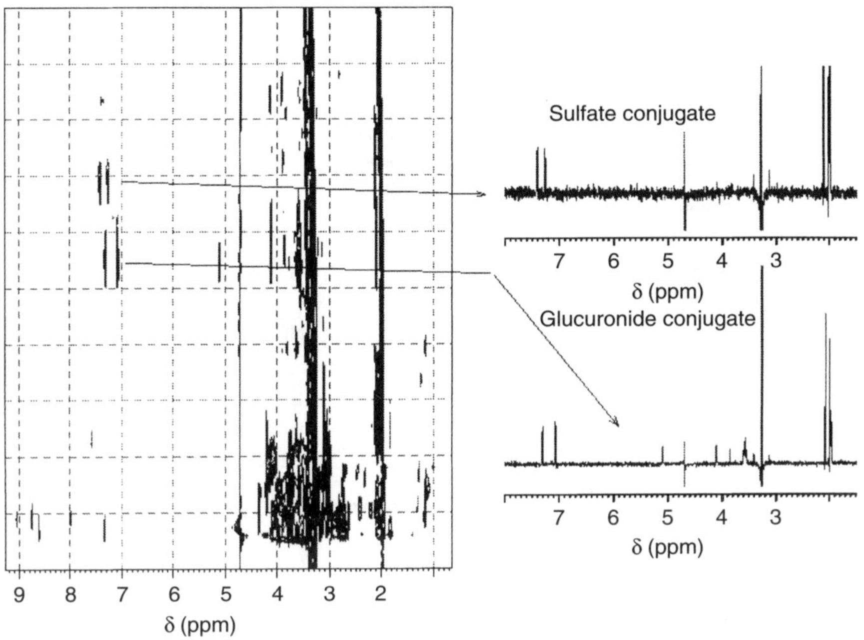

Figure 4.1 Pseudo-2D plot of continuous-flow ^{1}H LC–NMR data obtained on human urine after dosing with paracetamol (**1**). The resonances from the glucuronide (**2**) and sulfate (**3**) conjugate metabolites of paracetamol and their 1D slices are shown

The continuous-flow data for paracetamol was obtained with 16 scans per spectra which allowed detection of only the two major metabolites. Minor metabolites detected by mass spectrometry on this sample, such as the *N*-acetylcysteine conjugate, are below the NMR limit of detection in the continuous-flow mode. This lack of sensitivity severely limits the utility of continuous-flow ^{1}H NMR to the study of drug metabolites. However, continuous-flow ^{19}F can be a very powerful method for the study of the metabolism of fluorinated compounds. This is illustrated in the analysis of the metabolites of BW935U83 (**4**). [34]. The ^{1}H NMR spectrum of whole human urine from HIV-infected clinical trial volunteers following a single 1000 mg dose of BW935U83 is shown in Figure 4.2(b) and indicates the degree of difficulty in detecting drug metabolite resonances in the presence of the many endogenous compounds found in human urine. However, a remarkable degree of simplification is possible by measuring the ^{19}F NMR spectrum of the same urine sample (Figure 4.2(a)). Since there are no endogenous fluorine-containing compounds in the control urine, the observed resonances must be from drug-related substances. The two large peaks correspond to the parent drug at −178.10 ppm and its glucuronide conjugate (**5**), at −176.90 ppm, along with a second previously unidentified metabolite (**6**) at −191.25 ppm (see Scheme 4.2).

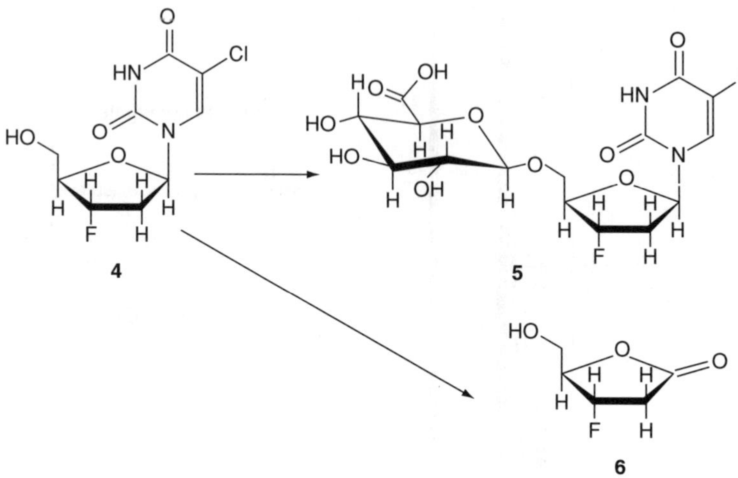

Scheme 4.2

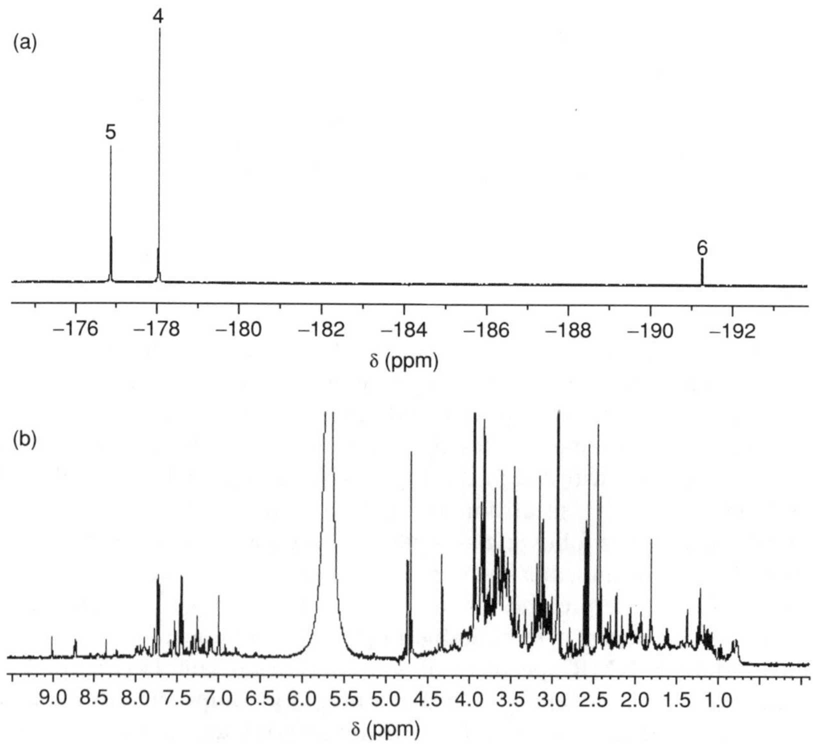

Figure 4.2 (a) The ^{19}F NMR spectrum of a sample of human urine collected from 0–4 h after dosing with 1000 mg BW935U83 and (b) the corresponding ^{1}H NMR spectrum of the same sample

The continuous-flow ^{19}F LC–NMR spectrum was measured on the urine sample to determine the retention times of the parent drug and its metabolites. The pseudo-two dimensional contour plot of the continuous-flow ^{19}F NMR spectra (Figure 4.3) shows responses for the three largest resonances and provides their retention times from the vertical axis. The retention times for BW935U83, its glucuronide conjugate and the unknown metabolite are 27.5, 12.2, and 3.5 min, respectively. Figure 4.4 shows the UV chromatogram obtained during this experiment. With this information, it was possible to obtain ^1H NMR and positive-ion electroscopy ionization (ESI) mass spectral data in a single LC–NMR–MS experiment which provided the data necessary to identify the unknown metabolite as 3-fluoro-ribolactone.

4.2.2 TIME-SLICE LC–NMR

Stopping the flow at short intervals over a chromatographic peak and collecting NMR data is referred to as 'time-slicing'. The time-slicing method may be useful if there is poor chromatographic separation, if the compounds under study have weak or no UV chromophores, or if the exact chromatographic retention time is unknown. It is also possible to time-slice through an entire

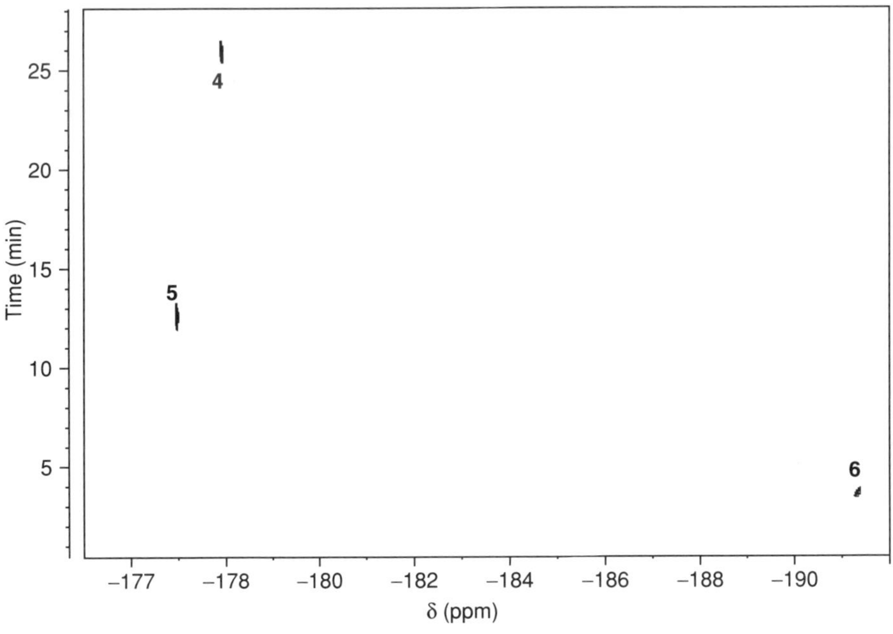

Figure 4.3 Pseudo-2D contour plot of the continuous-flow ^{19}F NMR data obtained on a 50 μl injection of human urine collected from 0–4 h after a 1000 mg dose of BW935U83

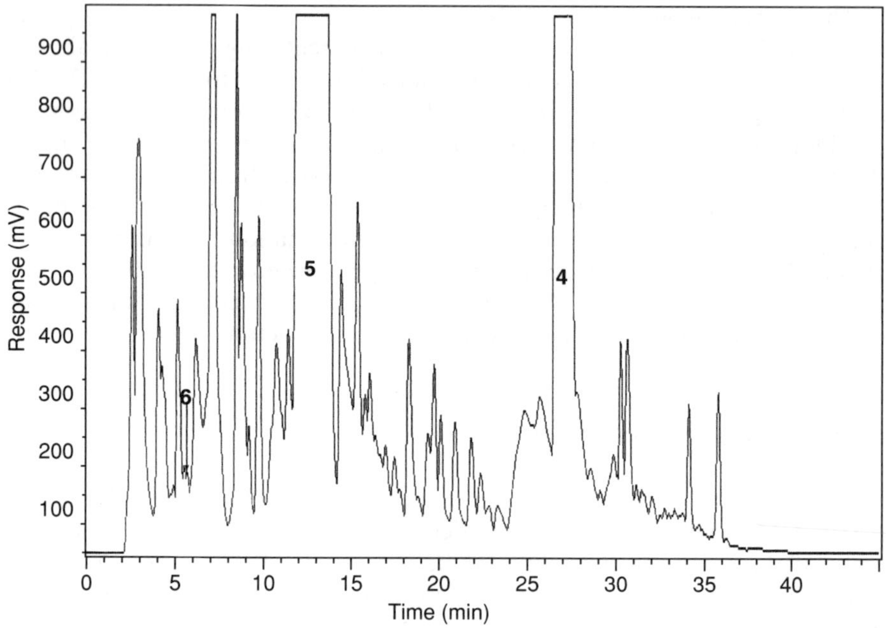

Figure 4.4 HPLC chromatogram obtained on a 50 μl injection of human urine collected from 0–4 h after a 1000 mg dose of BW935U83. The locations for BW935U83 (**4**), its glucuronide conjugate (**5**), and 3-fluoro-ribolactone (**6**) are indicated

chromatographic run, thus producing the equivalent of a continuous-flow experiment with a higher signal-to-noise ratio. The data from such a time-slicing experiment has been referred to as a *total NMR chromatogram* (tNMRc). Figure 4.5 shows the result of a time-slicing experiment on a sample of rat bile after dosing with an anti-viral compound. These spectra were obtained by stopping flow at 30 s intervals through a chromatographic run from 5.0 min into the run through 14.0 min. Each spectrum was acquired for 128 transients. The entire run was performed overnight under automation. Three of the thirteen metabolites structures elucidated from this rat bile sample were obtained directly from this experiment.

4.2.3 STOP-FLOW LC–NMR

If the retention times of the compounds to be separated are known, or if they can be detected by using UV (including diode arrays), radiochemical or fluorescence detectors, stop-flow LC–NMR becomes an option. Upon detection, the PC controlling the liquid chromatograph allows the pumps to continue running, moving the peak of interest into the NMR probe. Once the pumps have stopped, normal high-resolution NMR spectroscopy is possible. It could be

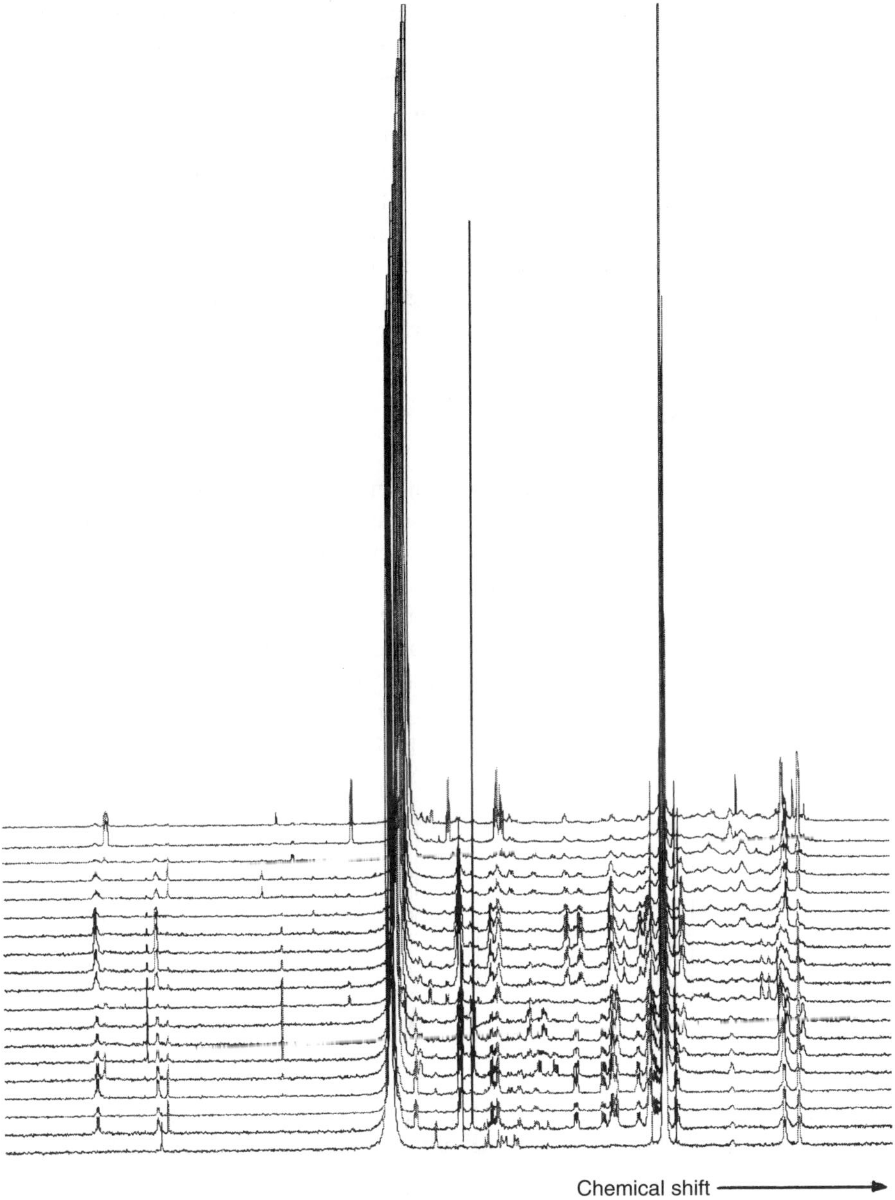

Figure 4.5 Typical ^{1}H LC–NMR data obtained from a time-slice experiment. These data were obtained by stopping flow at 30 s intervals through a chromatographic run from 5.0 min into the run through 14.0 min. Each spectrum was acquired for 128 transients

argued that the long length of capillary tubing connecting the LC system to the NMR probe causes significant loss of resolution in the separation. In fact, this is not a problem. The NMR detection cell volume is typically 60–120 μL, and this represents the limiting factor in the chromatographic resolution. The practicality of the stop-flow approach has been amply demonstrated and although several separate stops are often made in each chromatographic run, the quality of the resulting NMR spectra are such that good structural information can be obtained (Figure 4.6). Even long 2D experiments, which provide correlation between NMR resonances, based on mutual spin–spin coupling, such as COSY or TOCSY (Figure 4.7), and heteronuclear correlation studies, such as HMQC (Figure 4.8) or HSQC, can be performed.

4.2.4 LOOP–STORAGE

Most LC–NMR hardware offers the option of moving eluting fractions to capillary loops for storage and later off-line NMR study. This technique is often referred to as 'peak-picking'. When there are several fractions of interest in a chromatographic run it can be useful to move the first fraction in the NMR flow-probe without interrupting the chromatography and place the remaining fractions of interest in storage loops for subsequent analysis. Another advantage of loop-storage is that it avoids contamination of the flow-cell that often occurs in stop-flow mode when a high-concentration component elutes prior to the peak of interest. The high-concentration component will not be fully flushed from

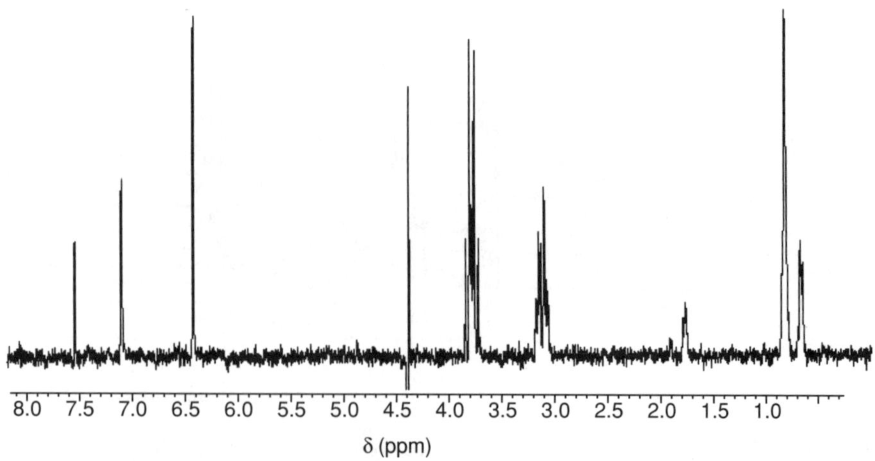

Figure 4.6 Typical stop-flow ^{1}H LC–NMR data obtained showing the cysteinylglycine adduct of a non-nucleoside reverse-transcriptase inhibitor from a sample of rat bile

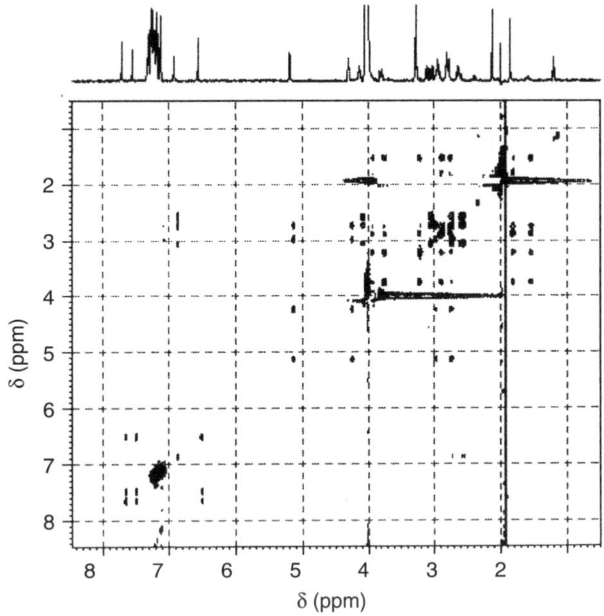

Figure 4.7 Typical 2D LC–NMR stop-flow data, showing a TOCSY spectrum obtained on the glucuronide conjugate of Naproxen from a sample of human urine

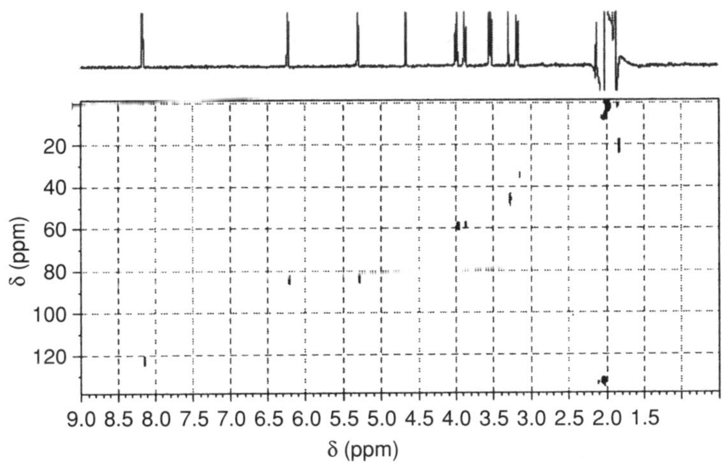

Figure 4.8 Typical 2D LC–NMR stop-flow data, showing HMQC data obtained on approximately 100 μg of a nucleoside reverse-transcriptase inhibitor from a sample of human urine

the NMR flow-cell, thus resulting in contamination of the peak of interest. In loop-storage mode, the peak of interest is directed to a clean storage-loop and the NMR flow-cell can be thoroughly washed prior to transfer. Typically, the loop temperature can be controlled to ensure that peaks do not decompose while awaiting transfer for analysis. This method is now the most commonly used technique for metabolism studies and is particularly useful when LC–NMR–MS is being employed.

4.2.5 LC–NMR–MS

The extension of an LC–NMR system to include mass spectrometry has been in application for several years [35,36]. By directly coupling a mass spectrometer to an LC–NMR system it is possible to obtain valuable mass spectral data. Configuring the system to have the sample reach the mass spectrometer before it reaches the NMR flow-cell enables the mass spectrometer to be employed as an experimental control device for analysing complex mixtures. Mass spectrometry is an ideal detector, provided that the molecules of interest are ionizable. It provides data rapidly and can thus yield valuable information on parent or daughter ion masses prior to initiating time-consuming NMR experiments. This synergy is not possible when the instruments are not directly coupled.

4.3 APPLICATION OF LC–NMR–MS TO DRUG METABOLISM: THE STRUCTURE ELUCIDATION OF RAT URINARY METABOLITES OF EFAVIRENZ BY LC–NMR–MS

An example of the utility of LC–NMR–MS is the study of the metabolism of the non-nucleoside HIV-1 reverse-transcriptase inhibitor (NNRTI) efavirenz (Sustiva®) (**7**, in Scheme 4.3 on page 104). The effective treatment of HIV infection and AIDS is still difficult, despite tremendous advances in our understanding of the pathogenesis of the disease and the arrival of potent drugs aimed at different, critical targets in the life cycle of the virus [37]. Efavirenz is a potent NNRTI that has been approved by regulatory agencies worldwide to treat AIDS. Extensive use of efavirenz has demonstrated a durable, long-lasting reduction in HIV RNA after once-a-day dosing in combination with other drugs such as AZT and 3-TC. Efavirenz is metabolized extensively by rats, cynomolgus monkeys and humans with the major metabolite being the 8-OH efavirenz glucuronide. LC–NMR–MS analyses of bile and urine samples from rats showed the presence of glutathione-derived adducts formed by enzymatic addition of glutathione across the triple bond of the acetylene moiety [38,39]. The formation of such conjugates has also been linked to the species-specific nephrotoxicity observed in rats [40].

4.3.1 EXPERIMENTAL

Sample Preparation

The sample for analysis was prepared by solid-phase extraction of 5 ml of 0–24 h urine obtained from rats after dosing with 800 mg/kg of efavirenz. The extract was dried and reconstituted with 100 ml of 80 % D_2O and 20 % acetonitrile-d_3.

Hyphenation

The control of the hyphenated system, comprised of an Agilent 1100 HPLC, a Bruker Avance 600 MHz NMR spectrometer and a Bruker Daltonics Esquire ion-trap mass spectrometer, was achieved by using HYSTAR software from Bruker Analytik. Peaks were selected for analysis using both UV and mass detection and were stored in the Bruker BPSU-36 loop storage unit prior to NMR analysis. A schematic diagram of the system is shown in Figure 4.9.

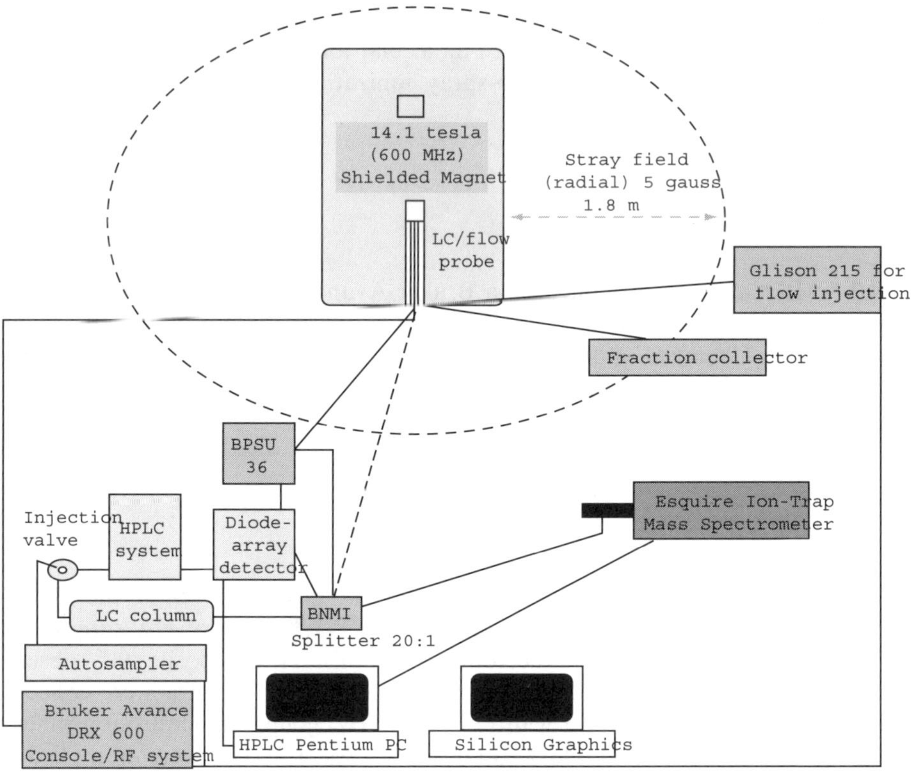

Figure 4.9 A schematic diagram of an LC–NMR–MS system

Liquid Chromatography

An LC system was used with a Bruker diode-array detector set at 254 nm. A 3.9 $\times$ 150 mm Waters Symmetry® C_{18} column was used. A gradient from 75 % D_2O and 25 % acetonitrile-d_3 to 50 % D_2O and 50 % acetonitrile-d_3 over 20 min at a flow rate of 0.8 ml/min was employed for separation. Both solvents contained 0.05 % trifluoroacetic acid (TFA).

Mass Spectrometry

The mass spectrometry portion of the analysis was carried out by coupling a Bruker Esquire ion-trap mass spectrometer to the LC–NMR system with a 20:1 splitter. The major portion of the flow was directed to the NMR system while the minor fraction went to the mass spectrometer. The system was plumbed such that the sample reached the mass spectrometer and the UV detector at the same time. In this configuration, it is possible to use the mass spectrometer as an intelligent detector, thus allowing stop-flow experiments to be initiated on the basis of observed molecular ions or daughter ion fragments. Data were acquired with electro-spray ionization (ESI) in the positive-ion mode.

NMR Spectroscopy

All NMR data were obtained on a Bruker Avance 600 MHz NMR spectrometer equipped with a 4 mm $^{1}H/^{19}F$ dual LC–NMR flow-probe, (cell volume of 120 μl). Chemical shifts were referenced to acetonitrile at δ 2.0 ppm for ^{1}H NMR, while ^{19}F NMR chemicals shifts were referenced to TFA at −78.5 ppm.

^{1}H *LC–NMR*

The ^{1}H LC–NMR spectra were obtained on peaks stored in the BPSU-36 storage loops. Data were acquired with WET [41] solvent suppression on the residual water and acetonitrile signals. A composite 90° observe pulse, $(\pi/2)_y$–$(\pi/2)_{-x}$–$(\pi/2)_{-y}$–$(\pi/2)_x$, was employed. Spectra were collected into 32K data points over a width of 12 019 Hz, giving an acquisition time of 1.36 s, with an additional relaxation delay of 1.5 s. The data were multiplied by a line-broadening function of 1 Hz to improve the signal-to-noise ratio and zero-filled by a factor of two before Fourier transformation.

^{19}F *Continuous-Flow LC–NMR*

Continuous-flow ^{19}F LC–NMR spectra were acquired for 16 transients using 60° pulses into 8192 data points over a spectral width of 11 364 Hz, giving an acquisition time of 0.36 s. A relaxation delay of 0.64 s was added to give a total acquisition time for each spectrum of 16 s. The data were multiplied by a line-broadening function of 3 Hz to improve the signal-to-noise ratio and zero-filled by a factor of two before Fourier transformation. The results are presented as a contour plot with ^{19}F NMR chemical shift on the horizontal axis and chromatographic retention time on the vertical axis.

TOCSY LC–NMR

TOCSY (^{1}H–^{1}H total correlation) LC–NMR spectra were measured using an acquisition time of 211 ms with a spectral width of 4854 Hz and the same saturation method as described above. The data were acquired using 2K complex points in the acquisition domain (F_2) with 256 increments in the orthogonal domain (F_1). Data were acquired in the phase-sensitive mode (using time proportional phase incrementation (TPPI)) with WET solvent suppression on the residual water and acetonitrile signals. The spin-lock period was 50 ms and the relaxation delay 1.0 s. The data were zero-filled to 1K points in F_1 and apodized using a cosine-squared function in F_1 and F_2 prior to the double Fourier transformation.

4.3.2 RESULTS

The initial experiment, a continuous-flow ^{19}F LC–NMR–MS run, takes advantage of the CF_3 group present in the parent drug. Since there were no endogenous fluorinated compounds in the control urine, responses in the spectrum must arise from metabolites of efavirenz. This experiment provided the retention times of the metabolites that were of sufficient concentration to be detected. The ^{19}F chemical shift and the mass of the metabolites were also obtained from this single experiment. Figure 4.10 shows the data as a pseudo-2D plot. The peaks of interest are labeled with their molecular weights. The M^+D masses which were observed in the continuous-flow ^{19}F LC–NMR–MS experiment and the M^+H masses are also shown. Since LC–NMR and LC–NMR–MS experiments are typically run with D_2O as one of the solvents, an M^+D ion is observed in positive-ion mode instead of an M^+H and exchangeable protons on a molecule are typically replaced with deuterium. Thus the ions observed will have a molecular weight higher than the M^+H ion. This might at first be thought of as a problem for structure elucidation. However, by obtaining

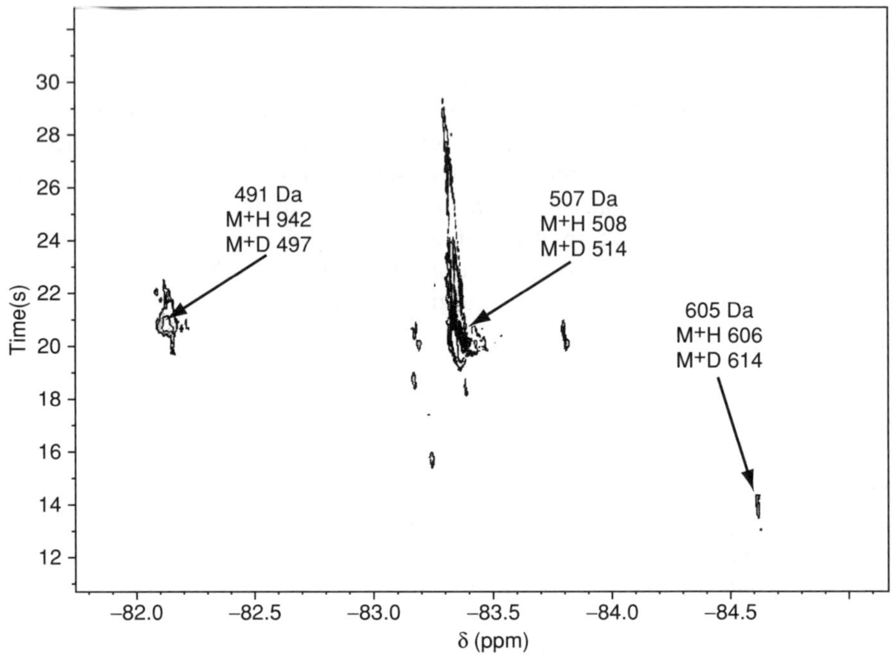

Figure 4.10 Continuous-flow ^{19}F LC–NMR–MS data displayed as a pseudo-2D plot with ^{19}F chemical shift on the horizontal axis and retention time on the vertical axis. The peaks of interest are labelled with the relevant molecular weights

additional LC–MS data in protonated solvents and comparing these data to that obtained in deuterated solvents, it is possible to obtain a count of exchangeable protons in the compound being analysed. This information can prove to be extemely valuable in the structure elucidation process.

The next phase of the experimental procedure was to collect in storage loops the peaks of interest as determined by the continuous-flow ^{19}F LC–NMR–MS experiment. Storage in the loops was triggered by the UV response and desired M^+D ion.

The 1D and TOCSY spectra obtained on the minor metabolite with retention time 14.2 min, M^+D 614 and ^{19}F chemical shift −84.6 ppm are shown in Figures 4.11 and 4.12, respectively. The observed loss of 80 Da from the parent ion in the mass spectrum is indicative of SO_3 loss. The absence of the methine proton from the cyclopropyl ring in the ^{1}H NMR spectrum and the downfield shift of the cycloproyl methylene resonances support hydroxylation at the cycloproyl methine position. The spin-system observed between 3.2 and 4.9 ppm and the presence of seven exchangeable protons, determined by comparison of the M^+H and M^+D masses, is consistent with the formation of a cysteinylglycine conjugate. However, the α-methine resonance of the cysteine could not be observed in the ^{1}H NMR spectrum. The assignment of the cysteine α-methine

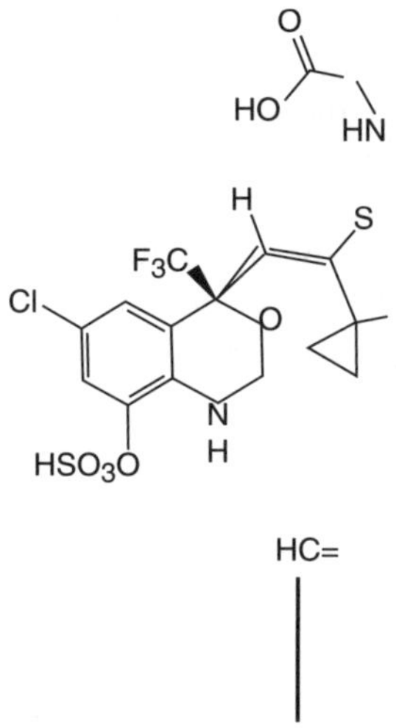

Figure 4.11 ^{1}H LC–NMR spectrum of the peak with retention time 14.2 min, M + D 614 and ^{19}F chemical shift −84.6 ppm. Assigned as structure 8

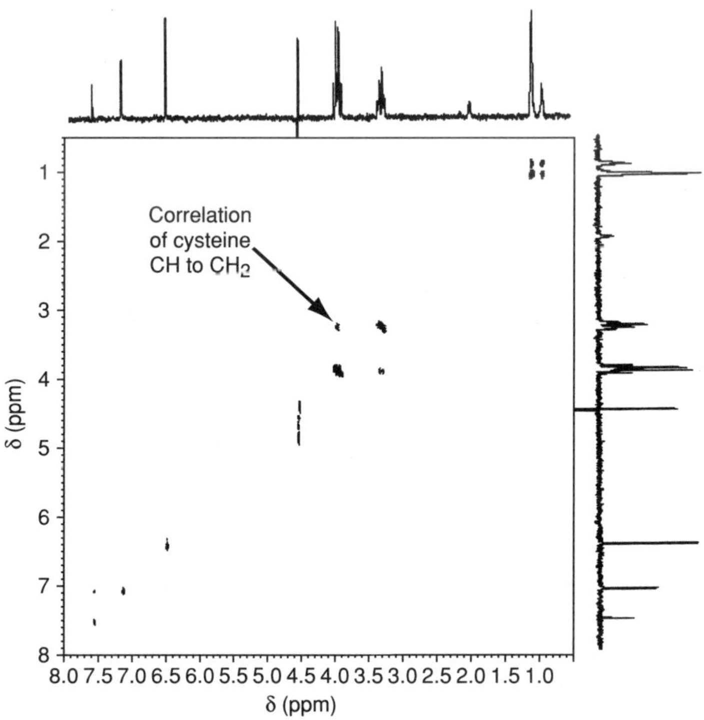

Figure 4.12 ^{1}H LC–NMR TOCSY spectrum of the peak with retention time 14.2 min, M + D 614 and ^{19}F chemical shift −84.6 ppm. Assigned as structure 8

is made from the TOCSY spectrum (see Figure 4.12), which clearly shows the correlation between the cysteine β-methylene protons centred at 3.25 ppm and the cysteine α-methine under the glycine methylene resonance at 4.9 ppm. The presence of an olefinic proton at 6.45 ppm indicates that the site of the initial glutathione conjugation occurred at the alkyne of the parent with formation of the cysteinylglycine occurring enzymatically via γ-glutamyltranspeptidase. These data are consistent with the formation of a hydroxylated cyclopropyl ring, 8-O-sulfate, cysteinylglycine di-conjugate (**8**) (see Scheme 4.3).

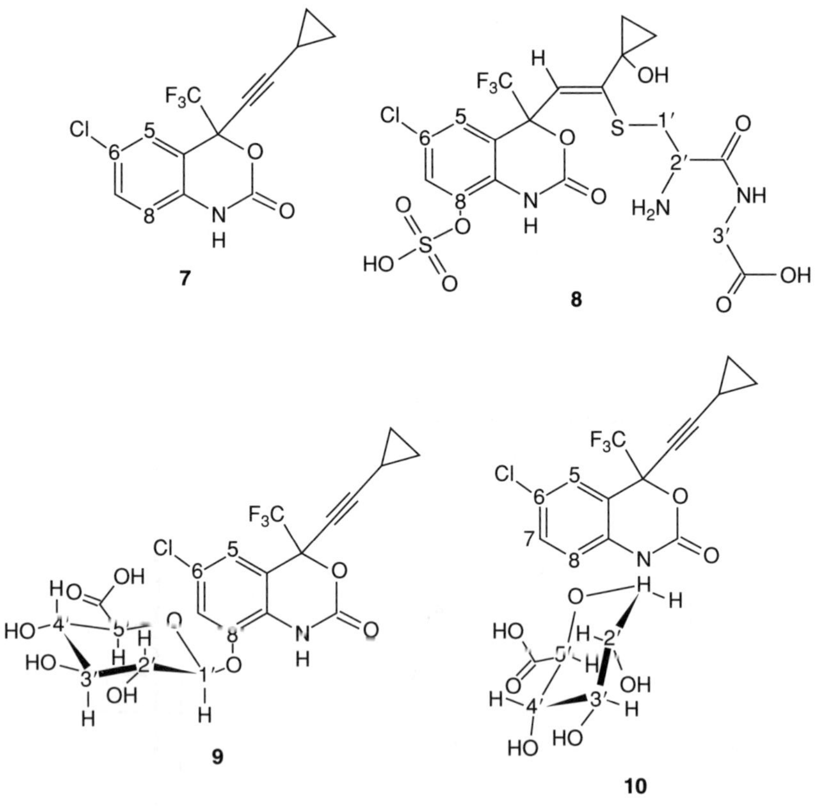

Scheme 4.3

The 1D and TOCSY spectra obtained on the major metabolite with retention time 20.5 min, M^+D 514 and ^{19}F chemical shift −83.3 ppm are shown in Figures 4.13 and 4.14. respectively. These spectra clearly show the presence of a glucuronide conjugate. The observed changes in the aromatic region of the spectrum and the presence of five exchangeable protons, determined by comparison of the M^+H and M^+D masses, are consistent with the 8-O-glucuronide conjugate of efavirenz (**9**)

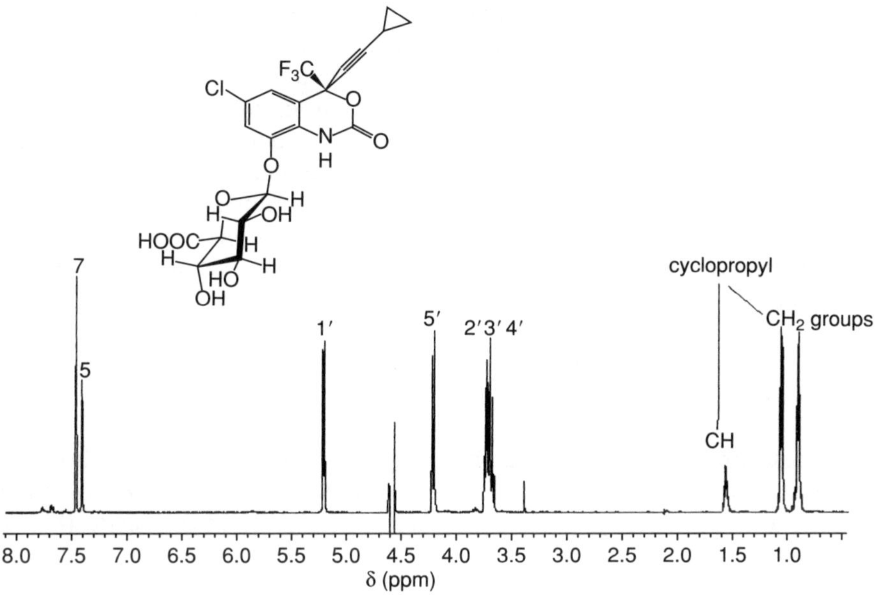

Figure 4.13 ^{1}H LC–NMR spectrum of the peak with retention time 20.5 min, M + D 514 and ^{19}F chemical shift −83.3 ppm. Assigned as structure 9

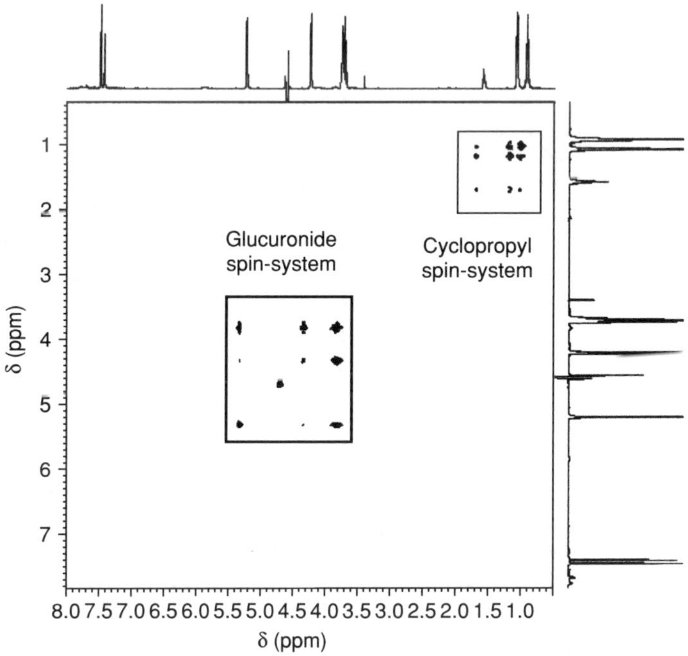

Figure 4.14 ^{1}H LC–NMR TOCSY spectrum of the peak with retention time 20.5 min, M + D 514 and ^{19}F chemical shift −83.3 ppm. Assigned as structure 9

Co-eluting with the major metabolite at 14.2 min is a component with mass 491 Da and ^{19}F chemical shift −82.1 ppm. Using the mass spectrometer as a detector, the loop storage was executed when the $M^{+}D$, 497 Da, was observed at its maximum intensity in the single-ion chromatogram. While the spectrum is dominated by the major metabolite, the 8-O-glucuronide conjugate of efavirenz, it is possible to assign several resonances to a second glucuronide conjugate, as shown on Figure 4.15. The assignment of this metabolite as the N-glucuronide conjugate (**10**) is consistent with the chemical shift of H1′ at 4.75 ppm, the observed molecular weight and the presence of four exchangeable protons, determined by comparison of the $M^{+}H$ and $M^{+}D$ masses.

4.4 CONCLUSIONS

These data provide unambiguous proof of structure for three metabolites of efavirenz, the major metabolite 8-O-glucuronide conjugate, (**9**), the novel hydroxylated cyclopropyl, 8-O-sulfate, cysteinylglycine di-conjugate (**8**), and

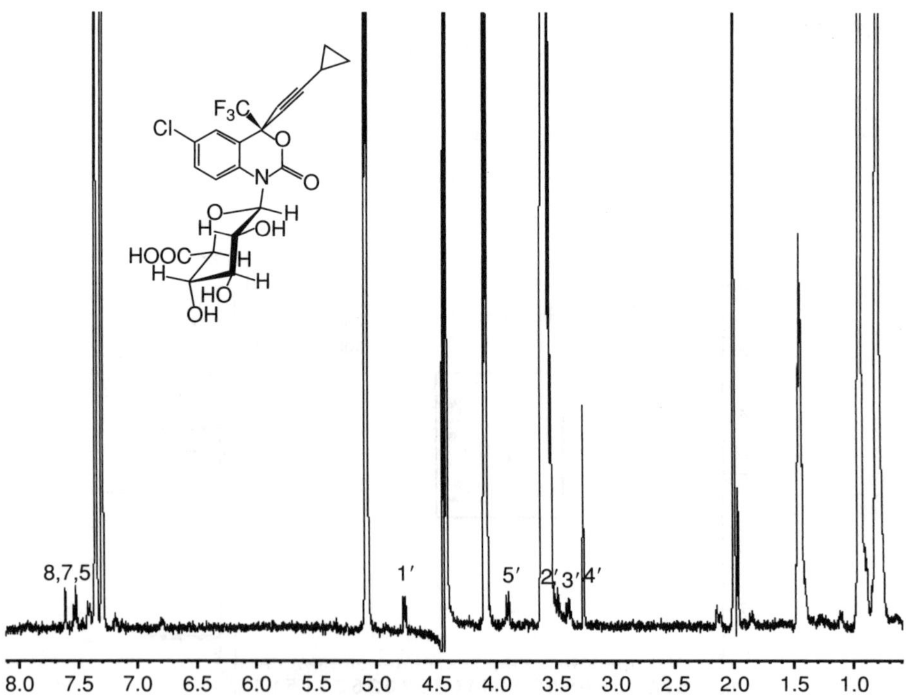

Figure 4.15 ^{1}H LC–NMR spectrum obtained when M + D, 497 Da, was at its maximum intensity showing the minor component at retention time 14.2 min with ^{19}F chemical at −82.1 ppm. Assigned as structure 10

the N-glucuronide conjugate (**10**). They also illustrate the power of hyphenated NMR methods for the study of drug metabolites. Exploiting the sensitivity of mass spectrometry and the selectivity of ^{19}F NMR spectroscopy by hyphenating the methods, the detection and analysis of a novel and previously undetected metabolite of efavirenz was achieved in a rapid and efficient manner. The utility of LC–NMR and LC–NMR–MS is still hampered by the inherently low sensitivity of NMR spectroscopy. It is hoped that in the near future, cryogenic LC–NMR probes will be available and with their increased sensitivity many of the limitations on use of hyphenated methods for metabolic studies should be eliminated.

REFERENCES

1. Lindon, J. C., Nicholson, J. K. and Wilson, I. D., *Adv. Chromatogr.*, 1995, **36**, 315.
2. Shockcor, J. P., Frick, L. W., Wurm, R. M., Sanderson, P. N., Farrant, R. D., Sweatman, B. C. and Lindon, J. C., *Xenobiotica*, 1996, **26**, 189.
3. Shockcor, J. P., Silver, I. S., Wurm, R. M., Sanderson, P. N., Farrant, R. D., Sweatman, B. C. and Lindon, J. C., *Xenobiotica*, 1996, **26**, 41.
4. Lindon, J. C., Nicholson, J. K. and Wilson, I. D., *Prog. Nucl. Magn. Reson. Spectrosc.*, 1996, **29**, 1.
5. Lindon, J. C., Nicholson, J. K., Sidelmann, U. G. and Wilson, I. D., *Drug Metab. Rev.*, 1997, **29**, 705.
6. Burton, K. I., Everett, J. R., Newman, M. J., Pullen, F. S., Richards, D. S. and Swanson, A. G., *J. Pharm. Biomed. Anal.*, 1997, **15**, 1903.
7. Farrant, R. D., Cupid, B. C., Nicholson, J. K. and Lindon, J. C., *J. Pharm. Biomed. Anal.*, 1997, **16**, 1.
8. Corcoran, O., Spraul, M., Hofmann, M., Ismail, I. M., Lindon, J. C. and Nicholson, J. K., *J. Pharm. Biomed. Anal.*, 1997, **16**, 481.
9. Ehlhardt, W. J., Woodland, J. M., Baughman, T. M., Vandenbranden, M., Wrighton, S. A., Kroin, J. S., Norman, B. H. and Maple, S. R., *Drug Metab. Disp.*, 1998, **26**, 42.
10. Scarfe, G. B., Wright, B., Clayton, E., Taylor, S., Wilson, I. D., Lindon, J. C. and Nicholson, J. K., *Xenobiotica*, 1998, **28**, 373.
11. Feng, N., Zhang, Z., An, D., Huang, W., Wang, G., Han, X. and Bao, X., *Eur. J. Drug Metab. Pharmacokin.*, 1998, **23**, 41.
12. Scarfe, G. B., Wright, B., Clayton, E., Taylor, S., Wilson, I. D., Lindon, J. C. and Nicholson, J. K., *Xenobiotica*, 1999, **29**, 77.
13. Abel, C. B. L., Lindon, J. C., Noble, D., Rudd, B. A. M., Sidebottom, P. J. and Nicholson, J. K., *Anal. Biochem.*, 1999, **270**, 220.
14. Mutlib, E., Chen, H., Nemeth, G., Gan, L. S. and Christ, D. D., *Drug Metab. Disp.*, 1999, **27**, 1045.
15. Nicholls, W., Lindon, J. C., Farrant, R. D., Shockcor, J. P., Wilson, I. D. and Nicholson, J. K., *J. Pharm. Biomed. Anal.*, 1999, **20**, 865.
16. Scarfe, G. B., Lindon, J. C., Nicholson, J. K., Wright, B., Clayton, E. and Wilson, I. D., *Drug Metab. Disp.*, 1999, **27**, 1171.
17. Ismail, M., Dear, G. J., Mutch, P. J., Davies, L. H., Plumb, R. S. and Sweatman, B. C., *Xenobiotica* 1999, **29**, 957.
18. Bailey, N. J. C., Cooper, P., Hadfield, S. T., Lenz, E. M., Lindon, J. C., Nicholson, J. K., Stanley, P. D., Wilson, I. D., Wright, B. and Taylor, S. D., *J. Agri. Food Chem.*, 2000, **48**, 42.

19. Bailey, N. J. C., Stanley, P. D., Hadfield, S. T., Lindon, J. C. and Nicholson, J. K., *Rapid Commun. Mass Spectrom.*, 2000, **14**, 679.
20. Scarfe, G. B., Lindon, J. C., Nicholson, J. K., Martin, P., Wright, B., Taylor, S., Lenz, E. and Wilson, I. D., *Xenobiotica*, 2000, **30**, 717.
21. Scarfe, G. B., Clayton, E., Wilson, I. D. and Nicholson, J. K., *J. Chromatogr. B*, 2000, **748**, 311.
22. Dear, G. J., Plumb, R. S., Sweatman, B. C., Parry, P. S., Roberts, A. D., Lindon, J. C., Nicholson, J. K. and Ismail, I. M., *J. Chromatogr. B*, 2000, **748**, 295.
23. Dear, G. J., Plumb, R. S., Sweatman, B. C., Ayrton, J., Lindon, J. C., Nicholson, J. K. and Ismail, I. M., *J. Chromatogr. B*, 2000, **748**, 281.
24. Bobzin, S. C., Yang, S. T. and Kasten, T. P., *J. Chromatogr. B*, 2000, **748**, 259.
25. Dachtler, M., Handel, H., Glaser, T., Lindquist, D., Hawk, R. M., Karson, C. N., Komoroski, R. A. and Albert, K., *Magn. Reson. Chem.*, 2000, **38**, 951.
26. Mutlib, E., Diamond, S., Shockcor, J., Way, R., Nemeth, G., Gan, L. and Christ, D. D., *Xenobiotica*, 2000, **30**, 1091.
27. Mutilb, E., Shockcor, J. P., Espina, J. R., Graciani, N., Du, A. and Gan, L. S., *J. Pharmacol. Exp. Theor.*, 2000, **294**, 735.
28. Sidelmann, U. G., Bjornsdottir, I., Shockcor, J. P., Hansen, S. H., Lindon, J. C. and Nicholson, J. K., *J. Pharm. Biomed. Anal.*, 2001, **24**, 569.
29. Webster, R., Beaumont, K., Ritzau, M. and Stachulski, A. V., *Biocatal. Biotransform.* 2001, **19**, 69.
30. Delort, M. and Combourieu, B., *J. Ind. Microbiol. Biotechnol.*, 2001, **26**, 2.
31. Zhang, K. Y. E., Hee, B., Lee, C. A., Liang, B. H. and Potts, B. C. M., *Drug Metab. Disp.*, 2001, **29**, 729.
32. Ruhl, R., Thiel, R., Lacker, T. S., Strohschein, S., Albert, K. and Nau, H., *J. Chromatogr. B*, 2001, **757**, 101.
33. Feng, W. Q., Liu, H. Y., Chen, G. D., Malchow, R., Bennett, F., Lin, E., Pramanik, B. and Chan, T. M., *J. Pharm. Biomed. Anal.*, 2001, **25**, 545.
34. Shockcor, J. P., Unger, S. E., Savina, P., Nicholson, J. K. and Lindon, J. C., *J. Chromatogr. B*, 2000, **748**, 269.
35. Shockcor, J. P., Unger, S. E., Wilson, I. D., Foxall, P. J. D., Nicholson, J. K. and Lindon, J. C., *Anal. Chem.*, 1996, **68**, 4431.
36. Burton, K. I., Everett, Jr, J. R., Newman, M. J., Pullen, F. S., Richards, D. S. and Swanson, A. G., *J. Pharm. Biomed. Anal.*, 1997, **15**, 1903.
37. Havlir, D, V. and Richman, D. D., *Ann. Int. Med.*, 1996, **124**, 984.
38. Mutlib, E., Chen, H., Nemeth, G., Markwalder, J., Seitz, S., Gan, L. S. and Christ, D. D., *Drug Metab. Disp.*, 1999, **27**, 1319.
39. Mutlib, A. E., Chen, H., Nemeth, G., Gan, L. S. and Christ, D. D., *Drug Metab. Disp.*, 1999, **27**, 1045.
40. Gerson, R. J., Mutlib, A. E., Meunier, P. C., Haley, P. J., Gan, L. S., Chen, H., Davies, M. H., Gemzik, B., Christ, D. D., Krahn, D. F., Markwalder, J. A., Seitz, S. P., Miwa, G. T. and Robertson, R. T., *Toxicol. Sci.*, 1999, **48**, 1833.
41. Smallcombe, S. H., Patt, S. L. and Keifer, P. A., *J. Magn. Reson., A*, 1995, **17**, 295.

5 LC–NMR for Natural Products Analysis

5.1 Application of LC–NMR and LC–NMR–MS Hyphenation to Natural Products Analysis

MARTIN SANVOSS
GlaxoSmithKline, Ware, Hertfordshire, UK

5.1.1 INTRODUCTION

5.1.1.1 GENERAL ASPECTS

Apart from pharmaceutical analysis and metabolism research, natural product analysis today is one of the most important fields of application for LC–NMR hyphenation. The reasons for this are obvious:

1. In search of new natural products, crude extracts are classically subjected to multi-step work-up and isolation procedures which include various separation methods (besides HPLC, for instance, column, gel or counter-current chromatography) in order to obtain pure compounds which are then structurally elucidated by using off-line spectroscopic methods such as nuclear magnetic resonance spectroscopy and mass spectrometry.

2. As natural product extracts often contain a large number of closely related and thus difficult to separate compounds, this classical approach may become very tedious and time-consuming. Thus, the direct hyphenation of an efficient separation technique with a powerful spectroscopic method bears great potential in order to speed up the analytical process in general.

3. Moreover, such conventional off-line approaches often result in the re-isolation ('replication') of already known compounds. Here, LC–NMR offers an unique insight into the composition of mixtures at an early stage of the analytical process for identifying ('dereplicating') unwanted or already known compounds and thus guide the targeted isolation of potentially new substances.

On-Line LC–NMR and Related Techniques
Edited by K. Albert.

4. In particular bioactivity guided screening approaches where unwanted false positive compounds (e.g. tanning agents), as well as the replication of known compounds, have to be excluded by a dedicated dereplication step (usually LC–diode array detection (DAD) or LC–MS) represent a valuable field of application for LC–NMR. In addition, with the trend towards high-throughput screening approaches, the demand for rapid methods for the structural elucidation of mixtures further increases [1–3].

Such applications require, on the one hand, an effective sample preparation (extraction, clean-up and separation) and, on the other hand, the ready and reliable availability of spectroscopic information of individual compounds at the earliest possible stage without necessarily providing the full structural information.

Unlike, e.g. in pharmaceutical metabolism studies, where the parent compound is known and thus some preliminary knowledge of the expected structures is available, the composition of a natural products extract is often completely unknown beforehand ('non-target analysis'). In such cases, NMR spectroscopy is especially well suited as a detection system since it does not discriminate any classes of compounds. (Sufficient relaxation delays provided, the NMR signal depends only on the number of nuclei in the active probe volume.)

Furthermore, in search of new natural products, NMR detection is particularly powerful for the differentiation of isomers such as different sugar configurations or different substitution patterns of cyclohexane or aromatic systems which other spectroscopic methods, including MS, cannot distinguish. However, the complementary mass spectroscopic information on molecular weight and functional groups which cannot be recognised by NMR (e.g. SO_4^-, NO_2, COOH) is in most cases needed for the identification of unknowns. Here, the hyphenation of an LC–NMR system to a mass spectrometer represents the next obvious step towards creating a comprehensive analytical system providing the complementary information of both NMR spectroscopy and MS in a single chromatographic run.

At this point it should be noted, that for a comprehensive structural elucidation of a novel natural product a preparative isolation still has to be performed, since in LC–NMR a part of the spectral region is lost due to solvent suppression (which may be overcome by a solvent change) and more importantly, even though the amount of sample available is generally not a limiting factor in natural products chemistry, LC–NMR does not provide ^{13}C NMR data yet, which is a prerequisite for the unambiguous assignment of a new structure.

However, the applications presented in the literature so far (see below) impressively show the capability of LC–NMR to reliably recognise a wide range of known compounds and to characterise the structures of new compounds which are then in most cases corroborated by off-line NMR spectroscopy.

5.1.1.2 APPLICATIONS

While important early publications demonstrated the suitability of LC–NMR to examine mixtures of isomerisation products of natural product standards that were exposed to light [4] or heat [5], the application of LC–NMR to natural products extracts was first presented with the characterisation of sesquiterpene lactones from the Mexican plant *Zaluzania grayana* [6]. In this chemotaxonomic investigation, a new lactone was already identified in on-flow mode.

It was about 1997 that LC–NMR hyphenation began to find wider use in natural products analysis. Since then, several applications to characterise more or less purified as well as crude natural product extracts have been presented. In particular, the group of J.-L. Wolfender [7] has investigated various plant species in which they characterised among other classes of compounds [8], prenylated flavonones [9], secoiridoids [10,11], naphthoquinones [12], pyrrolizidine alcaloids [13] and benzophenones [14]. Other groups dealt with naphthylisoquinoline alcaloids [15–17], sesquiterpene lactones [6,18,19], triterpene saponins [20] taxanes [21], lignanes [22], phenylphenalenones [23], polyhydroxy steroids [24], fasciculol triterpenes [25], tocopherols and tocotrienols [26], carotenoid isomers [27] and flavonoids [28,29], as well as hop bitter acids [30,31].

Remarkably, most applications published so far deal with the characterisation of plant-derived mixtures, while applications to secondary metabolites of micro-organisms [32,33] or marine natural products [2,34,35] are still rare.

In natural products analysis, most frequently the stop-flow mode is chosen to acquire ^{1}H spectra of the compounds of interest, or if further structural information is required to perform two-dimensional ^{1}H NMR spectra, such as COSY, TOCSY, NOESY or ROESY. In many cases an on flow NMR chromatogram (usually at flow rates between 0.3 and 1 ml min^{-1}) is recorded beforehand, either to screen for the presence of particular groups of compounds or to gain a general overview on the sample composition. (Heteronuclear LC–NMR experiments, such as HSQC and HMBC of a natural product, have been reported in the literature once [9]; however, this was of a highly enriched fraction.) More recently, 'time-sliced' stop flow [14,16] and on-flow approaches at low flow rates [34,35] have been applied to natural product extracts in order to combine the advantages of both on-flow (a ready overview on the entire sample) and stop-flow (sufficient acquisition time for minor compounds) modes.

The sensitivity of the NMR experiment is concentration-dependent, and thus at a given amount of sample, molecules with a lower molecular weight (MW) give a more intensive signal. On the other hand, larger molecules usually possess shorter relaxation times which allow more spectra to be accumulated per time increment by using shorter relaxation delays. The work published so far has focused mainly on the characterisation and identification of small molecules up to an MW of ca. 700. However, LC–NMR has been successfully applied to identify saponins of an MW up to 1400 [34,35] and to characterise larger biomolecules such as glycosphingolipids [36].

The first application of LC–NMR–MS to natural products analysis was presented in 1999 [37]. The additional mass spectroscopic information allowed the identification of a further ecdysteroid in an extract of *Silene otides* which could not be identified by LC–NMR alone [24]. Further applications of this double hyphenation dealt with the identification of napthodianthrones [38] and flavone glycosides [38,39] in natural products extracts. Taking the hyphenated technique one step further, the suitability of an integrated LC–UV–IR–NMR–MS system for natural products analysis has been assessed [40] – again using ecdysteroids as an example. Such systems still suffer from different requirements of the individual detectors (mainly in terms of sensitivity). However, the results obtained are promising.

Another application of LC–NMR in natural products chemistry concerns biosynthetic studies employing feeding experiments with stable isotope-labelled compounds. ^{1}H NMR spectra allow the determination of the amount of isotopic label incorporated into metabolites, e.g. by observing signals that arise from *J*-couplings of protons to ^{13}C-labelled nuclei [41,42].

A comprehensive recent review paper on the application of LC–NMR in phytochemical analysis is recommended for further reading [1].

In the following sections, some examples of the advantageous use of LC–NMR and LC–NMR–MS in natural products analysis are presented in order to demonstrate the possibilities and limitations of these hyphenated techniques.

5.1.2 APPLICATION OF LC–NMR–MS TO GLYCOSIDIC NATURAL PRODUCTS OF MARINE ORIGIN

5.1.2.1 INTRODUCTION – NEED FOR LC–NMR

In the search for biologically active natural products for a technical application, constituents of marine organisms were systematically investigated [34,35]. Biological tests on extracts of the Baltic starfish *Asterias rubens* led to a class of closely related glycosidic compounds, the so-called asterosaponins. These are $\Delta^{9(11)}$-3β, 6α-dioxygenated steroids with a sulphate group attached at C-3 and an oligosaccharide chain containing five or six sugar units at C-6. Individual compounds of this class differ in their steroidal side-chain or their sugar moiety, while the steroidal nucleus is common to all asterosaponins. Approximately 100 different combinations of sugar moiety and side-chain are described in the literature [43,44].

An initially pursued strategy consisting of an acetonitrile extraction, followed by a classical column chromatographic work-up, and subsequential comprehensive off-line NMR structural elucidation, proved to be unsuitable for two reasons: on the one hand, mostly sub-fractions of asterosaponins were obtained which required further (HPLC) chromatographic purification. On the other hand, those compounds that could instantly be obtained in pure form proved, after a time-consuming NMR examination (COSY, TOCSY, multiplicity-edited HSQC, HMBC and ROESY), to be without exception already described in the literature, i.e. a case of replication of known compounds.

In this context, LC–NMR and LC–NMR–MS proved to be powerful tools for speeding up the process of screening fractions of biologically active natural products for novel compounds, justifying the laborious isolation and comprehensive off-line NMR examination.

5.1.2.2 METHODOLOGY: ON-FLOW LC–NMR–MS SCREENING

On-flow ^{1}H NMR chromatograms, offer several advantages for the screening of mixtures of closely related natural products. First, they provide a straightforward overview of the sample and permit a rapid comparison of spectra of individual compounds. In addition, even signals of partial co-eluting constituents can be assigned to distinct compounds, which is more difficult in spectra obtained under stop-flow conditions or in off-line NMR spectra. Furthermore, minor chemical shift differences (e.g. the H_3-18 signals in the right-hand dotted frame in Figure 5.1.4 shown below) can be reliably recognised in isocratic on-flow mode, since each spectrum is recorded under identical conditions.

In order to gain sufficient accumulation time for NMR spectra to be acquired, on-flow LC–NMR–MS investigations can be performed as overnight experiments at eluent flow rates lower than usual, e.g. 0.05 ml min^{-1}. This allows a number of up to 128 scans to be recorded per spectrum. Preliminary tests showed that both for early- and late-eluting asterosaponins such a low flow rate did not result in peak broadening. On the contrary, a broad van-Deemter minimum is found and the actual eluted peak volume (peak width at half-height $\times$ flow rate), which is crucial for the sensitivity of the concentration-dependent NMR experiment, is lower than at the conventional flow rate of 0.8 ml min^{-1}. For a late-eluting asterosaponin (ruberoside D, MW 1225) it decreases from 400 μl at 0.8 ml min^{-1} over 360 μl (0.4 ml min^{-1}) to 280 μl at 0.1 and 0.05 ml min^{-1}. Furthermore, the peak broadening due to turbulences in the flow cell is reduced at low flow rates (see also Chapter 6 in this present volume). Thus, the flow rate should be taken into account when optimising on-flow parameters.

A similar approach for increasing the time available for data accumulation has been suggested recently [14] for natural products metabolite profiling purposes: in the time-slicing mode, the LC pump is stopped after certain defined intervals (e.g. every 60 s) to allow the acquisition of up to 1024 scans per spectrum. While this technique enables longer accumulation times, it should in theory affect the chromatographic separation to a higher degree. A future comparison of both approaches should be most interesting.

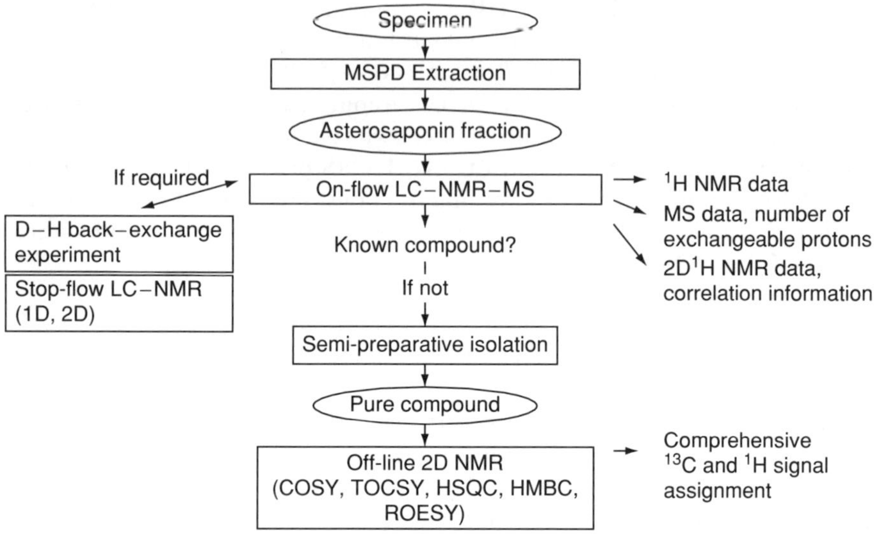

Figure 5.1.1 Extraction scheme and LC–NMR–MS screening approach for the targeted isolation of new natural products as applied to asterosaponins of *Asterias rubens*

In our laboratory, an on-flow LC–NMR–MS screening (Figure 5.1.1) was applied to both saponin fractions which were not separated into pure compounds by classical column chromatography and further to total asterosaponin fractions obtained by the micropreparative technique, matrix solid-phase dispersion (MSPD) extraction [45] (see Figure 5.1.2). The LC–NMR–MS hyphenation is set up in the widely used parallel configuration of NMR and mass spectrometer (Figure 5.1.3). Typically, absolute amounts of asterosaponin mixtures of about 500 μg – 1 mg are injected onto the column.

5.1.2.3 NMR – STRUCTURAL INFORMATION

Figure 5.1.4 shows examples of what kind of structural information can be obtained from such an LC–NMR–MS run. This figure shows the NMR chromatogram of a total asterosaponin fraction obtained from MSPD extraction. For clarity, the regions of the methyl and olefinic/anomeric resonances are shown.

As a complete structural elucidation of complex glycosidic compounds cannot be achieved by one-dimensional ^{1}H–NMR spectra, which are provided by on-flow LC–NMR (^{13}C–NMR data are still required for an unambiguous assignment of resonances), two approaches have proven to be effective in the case of asterosaponins in order to identify new compounds for a subsequent semi-preparative isolation:

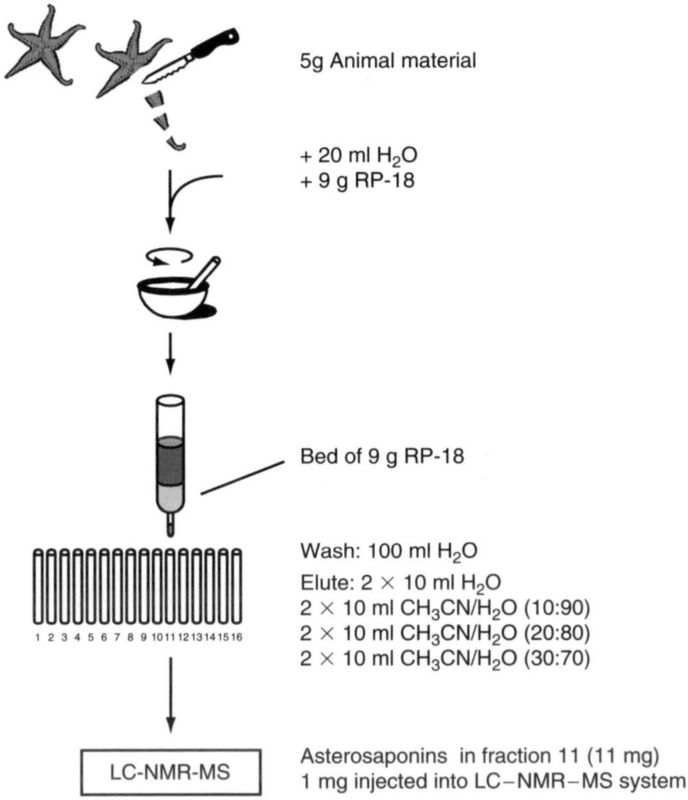

Figure 5.1.2 Matrix solid-phase dispersion (MSPD) extraction as a micro-preparative extraction technique for an on-flow LC–NMR–MS screening. Since the latter requires only sample amounts in the 0.5–2 mg range, the sample preparation can be achieved by fast small-scale extraction procedures, such as MSPD. This is a sample preparation technique that combines both sample homogenisation and extraction of compounds of interest in one single step starting from the intact sample material. Thus, it simplifies the extraction and clean-up steps, reduces the sample manipulation and is much faster than conventional techniques. It is therefore very well suited for a rough separation of extracts into classes of compounds of similar polarities, which can then be submitted to LC–NMR–MS analysis

1. *The analysis of easy-to-assign resonances characteristic of certain structural features (e.g. olefinic protons, methyl groups or anomeric protons of sugar units).* A good example of this category is the study of the resonances of the steroidal H_3-18 methyl groups. Most asterosaponins possess a hydroxy function at C-20. The corresponding H_3-18 methyl group resonates at ca. 0.7 ppm (see Figure 5.1.4, dotted frames). The significant upfield shift of this resonance in ruberosides A, B and G clearly indicates a lack of this C-20 hydroxy function in the proximity of H_3-18. The resonances of the actual H_3-21 methyl group corroborate this assignment: in asterosaponins with a C-20 OH group, a singlet at about

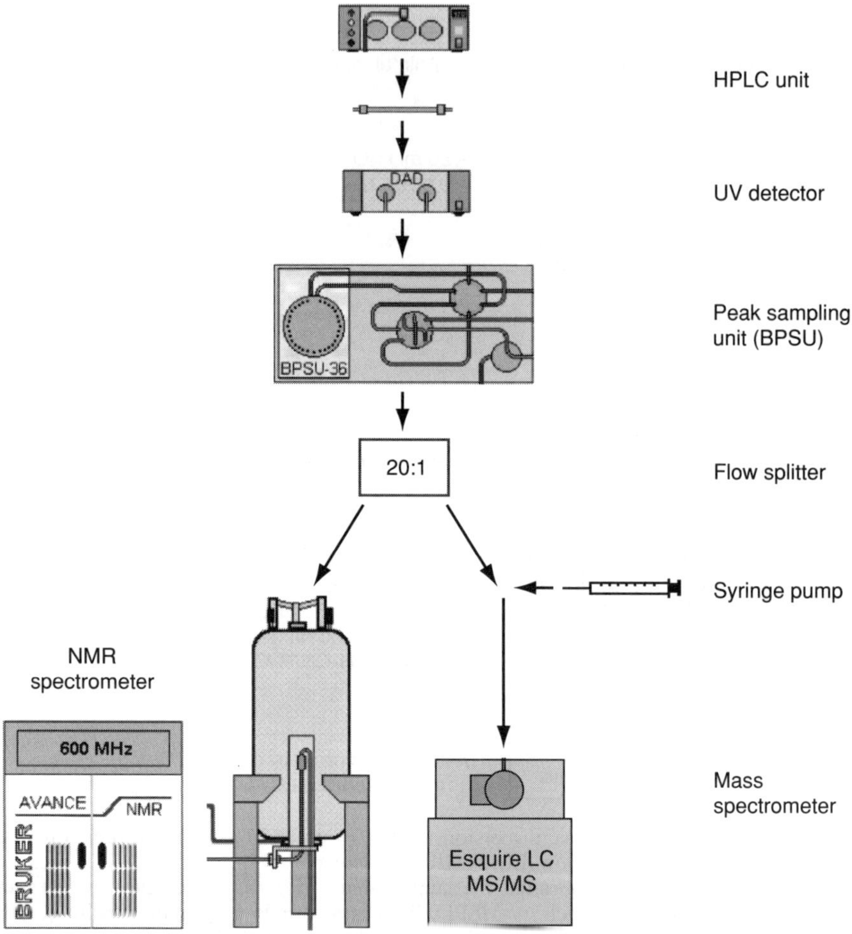

Figure 5.1.3 Schematic of an LC–NMR–MS set-up

1.25 ppm appears, while in those without one a doublet between 0.8 and 0.9 ppm appears. However, the H_3-21 methyl resonances sometimes overlap with other signals.

A further example is the ready assignment of an olefinic resonance which appears in the spectrum of ruberoside G in addition to the H-11 signal at 5.3 ppm, common to asterosaponins. Its low-field chemical shift of 6.2 ppm (solid frame in Figure 5.1.4, and Figure 5.1.5) indicates the presence of a double bond in the steroidal side-chain in proximity to a deshielding functional group such as a carbonyl.

2. *The comparison of sub-spectra of unknowns with those of known compounds or with those of other unknowns in order to recognise structural analogies or differ-*

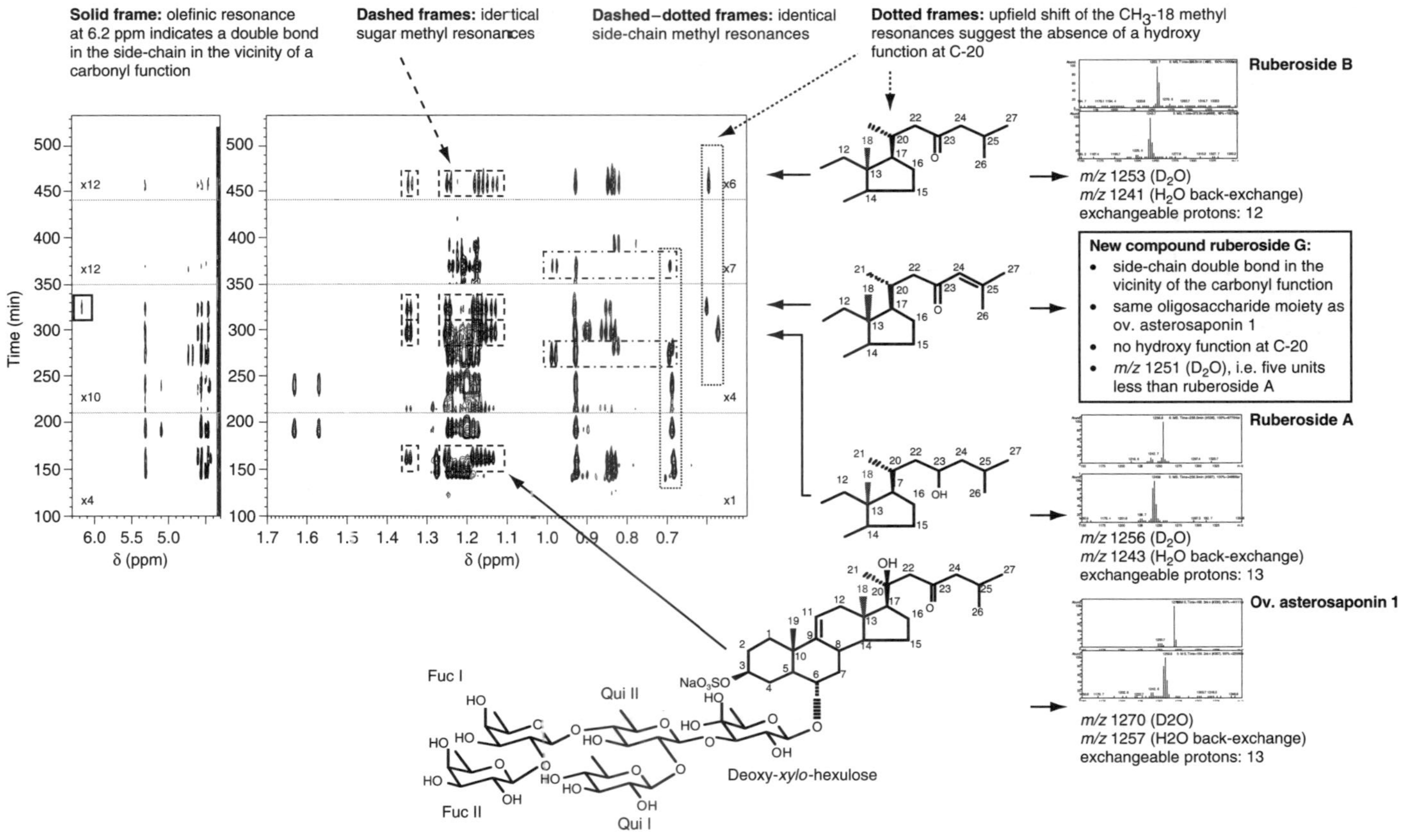

Figure 5.1.4 On-flow LC–NMR–MS chromatogram of a total asterosaponin fraction obtained from MSPD extraction. Individual segments of the chromatogram are scaled-up for clarity

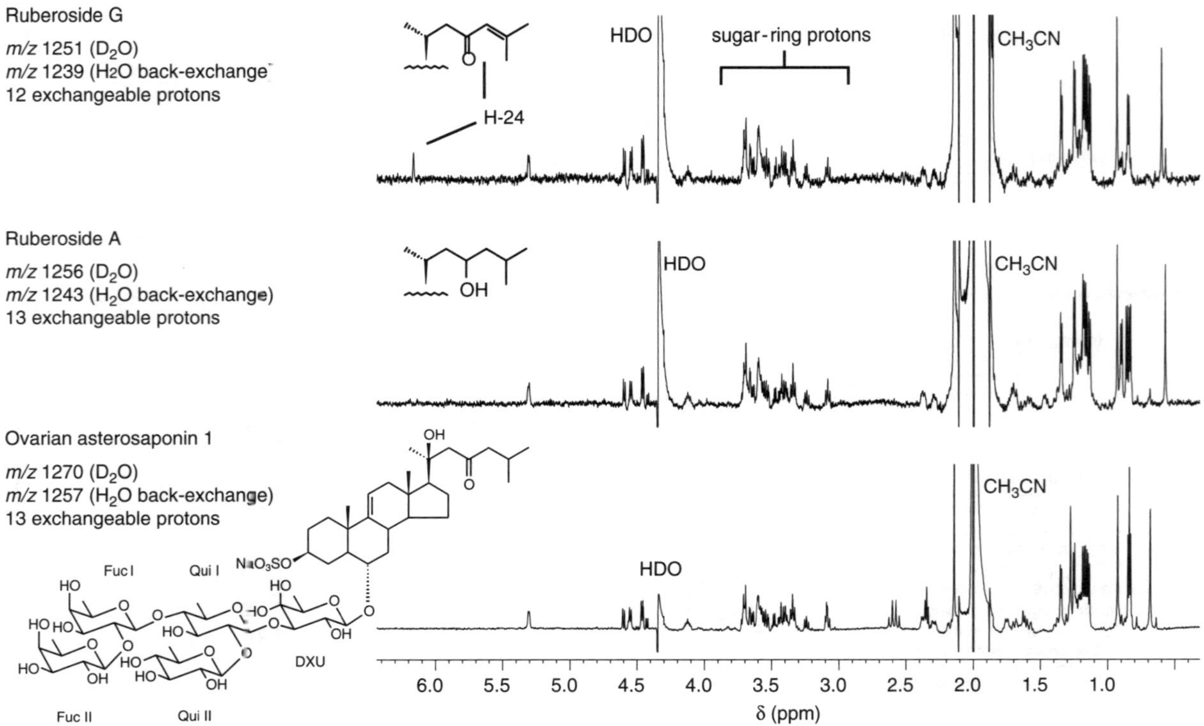

Figure 5.1.5 One-dimensional ^{1}H NMR spectra of ruberosides A, B and G, extracted from the on-flow NMR chromatogram shown in Figure 5.1.4. The conformity of the sugar-ring proton signals indicates identical oligosaccharide moieties in all three compounds. The additional olefinic resonance, at 6.16 ppm, of ruberoside G suggests a double bond in proximity to a carbonyl function. The ESI MS data and information from D–H back-exchange experiments support this assignment. Reprinted from Sandvoss, M., Weltring, A., Preiss, A., Levsen, K. and Wuensch, G., 'Combination of matrix solid-phase dispersion extraction and direct on-line liquid chromatography–nuclear magnetic resonance spectroscopy–tandem mass spectrometry as a new efficient approach for the rapid screening of natural products: Application to the total asterosaponin fraction of the starfish *Asterias rubens*', J. Chromatogr., A, 917, 75–86, Copyright (2001), with permission of Elsevier Science

ences. For instance, the conformity of both the carbohydrate methyl signals and anomeric proton resonances of ruberoside A, B and G with those of the previously known compound ovarian asterosaponin 1 (see Figure 5.1.4, dashed frames) suggests that all four compounds possess the same glycoside moiety carrying a 6-deoxy-*xylo*-hex-4-ulose unit directly attached to the aglycone. This is further corroborated by the conformity of the corresponding sugar-ring proton signals between $\delta_H = 3.00$ and 3.90 (see Figure 5.1.5).

In the same manner, conformities of aglycone side-chain signals can be recognised from the chromatogram as well (see Figure 5.1.4, dashed-dotted frames; for another example, see Section 5.1.2.6 below).

Since asterosaponin mixtures consist of a variety of compounds with different combinations of reoccurring structural features (oligosaccharide or steroidal side-chains), this approach has already yielded valuable information for identifing known constituents and for selecting compounds of interest for further investigations, such as 2D stop-flow NMR spectra (see Section 5.1.2.5 below).

5.1.2.4 MASS SPECTROMETRY AND D–H BACK-EXCHANGE EXPERIMENTS

Today, deuterium oxide and (undeuterated) acetonitrile or methanol are by far the most frequently used eluent constituents employed in LC–NMR unless normal-phase material is used or NH protons [46] are studied. The use of these eluents leads to shifted masses in the mass spectrum, since all exchangeable protons of the analytes are exchanged for deuterons.

However, by means of an additional D–H back-exchange (performed by adding water or aqueous buffer via a syringe pump to the MS eluent flow in a ratio of 1:4), information on the actual molecular ion, as well as the number of exchangeable protons, is available.

In Figure 5.1.4, electrospray ionization (ESI) mass spectra of selected asterosaponins in both deuterated solvent and under D–H back-exchange conditions are shown. For instance, the molecular ion of ruberoside A in deuterated solvent appears at *m/z* 1256. Under D–H back-exchange conditions, the most abundant mass is 1245 $[M + D2]^-$. However, the molecular ion at *m/z* 1243 can be clearly distinguished. Thus, the number of exchangeable protons of this molecule is 13. Such additional information turned out to be particularly useful in the case of these saponins. In comparison, the similar compound, ruberoside B, showed only 12 exchangeable protons. As the on-flow NMR chromatogram had already indicated that both compounds carry the same oligosaccharide chain and no hydroxy function at C-20, this suggested, in conjunction with the mass spectral data, the presence of an additional hydroxy group in its steroidal side-chain, which was then corroborated by a 2D stop-flow NMR experiment (see Section 5.1.2.5 below).

Moreover, the information on the number of exchangeable protons provided by back-exchange experiments allows the further differentiation of hydrogen equivalents. For example, the molecular ion of ruberoside G after H–D exchange is found to be *m/z* 1251, i.e. five mass units less than that of ruberoside A, which elutes directly prior to it (see Figure 5.1.4). The loss of two mass units corresponds to the additional double bond in the side-chain of ruberoside G, while the remaining difference of three units is in conformity with the exchange of a hydroxy function for a carbonyl function in the side-chain of ruberoside G.

In principle, a D–H back-exchange experiment does not necessarily need to be performed since the required structural information can also be deduced from the mass spectrum recorded under H–D exchange conditions (deuterated eluent), which is of advantage in terms of the ready distinction of hydrogen equivalents. However, it has to be noted that this approach is not compatible with the extensive use of existing MS data libraries (data bases) and it thus depends on the individual problem as to whether such spectra can be used to advantage [1].

5.1.2.5 STOP-FLOW EXPERIMENTS

Two-dimensional ^{1}H NMR experiments acquired in stop-flow mode represent to some extent a link between on-flow LC–NMR screening and the detailed structural elucidation of isolated compounds employing two-dimensional NMR techniques in conventional off-line probe-heads. As the LC–NMR–MS hyphenation offers the possibility of triggering the stopping of the LC pump by the MS signal, this technique is particularly well suited for the reliable detection of compounds showing only weak UV absorbances.

MS triggering of stop-flow spectra even provides an additional feature – it is now possible to monitor the peak purity as a triggering criterion, instead of just the peak maximum in conventional UV-triggered experiments. Thus, in the case of partial co-elution, the stopping of the LC pump can be delayed until a peak elutes without major impurities.

Such MS-triggered stop-flow experiments are preferably performed in the second LC–NMR–MS run, required for the determination of the number of exchangeable protons by D–H back-exchange.

As an example, Figure 5.1.6 shows a 2D WET-TOCSY spectrum of ruberoside A. In this figure, a correlation between a downfield methine proton at 3.67 ppm and both the aglycone methyl groups A-H_3-26/27 (0.84/0.86 ppm) and A-H_3-21 (0.90 ppm) confirms the 23-hydroxy function in the side-chain which was suggested by the on-flow NMR and MS-back-exchange data (see above). Ruberoside A was thus identified as a new asterosaponin and subjected to semi-preparative isolation.

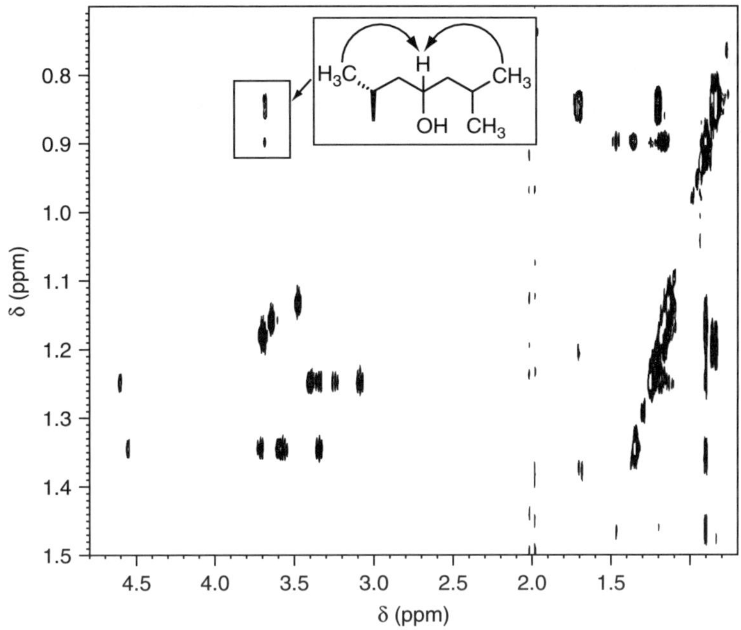

Figure 5.1.6 Stop-flow 2D WET-TOCSY data of ruberoside A. Magnetisation transfer from the methyl groups A-H_3-21 and A-H_3-26/27 to a resonance at 3.67 ppm indicates a hydroxy function in the aglycone side-chain

5.1.2.6 COMPLEMENTARY STRUCTURAL INFORMATION OF NMR AND MS

The complementary structural information of both NMR spectroscopy and MS detection is particularly valuable for the analysis of closely related glycosidic natural products, where compounds of the same molecular mass (isobars, see Table 5.1.1), and even with identical MS/MS fragmentation patterns, frequently occur, as this example demonstrates.

Modern ion-trap spectrometers are capable of performing MS^n experiments which yield unique information on the molecular weight of individual sugar units in the oligosaccharide chain of saponins. While fragmentations up to MS^5 could be performed in off-line mode, MS/MS spectra were recorded in on-flow LC–NMR–MS runs by default (auto-MS/MS mode).

Figure 5.1.7 shows the ESI-MS/MS spectra of the compounds solasteroside A and ruberoside E. Both compounds have the molecular mass, *m/z* 1223, in deuterated solvents and show identical MS/MS fragments at *m/z* 1075, 927, 779 and 645, corresponding to the sequential loss of three methyl pentoses and one pentose. Nevertheless, mass spectrometry yields the information

Table 5.1.1 Asterosaponins identified in the total asterosaponin fraction obtained from MSPD ex-traction (see Figure 5.1.4). The molecular ions in bold typeface indicate compounds with identical molecular weights (isobars).

Compound	Retention time (min)	*m/z*[a] (D_2O)	*m/z*[b] (D–H back-exchange)	Saccharide chain[c]	Steroidal side-chain
Versicoside A, forbeside A	128.5	1421	1405	G–$_3$F–$_2$G–$_4$*X*$_2$–$_3$Q– *Q*	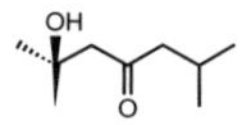
Glycoside B$_2$, Forbeside B	150.7	**1256**	**1243**	Q–$_2$G–$_4$*X*$_2$–$_3$Q– *Q*	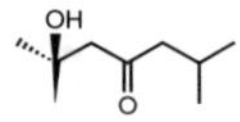
Ov. asterosaponin 1, Forbeside C	164.0	**1270**	**1257**	F–$_2$F–$_4$*Q*$_2$–$_3$D– *Q*	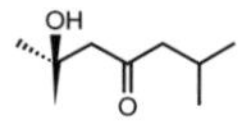
—[d]	176.8	1254	**1241**	F–$_2$F–$_4$*Q*$_2$–$_3$D– *Q*	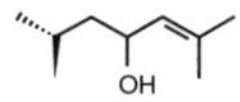
Asteriidoside C	192.7	1388	1373	G–$_3$F–$_2$F–$_4$*X*$_2$–$_3$Q– *Q*	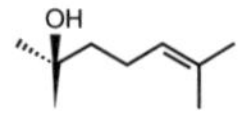
Asteroside C	"	**1270**	**1257**	Q–$_2$G–$_4$*X*$_2$–$_3$Q– *Q*	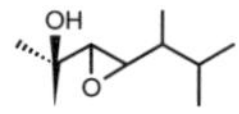
Ruberoside F	"	"	"	"	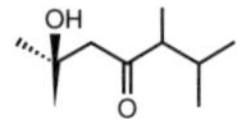
—[d]	220.9	1284	1271	F–$_2$F–$_4$*Q*$_2$–$_3$D– *Q*	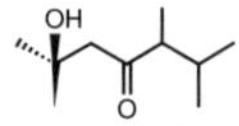
Solasteroside A	240.9	**1223**	**1211**	F–$_2$F–$_4$*X*$_2$–$_3$Q– *Q*	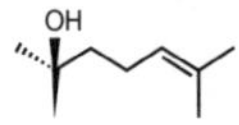
Ruberoside E	251.3	**1223**	**1211**	Q–$_2$F–$_4$*X*$_2$–$_3$Q– *Q*	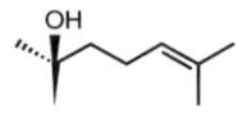
Asteriidoside B	272.8	1402	1387	G–$_3$F–$_2$F–$_4$*X*$_2$–$_3$Q– *Q*	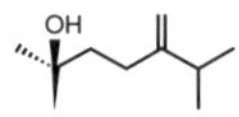
Ruberoside C	283.3	1390	1375	G–$_3$F–$_2$F–$_4$*X*$_2$–$_3$Q– *Q*	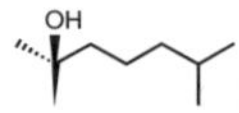
Ruberoside A	300.1	**1256**	**1243**	F–$_2$F–$_4$*Q*$_2$–$_3$D– *Q*	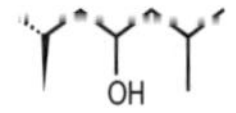
Ruberoside G	326.3	1251	1239	F–$_2$F–$_4$*Q*$_2$–$_3$D– *Q*	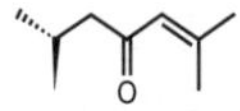
Ruberoside D	374.9	1237	1225	Q–$_2$F–$_4$*X*$_2$–$_3$Q– *Q*	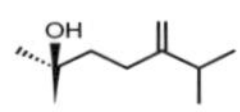
—[d]	399.6	1225	1213	Q–$_2$F–$_4$*X*$_2$–$_3$Q– *Q*	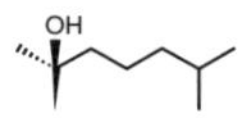
Ruberoside B	463.9	1253	**1241**	F–$_2$F–$_4$*Q*$_2$–$_3$D– *Q*	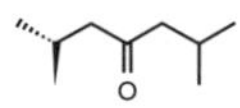

[a] Molecular anion after H–D exchange.
[b] Molecular anion after D–H back-exchange.
[c] Abbreviated representations, where all glycosidic linkages are (1—X)-linked: Q, quinovase; X, xylose; F, fucose; D, deoxy-xylo-hex-ulose; G, galactose.
[d] Compound tentatively assigned.

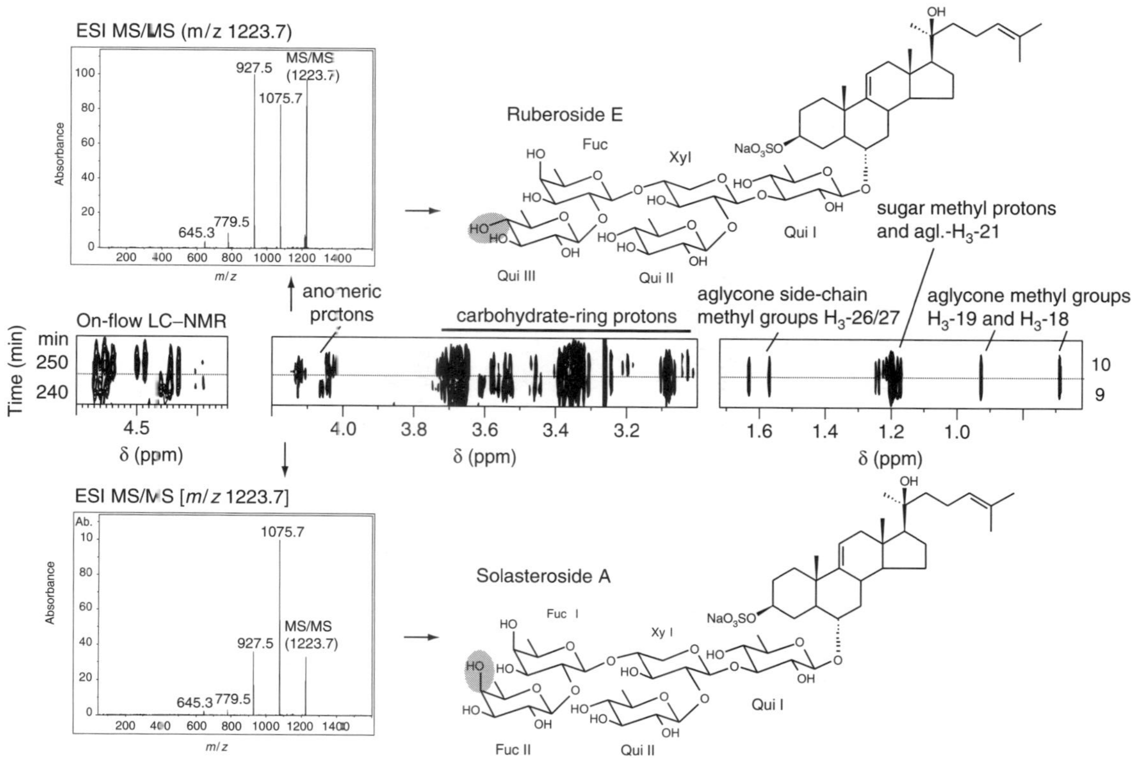

Figure 5.1.7 The MS/MS spectra of solasteroside A and ruberoside E. Both MS/MS spectra show identical mass fragments. The corresponding expansion of the LC–NMR chromatogram indicates that both compounds give the same steroidal side-chain resonances, but differ in their carbohydrate signals. Thus, ruberoside E was readily identified as a novel asterosaponin, which differs from solasteroside A only in the arrangement of the hydroxy group at C-4 of the terminal sugar unit. Reprinted from Sandvoss, M., Weltring, A., Preiss, A., Levsen, K. and Wuensch, G., 'Combination of matrix solid-phase dispersion extraction and direct on-line liquid chromatography–nuclear magnetic resonance spectroscopy–tandem mass spectrometry as a new efficient approach for the rapid screening of natural products: Application to the total asterosaponin fraction of the starfish *Asterias rubens*', J. Chromatogr., A, 917, 75–86, Copyright (2001), with permission of Elsevier Science

that the two compounds are distinct from one another, since the intensities in the MS/MS spectra, recorded under identical conditions, are different. However, for further information on the two isomers, on-flow NMR data are needed (see Figure 5.1.7, middle section).

Comparison of side-chain methyl resonances from the NMR chromatogram readily indicated that both compounds possess the same steroidal side-chain as an already known asterosaponin (asteriidoside C), while analysis of the sugar resonances indicated that the oligosaccharide chain of ruberoside E is identical to that of ruberoside D, which had been isolated before. Thus, it was clear that ruberoside E possesses a combination of oligosaccharide and steroidal side-chains not previously reported, i.e. it was identified as a new asterosaponin. Preparative isolation and extensive two-dimensional NMR examination resulted in corroboration of the structure deduced from LC–NMR [35]. The other compound, solasteroside A, differs from ruberosid E only in the terminal sugar unit of the oligosaccharide side-chain, i.e. a fucose in solasteroside A instead of a quinovose in ruberoside E. That is to say that both compounds differ only from each other in the arrangement of the hydroxy group at C-4 of the terminal sugar unit.

5.1.2.7 CONCLUSIONS

In addition to the ruberosides A, B and G discussed in some detail here, the application of LC–NMR–MS hyphenation has resulted so far in the targeted isolation and off-line structural elucidation of a total of seven new asterosaponins. In a single chromatogram of an asterosaponin fraction, up to 17 individual constituents could be characterised (see Table 5.1.1).

In particular, in conjunction with fast and simple sample preparation methods such as MSPD it is now possible to perform an LC–NMR–MS screening of a fraction of interest (e.g. one that shows biological activity or contains a desired class of compounds) before investing time and effort in the preparative isolation and detailed structural elucidation.

Apart from the gain in time and sample efficiency, the most important advantage of the doubly hyphenated LC–NMR–MS set-up over stand-alone LC–MS and LC–NMR set-ups is the unequivocal assignment of the MS data to the NMR peaks as LC–MS and LC–NMR chromatograms of the same sample are sometimes difficult to correlate because slightly different chromatograms are obtained in the two systems. This may be due to effects arising from the higher amounts of injected sample in LC–NMR, the different chromatographic behaviour of deuterium oxide compared to water, or different eluent gradient-forming units in the respective chromatographic systems.

5.1.3 ACKNOWLEDGEMENTS

The work presented here has been performed at the Fraunhofer Institute of Toxicology and Aerosol Research, Hannover, Germany, under the supervision of Professor G. Wuensch, Professor C. Levsen and Dr A. Preiss. The contribution of Ms A. Weltring to the development of the MSPD method is kindly acknowledged.

REFERENCES

1. Wolfender, J.-L., Ndjoko, K. and Hostettmann, K., *Phytochem. Anal.*, 2001, **12**, 2.
2. Hostettmann, K., Potterat, O. and Wolfender, J.-L., *Chimia*, 1998, **52**, 10.
3. Bobzin, S. C., Yang, S. and Kasten, T. P., *J. Chromatogr.*, B, 2000, **748**, 259.
4. Johnson, S., Morgan, E. D., Wilson, I. D., Spraul, M. and Hofmann, M., *J. Chem. Soc., Perkin Trans.*, 1, 1994, 1499.
5. Albert, K., Schlotterbeck, G., Braumann, U., Händel, H., Spraul, M. and Krack, G., *Angew. Chem.*, 1995, **107**, 1102.
6. Spring, O., Buschmann, H., Vogler, B., Schilling, E. E., Spraul, M. and Hofmann, M., *Phytochemistry*, 1995, **39**, 609.
7. Wolfender, J.-L., Rodriguez, S. and Hostettmann, K., *J. Chromatogr., A*, 1998, **794**, 299.
8. Cavin, A., Potterat, O., Wolfender, J.-L., Hostettmann, K. and Dyatmyko, W., *J. Nat. Prod.*, 1998, **61**, 1497.
9. Garo, E., Wolfender, J.-L., Hostettmann, K., Hiller, W., Antus, S. and Mavi, S., *Helv. Chim. Acta*, 1998, **81**, 754.
10. Wolfender, J.-L., Rodriguez, S., Hostettmann, K. and Hiller, W., *Phytochem. Anal.*, 1997, **80**, 97.
11. Rodriguez, S., Wolfender, J.-L., Stoeckli-Evans, H., Gupta, M. P. and Hostettmann, K., *Helv. Chim. Acta*, 1998, **81**, 1393.
12. Ioset, J. R., Wolfender, J.-L., Marston, A., Gupta, M. P. and Hostettmann, K., *Phytochem. Anal.*, 1999, **10**, 137.
13. Ndjoko, K., Wolfender, J.-L., Röder, E. and Hostettmann, K., *Planta Med.*, 1999, **65**, 562.
14. Ferrari, J., Terreaux, C., Wolfender, J.-L. and Hostettmann, K., *Chimia*, 2000, **54**, 406.
15. Bringmann, G., Günther, C., Schlauer, J. and Rückert, M., *Anal. Chem.*, 1998, **70**, 2805.
16. Bringmann, G., Rückert, M., Messer, K., Schupp, O. and Louis, A. M., *J. Chromatogr., A*, 1999, **837**, 267.
17. Bringmann, G., Messer, K., Wohlfahrt, M., Kraus, J., Dumbuya, K. and Rückert, M., *Anal. Chem.*, 1999, **71**, 2678.
18. Spring, O., Heil, N. and Vogler, B., *Phytochemistry*, 1997, **46**, 1369.
19. Vogler, B., Klaiber, I., Roos, G., Walter, C. U., Hiller, W., Sandor, P. and Kraus, W., *J. Nat. Prod.*, 1998, **61**, 175.
20. Renupka, T., Roos, G., Klaiber, I., Vogler, B. and Kraus, W., *J. Chromatogr., A*, 1999, **847**, 109.
21. Schneider, B., Zhao, Y., Blitzke, T., Schmitt, B., Nookandeh, A., Sun, X. and Stöckigt, J., *Phytochem. Anal.*, 1998, **9**, 237.
22. Zhao, Y., Nookandeh, A., Schneider, B., Sun, X., Schmitt, B. and Stöckigt, J., *J. Chromatogr., A*, 1999, **837**, 83.

23. Hölscher, D. and Schneider, B., *Phytochemistry*, 1999, **50**, 155.
24. Wilson, I. D., Morgan, E. D., Lafont, R. and Wright, B., *J. Chromatogr., A*, 1998, **799**, 333.
25. Kleinwächter, P., Luhmann, U., Schlegel, B., Heinze, S., Härtl, A., Kiet, T. T. and Gräfe, U., *J. Basic Microbiol.*, 1999, **39**, 345.
26. Strohschein, S., Rentel, C., Lacker, T., Bayer, E. and Albert, K., *Anal. Chem.*, 1999, **71**, 1780.
27. Dachtler, M., Glaser, T., Kohler, K. and Albert, K., *Anal. Chem.*, 2001, **73**, 667.
28. Vilegas, W., Vilegas, J. H., Dachtler, M., Glaser, T. and Albert, K., *Phytochem. Anal.*, 2000, **11**, 317.
29. Santos, L. C., Dachtler, M., Andrade, F. D. P., Albert, K. and Vilegas, W., *Fresenius' J. Anal. Chem.*, 2000, **368**, 540.
30. Höltzel, A., Schlotterbeck, G., Albert, K. and Bayer, E., *Chromatographia*, 1996, **42**, 499.
31. Pusecker, K., Albert, K. and Bayer, E., *J. Chromatogr., A*, 1999, **836**, 245.
32. Abel, C. B. L., Lindon, J. C., Noble, D., Rudd, B. A. M., Sidebottom, P. J. and Nicholson, J. K., *Anal. Biochem.*, 1999, **270**, 220.
33. Kleinwächter, P., Martin, K., Groth, I. and Dornberger, K., *J. High Resol. Chromatogr.*, 2000, **23**, 609.
34. Sandvoss, M., Pham, L. H., Preiss, A., Levsen, K., Mügge, C. and Wünsch, G., *Eur. J. Org. Chem.*, 2000, 1253.
35. Sandvoss, M., Weltring, A., Preiss, A., Levsen, K. and Wuensch, G., *J. Chromatogr., A*, 2001, **917**, 75.
36. Bäcker, A. E., Thorbert, S., Rakotonirainy, O., Hallberg, E. C., Olling, A., Gustavsson, M., Samuelsson, B. E. and Soussi, B., *Glycoconj. J.*, 1999, **16**, 45.
37. Wilson, I. D., Morgan, E. D., Lafont, R., Shockor, J. P., Lindon, J. C., Nicholson, J. K. and Wright, B., *Chromatographia*, 1999, **49**, 374.
38. Hansen, S. H., Jensen, A. G., Cornett, C., Bjornsdottir, I., Taylor, S., Wright, B. and Wilson, I. D., *Anal. Chem.*, 1999, **71**, 5235.
39. Lommen, A., Godejohann, M., Venema, D. P., Hollmann, P. H. C. and Spraul, M., *Anal. Chem.*, 2000, **72**, 1793.
40. Louden, D., Handley, A., Taylor, S., Lenz, E., Miller, S., Wilson, I. D., Sage, A. and Lafont, R., *J. Chromatogr., A*, 2001, **910**, 237.
41. Schmitt, B. and Schneider, B., *Phytochemistry*, 1999, **52**, 45.
42. Schmitt, B. and Schneider, B., *Phytochem. Anal.*, 2001, **12**, 43.
43. D'Auria, M. V., Minale, L. and Riccio, R., *Chem. Rev.*, 1993, 1839.
44. Kornprobst, J. M., Sallenave, C. and Barnathan, G., *Comp. Biochem. Physiol., B*, 1998, **119**, 1.
45. Barker, S. A., *J. Chromatogr., A*, 2000, **885**, 115.
46. Bringmann, G., Wohlfahrt, M. and Heubes, M., *J. Chromatogr., A*, 2000, **904**, 243.

5.2 Hyphenation of Modern Extraction Techniques to LC–NMR for the Analysis of Geometrical Carotenoid Isomers in Functional Food and Biological Tissues

TOBIAS GLASER and KLAUS ALBERT
Institut für Organische Chemie, Universität Tübingen, Tübingen, Germany

5.2.1 INTRODUCTION

The hyphenation of modern extraction techniques to LC–NMR is of general interest for natural products analysis. Apart from being the most tedious and time-consuming step, the isolation of bioactive agents is the most common source of error. Especially in carotenoid research, where such materials act as antioxidants and free-radical scavengers, a non-mild extraction technique leads to the formation of artifacts and, therefore, to misleading nutrition studies. Carotenoids are a common nutrition supplement because of their vitamin A activity and, according to several nutrition studies, their health benefit [1–5]. For instance, the carotenes β-carotene and lycopene as well as the xanthophylls lutein and zeaxanthin, are ascribed as having preventive effects against eye diseases, such as age-related macular degeneration (AMD) [6–8] or cataracts [9,10], heart diseases [11,12] and certain cancers [13–15].

Due to their antioxidant status, carotenoids are extremely sensitive to UV light, as well as air and temperature [16]. In nature, each carotenoid occurs with several geometrical *Z*/*E* stereoisomers, which can isomerise or oxidise easily [17]. These carotenoid stereoisomers can differ considerably in their biological effectiveness, for instance in bioavailability [18,19], in quenching free radicals, or in the prevention of diseases [20]. Therefore, the unambiguous and quantitative analysis of the pattern of carotenoid stereoisomers from biological matrices is indispensable, especially with regard to dietary supplementation.

On-Line LC–NMR and Related Techniques
Edited by K. Albert.

The requirements for an artifact-free analysis of geometrical isomers includes a mild and quick extraction method allowing enrichment of the analyte, as well as the on-line coupling of the separation system with the identification system. Hereby, LC–MS gives good results in the identification of carotenoids in the picogram range of all kind of matrices [21–23]. Unfortunately, *Z/E* stereoisomers lead to the same mass spectra and fragmentation patterns, even with LC–MS^n, and therefore they are not distinguishable by using LC–MS [24]. LC–diode-array detection (DAD) gives a hint about the occurrence of *cis*-bonds in the analyte, as different geometrical isomers have characteristic wavelength shifts and absorption ratios [21]. However, the identification of unknown or multi*cis*-isomers is almost impossible. Therefore, LC–NMR is the only unambiguous proof for the structural elucidation of geometrical carotenoid isomers, as it gives information about the chemical shift and the coupling system of every proton [23,25].

5.2.2 ARTIFACT-FREE ISOLATION OF GEOMETRICAL CAROTENOID ISOMERS

Matrix solid-phase dispersion (MSPD) is the extraction method of choice for the analysis of solid samples, such as plant material, foodstuffs or tissue samples [26]. This method has been developed especially for solid or viscous matrices. MSPD is preferable to other extraction techniques, because the solid or viscous sample can be directly mixed with the sorbent material of the stationary phase [27]. As the carotenoid stereoisomers stay bound in their biological matrix until the elution step, they are protected against isomerisation and oxidation [28]. The extraction scheme of MSPD is shown in Figure 5.2.1.

After mixing the biological sample with the sorbent material, the homogenised blend is packed in a solid-phase extraction (SPE) column. After washing out polar interferences with water, the carotenoid stereoisomers are eluted in a concentrated fraction, requiring only small amounts of solvent, and thus leading to a complete and reproducible extraction.

Body fluids, such as serum, contain several different carotenoids in low amounts. The crucial point in the isolation and analysis of these samples is the enrichment factor. Serum samples can be directly analysed with hyphenated extraction–sample enrichment–separation systems, such as on-line SPE–HPLC employing tailored stationary phases [29]. By using special restricted access materials (RAMs) for sample enrichment, the carotenoids are retarded on the pre-column while the protein binding is broken and the macromolecules are eluted. The preparation of artifacts is hindered, as the whole analysis steps take place under conditions of light- and oxygen-exclusion. The scheme of on-line SPE–HPLC is presented in Figure 5.2.2.

The serum sample is transferred into the pre-column. Thereby, a small amount of organic solvent is required for breaking the protein binding. The

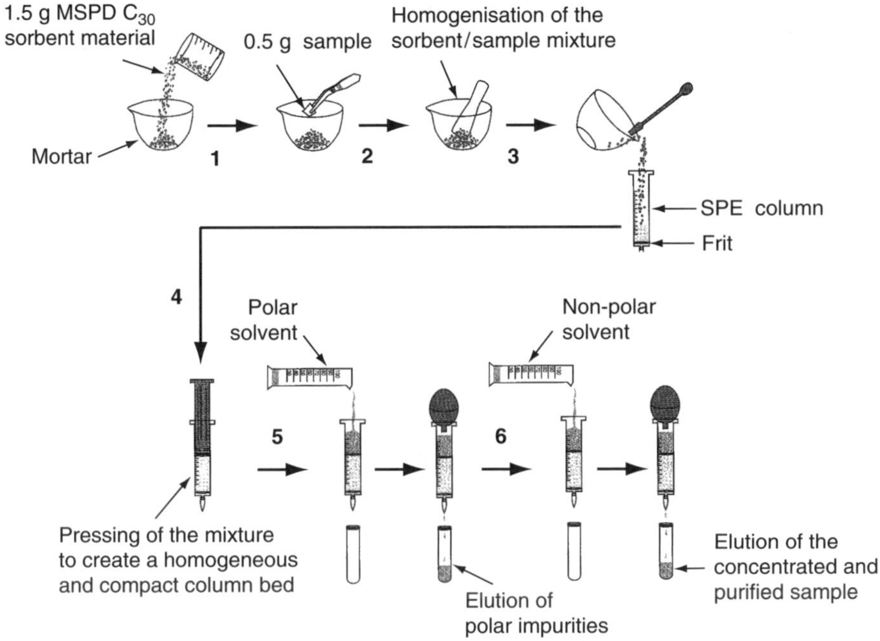

Figure 5.2.1 Scheme for the extraction of geometrical carotenoid isomers from solid biological samples employing matrix solid phase dispersion (MSPD)

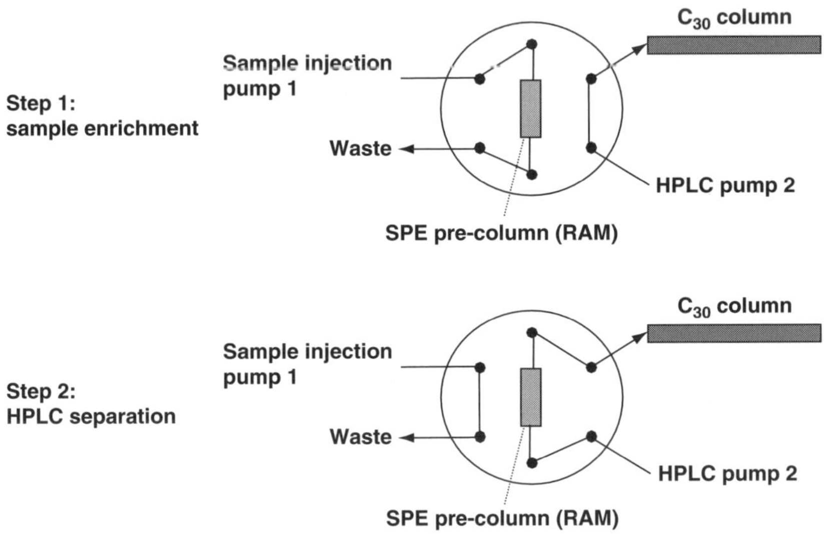

Figure 5.2.2 Scheme for the isolation and separation of geometrical carotenoid isomers from serum samples employing on-line solid phase extraction (SPE)–HPLC coupling

macromolecules are washed out while the carotenoids stay retained on the pre-column. By switching the valves, the pre-column is then connected to the analytical column. As the mobile phase consists mainly of organic solvent, the carotenoids are transferred immediately to the separation system.

5.2.3 ANALYSIS OF LYCOPENE STEREOISOMERS IN TOMATO EXTRACTS AND HUMAN SERUM

Vegetables are the most important source of carotenoids for humans. Most vegetables contain a variety of carotenoids [30,31]. Only carrots (β-carotene) and tomatoes (lycopene) [32] contain one major carotenoid. However, even in these vegetables, several minor components can be identified, most of them geometrical isomers – structurally similar, but nevertheless totally different. For nutritional aspects, the level of these stereoisomers in nature is significant, because their biomedical properties can differ enormously.

Figure 5.2.3 depicts the HPLC chromatogram of a tomato peel extract monitored by UV absorbance at 469 nm. The separation was performed on a 150 × 4.6 mm C_{30} column (ProntoSil, 3 μm, 200 Å, Bischoff, Germany) at room temperature and a flow rate of 1 ml/min with a binary mixture of acetone/water, developed for LC–NMR experiments. The 50-min gradient elution was performed in four steps, i.e. (1) an initial 3 min with 75/25 (v/v) acetone/water, (2) a 24-min gradient to 100 % acetone, (3) an isocratic step from 27–45 min with 100 % acetone, and (4) a 2-min gradient back to the initial conditions.

As shown, tomatoes contain solely carotenes – lycopene (95 % of the total carotenoid content) and β-carotene (5 %). The good resolution and shape of the separation is due to the enhanced shape selectivity of C_{30} stationary phases in comparison to C_{18} phases [33,34]. Furthermore, C_{30} phases have a higher loading capacity, and therefore are preferably suitable for LC–NMR experiments [35–37].

The main peak at 35 min belongs to all-*E* lycopene (82 %, peak 6). Furthermore, peak 1 can be assigned to all-*E* β-carotene, whereas peaks 2–5 belong to various *Z* and *ZZ* lycopene stereoisomers, as summarised in Table 5.2.1. Thereby, the peak assignment can be proven by recording stopped-flow LC–NMR spectra as shown in Figures 5.2.5 and 5.2.6 (see below).

In comparison to the tomato vegetable extract, the HPLC separation of a human serum sample is shown in Figure 5.2.4. The HPLC system used consisted of a 150 × 4.6 mm C_{30} column (3 μm, 200 Å, YMC, USA) with a ternary mixture of methanol/TBME/water (83/15/2 v/v/v for solvent A, and 8/90/2 v/v/v for solvent B, respectively). The 45-min gradient separation at a flow rate of 1 ml/min and 16°C was as follows: (1) 100 % A for 1 min, (2) a 7-min gradient to 70 % A, (3) a 5-min hold at 70 % A, (4) a 9-min gradient to 45 % A, (5) a 2-min hold at 45 % A, (6) a 10-min gradient to 5 % A, (7) a 4-min hold at 5 % A, and (8) a 2-min gradient back to the initial conditions.

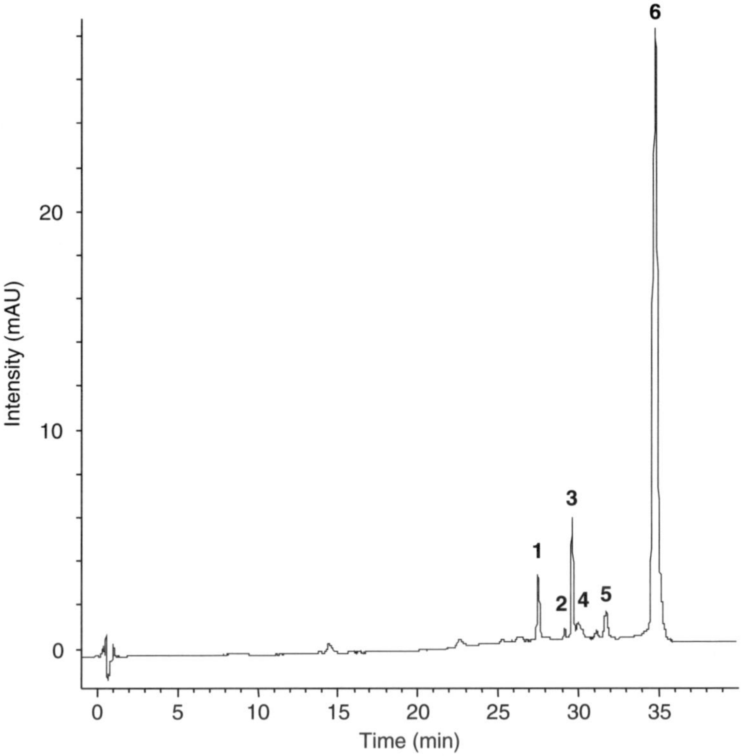

Figure 5.2.3 HPLC chromatogram of a raw tomato peel sample: 150 × 4.6 mm C_{30} column (ProntoSil, 3 μm, 200 Å, Bischoff, Germany), flow rate 1 ml/min, 22 °C, 469 nm, eluent acetone/water. Gradient steps: (1) 3 min with 75/25 (v/v) acetone/water; (2) a 24-min gradient to 100 % acetone; (3) an 18-min hold at 100 % acetone; (4) a 2-min gradient back to the initial conditions

Table 5.2.1 HPLC data obtained for a tomato peel extract.

Peak	Retention time (min)	Peak area (%)	Identification
1	27.7	5.0	All-*E* β-carotene
2	29.5	0.5	9,13′-*ZZ* lycopene
3	29.9	9.5	13-*Z* lycopene
4	30.3	0.5	9,13-*ZZ* lycopene
5	32.0	2.5	9-*Z* lycopene
6	35.0	82.0	All-*E* lycopene

The human body stores a variety of carotenes (lycopene, α- and β-carotenes), as well as xanthophylls (lutein, zeaxanthin and cryptoxanthin) [38]. Besides the main naturally occurring all-*E* configuration, there also exist some *Z*-stereoisomers of β-carotene and lycopene in the human serum at remarkable levels, as shown in Table 5.2.2 [39,40].

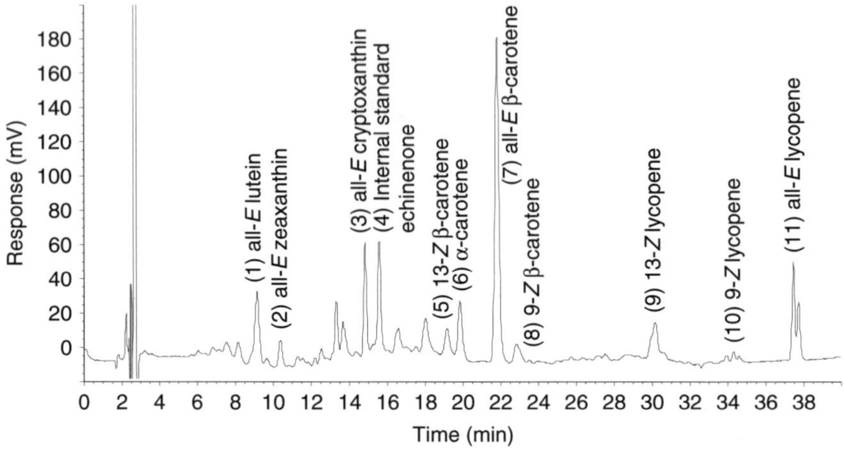

Figure 5.2.4 HPLC chromatogram of a human serum sample: 150 × 4.6 mm C_{30} column (3 μm, 200 Å, YMC, USA), flow rate 1 ml/min, 16 °C, 450 nm, eluent methanol/TBME/water (solvent A 83/15/2 v/v/v, and solvent B 8/90/2 v/v/v, respectively). Gradient steps: (1) 100 % A for 1 min; (2) a 7-min gradient to 70 % A; (3) a 5-min hold at 70 % A; (4) a 9-min gradient to 45 % A; (5) a 2-min hold at 45 % A; (6) a 10-min gradient to 5 % A; (7) a 4-min hold at 5 % A; (8) a 2-min gradient back to the initial conditions

Table 5.2.2 HPLC data obtained for a human serum sample.

Peak	Retention time (min)	Peak area (%)	Identification
1	9.2	9.9	All-*E* lutein
2	10.4	3.2	All-*E* zeaxanthin
3	14.9	7.6	All-*E* cryptoxanthin
4	15.7		Internal standard – echinenone
5	19.3	2.7	13-*Z* β-carotene
6	20.0	4.4	All-*E* α-carotene
7	21.9	30.3	All-*E* β-carotene
8	23.0	3.6	9-*Z* β-carotene
9	30.4	12.4	13-*Z* lycopene
10	34.6	5.5	9-*Z* lycopene
11	37.7	20.4	All-*E* lycopene

The ratio of all lycopene *Z*-isomers to the total lycopene content in human serum is 0.47, and is therefore more than threefold higher in comparison to that of tomato peel (ratio of lycopene *Z*-isomers to the total lycopene content is 0.14), which contains mainly all-*E* lycopene. This concludes that these geometrical isomers are produced in the human body, and therefore have a special role to play within human organisms.

5.2.4 IDENTIFICATION OF LYCOPENE STEREOISOMERS IN TOMATO EXTRACTS EMPLOYING LC–NMR

The lycopene stereoisomers were identified by using a Bruker AMX 600 spectrometer and, as a control unit, a Bruker peak sampling unit (BPSU 12). In general, the assignments of the *Z*/*E*-isomers are available, adapted from the Karplus equation, by taking the differences of the coupling constants into consideration.

Nevertheless, carotenoids contain a system of conjugated olefinic protons (up to 11 conjugated double bonds). In this case, the differences in the chemical shift values are more expressive – they can amount up to 0.6 ppm. Therefore, geometrical carotenoid isomers are identified by their signals in the olefinic region. In comparison to the respective all-*E* configuration, the protons near to the *cis*-bond are shifted, the 'outer' protons experience a shift to higher field, while the 'inner' protons shift to lower field.

The acquisition of a two-dimensional (2D) NMR spectrum gives the advantage of obtaining information about both the chemical shifts values δ and the coupling system of each proton (the 2D plot allows us to establish the assignments of such coupling systems). Figure 5.2.5 depicts the stopped-flow 2D COSY NMR spectrum of all-*E* lycopene from a tomato extract. This was recorded within 24 h with a total of 280 increments and 256 transients.

All-*E* lycopene is a centrosymmetric molecule, and, therefore contains nine distinguishable olefinic protons. As demonstrated in the 2D plot, there exist three sets of coupled protons. Proton H_7 at 6.58 ppm (t) couples with H_6 (d, 5.99 ppm) and H_8 (d, 6.29 ppm), while proton H_{11} at 6.75 ppm (t) couples with H_{10} (d, 6.24 ppm) and H_{12} (d, 6.42 ppm). Proton H_{15} (d, 6.75 ppm) shows only a coupling to H_{14} (d, 6.35 ppm), as it is identical to proton $H_{15'}$.

The differences in the chemical shift values in the NMR spectra of carotenoid stereoisomers are demonstrated in Figure 5.2.6. In this figure, the NMR spectra (olefinic range) of all of the five lycopene strereoisomers identifiable in tomato peel extract are depicted.

Due to the *Z*-configuration, peaks 2–5 have an unsymmetrical structure, thus leading to different chemical shifts for the quoted and unquoted protons. For example, the unsymmetrical 9-*Z* lycopene (peak 5, Figure 5.2.6(b)) shows a strong low-field shift for the 'inner' protons H_8 ($\delta = 6.92$ ppm, $\Delta\delta = 0.63$ ppm) and H_{11} ($\delta = 6.95$ ppm, $\Delta\delta = 0.20$ ppm), and a weak one for H_6 ($\delta = 6.06$ ppm, $\Delta\delta = 0.06$ ppm). The 'outer' protons H_{10} ($\delta = 6.09$ ppm, $\Delta\delta = 0.15$ ppm) and H_{12} ($\delta = 6.36$ ppm, $\Delta\delta = 0.06$ ppm) are shifted to higher field. The *cis*-bond at C-9 has no influence on the chemical shift of all quoted protons as these are too far away from the *cis*-bond. The assignment of 13-*Z* lycopene (peak 3, Figure 5.2.6(c)) is similar. The 'inner' protons H_{12} ($\delta = 7.03$ ppm, $\Delta\delta = 0.61$ ppm) and H_{15} ($\delta = 6.95$ ppm, $\Delta\delta = 0.20$ ppm) show a strong shift to lower field, whereas H_{10} ($\delta = 6.29$ ppm, $\Delta\delta = 0.05$ ppm) is weakly shifted to lower field. The 'outer'

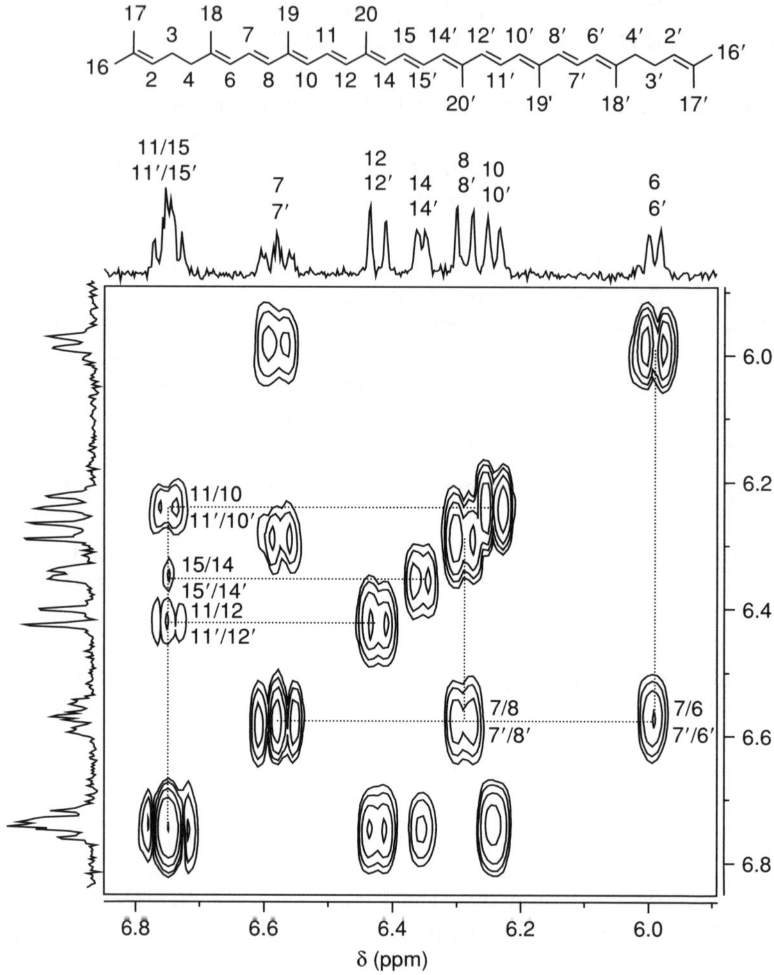

Figure 5.2.5 Stopped-flow COSY NMR spectrum of all-*E* lycopene from a tomato peel extract

protons H_{14} ($\delta = 6.18$ ppm, $\Delta\delta = 0.17$ ppm) and $H_{15'}$ ($\delta = 6.66$ ppm, $\Delta\delta = 0.09$ ppm) are shifted markedly to higher field.

The other two lycopene stereoisomers identified in the extract can be assigned to 9,13-*ZZ* lycopene and 9,13′-*ZZ* lycopene. These NMR spectra summarise the effects of the NMR spectra of both 9-*Z* and 13-*Z* lycopene. The assignment of 9,13′-*ZZ* lycopene (peak 2, Figure 5.2.6(d)) can be easily achieved because the two *cis*-bonds do not interfere, and so the resulting NMR spectrum looks like the addition of the NMR spectra of 9-*Z* and 13-*Z* lycopene.

The NMR spectrum of 9,13-*ZZ* lycopene (peak 4, Figure 5.2.6(e)) is more difficult to evaluate, as the two *cis*-bonds affect one another. This can be

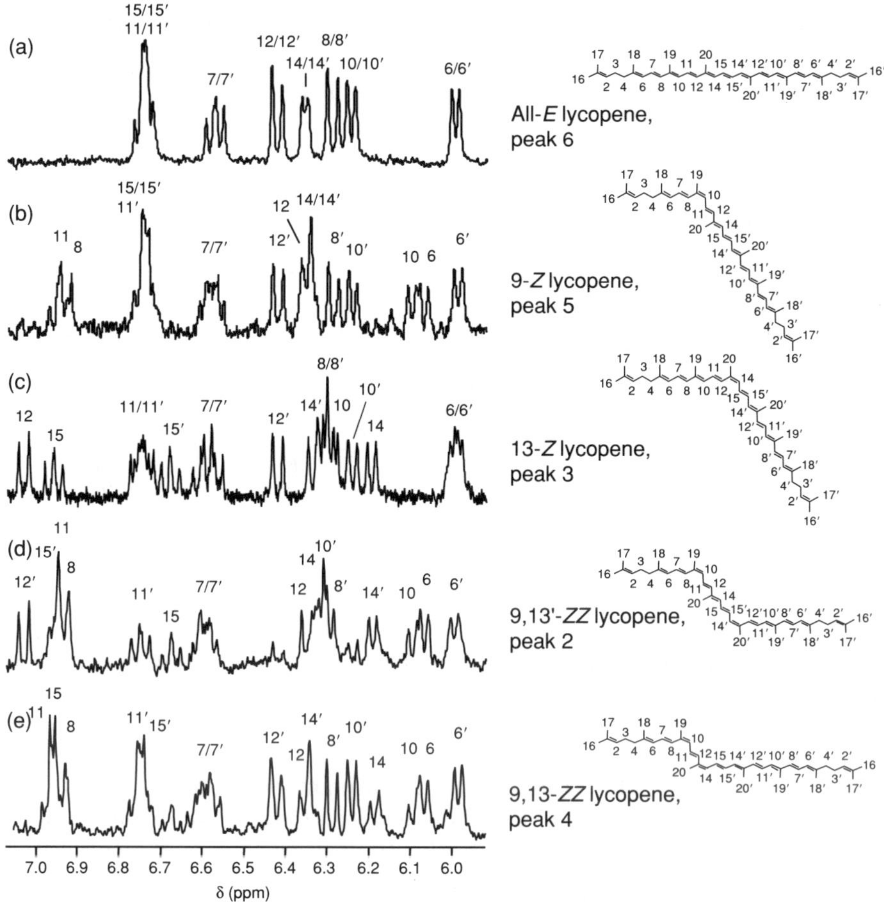

Figure 5.2.6 Stopped-flow ^{1}H NMR spectra of various geometrical lycopene stereoisomers: (a) all-*E*; (b) 9-*Z*; (c) 13-*Z*; (d) 9,13′-*ZZ*; (e) 9,13-*ZZ*

demonstrated with the chemical shift of the protons H_{10} and H_{12}, located between the two *cis*-bonds. Proton H_{10} ($\delta = 6.09$ ppm, $\Delta\delta = 0.15$ ppm) obtains a strong high-field shift from the *cis*-bond at the C-9 position and a weak low-field shift from the bond at the *cis*-13 position. Proton H_{12} ($\delta = 6.36$ ppm, $\Delta\delta = 0.06$ ppm) is more affected by the *cis*-bond at the C-9 position and therefore results in a weak high-field shift. The assignment of the olefinic protons of all identified lycopene stereoisomers is summarised in Table 5.2.3

5.2.5 CONCLUSIONS

Human serum contains a variety of carotenes and xanthophylls. These carotenoids are solely ingested from nutritional sources, for instance lycopene is

Table 5.2.3 ^{1}H NMR data (in ppm) of the olefinic protons for the identified lycopene stereoisomers

	all-E lycopene	9-Z lycopene	13-Z lycopene	9,13'-ZZ lycopene	9,13-ZZ lycopene
H-6	5.99	6.06	5.99	6.06	6.06
H-6'	5.99	5.99	5.99	5.99	5.99
H-7	6.58	6.58	6.58	6.58	6.58
H-7'	6.58	6.58	6.58	6.58	6.58
H-8	6.29	6.92	6.29	6.92	6.92
H-8'	6.29	6.29	6.29	6.29	6.29
H-10	6.24	6.09	6.29	6.09	6.09
H-10'	6.24	6.24	6.24	6.29	6.24
H-11	6.75	6.95	6.75	6.95	6.95
H-11'	6.75	6.75	6.75	6.75	6.75
H-12	6.42	6.36	7.03	6.36	6.36
H-12'	6.42	6.42	6.42	7.03	6.42
H-14	6.35	6.35	6.18	6.35	6.18
H-14'	6.35	6.35	6.35	6.18	6.35
H-15	6.75	6.75	6.95	6.66	6.95
H-15'	6.75	6.75	6.66	6.95	6.75

mainly ingested from tomatoes or tomato products. However, tomato peel extracts contain not just natural all-*E* lycopene, as various *Z*-stereoisomers of lycopene, at a content of 13 % of the total carotenoid content, can be identified.

Therefore, a mild and quick extraction technique is necessary to exclude the preparation of artifacts. The carotenoid stereoisomers can be quantitatively analysed, employing MSPD extraction, from plant material, as well as from serum samples, using on-line SPE without any isomerisation or oxidation of the carotenoids. The extraction step is coupled to the separation and identification steps. Here, LC–NMR hyphenation, employing C_{30} stationary phases, is suitable for unambiguous distinction between all of these stereoisomers.

We were able to identify five geometrical isomers of lycopene in tomato peel extracts (all-*E*, 9-*Z*, 13-*Z*, 9,13-*ZZ* and 9,13′-*ZZ* lycopenes) by recording LC–NMR spectra. In human serum, we have identified three of these isomers (all-*E*, 9-*Z* and 13-*Z* lycopenes). In comparison to the nutritional source (tomato), the two identified lycopene *Z*-isomers are enriched in the human serum sample, which indicates a specific role of these geometrical isomers within human organisms.

REFERENCES

1. Omenn, G. S., Goodman, G. E., Thornquist, M. D., Balmes, J., Cullen, M. R., Glass, A., Keogh, J. P., Meyskens, F. L., Jr, Valanis, B., Williams, J. H., Barnhart, S. and Hammar, S., 'The CARET study', *New Engl. J. Med.*, 1996, **334**, 1150.

2. Klipstein-Grobusch, K., Launer, L. J., Geleijnse, J. M., Boeing, H., Hofman, A. and Witteman, J. C. M., 'The Rotterdam study', *Atherosclerosis*, 2000, **148**, 49.
3. Seddon, J. M., Ajani, U. A., Sperduto, R. D., Hiller, R., Blair, N., Burton, T. C., Farber, M. D., Gragoudas, E. S., Haller, J., Miller, D. T., Yannuzzi, L. A. and Willett, W., 'The Eye Disease Case-Control Study Group', *J. Am. Med. Assoc.*, 1994, **272**, 1413.
4. Hennekens, C. H., *Pure Appl. Chem.*, 1997, **69**, 2141.
5. Lyle, B. J., Mares-Perlman, J. A., Klein, B. E. K., Klein, R. and Greger, J. L. 'The Beaver Dam eye study', *Am. J. Epidemiol.*, 1999, **149**, 801.
6. Hammond, Jr, B. R., Wooten, B. R. and Curran-Celentano, J., *Arch. Biochem. Biophys.*, 2001, **385**, 41.
7. Landrum, J. T. and Bone, R. A., *Arch. Biochem. Biophys.*, 2001, **385**, 28.
8. Jantzi, J. D. and Jantzi, J. A., *Prac. Optom.*, 2000, 11.
9. Chaban-Taber, L., Willett, W. C., Seddon, J. M., Stampfer, M. J., Rosner, B., Colditz, G. A., Speizer, F. E. and Hankinson, S. E., *Am. J. Clin. Nutr.*, 1999, **70**, 509.
10. Brown, L., Rimm, E. B., Seddon, J. M., Giovannucci, E. L., Chaban-Taber, L., Spiegelman, D., Willett, W. C. and Hankinson, S. E., *Am. J. Clin. Nutr.*, 1999, **70**, 517.
11. Arab, L. and Stock, S., *Am. J. Clin. Nutr.*, 2000, **71**, 1691.
12. Hennekens, C. H., Buring, J. E., Manson, J. E., Stampfer, M. J., Rosner, B., Cook, N. R., Belanger, C., LaMotte, F., Gaziano, J. M., Ridker, P. M., Willett, W. C. and Peto, R., *New Engl. J. Med.*, 1996, **334**, 1145.
13. Cooper, D. A., Eldrigde, A. L. and Peters, J. C., *Nutr. Rev.*, 1999, **57**, 133, 201.
14. Zhang, S., Hunter, D. J., Forman, M. R., Rosner, B. A., Speizer, F. E., Colditz, G. A., Manson, J. E., Hankinson, S. E. and Willett, W. C., *J. National Cancer Inst.*, 1999, **91**, 547.
15. Nishino, H., Tokuda, H., Satomi, Y., Masuda, M., Bu, P., Onozuka, M., Yamaguchi, S., Okuda, Y., Takayasu, J., Tsuruta, J., Okuda, M., Ichiishi, E., Murakoshi, M., Kato, T., Misawa, N., Narisawa, T., Takasuka, N. and Yano, M., *Pure Appl. Chem.*, 1999, **71**, 2273.
16. Rice-Evans, C. A., Sampson, J., Bramley, P. M. and Holloway, D. E., *Free Radical Res.*, 1997, **26**, 381.
17. Müller, R. K., Bernhard, K., Giger, A., Moine, G. and Hengartner, U., *Pure Appl. Chem.*, 1997, **69**, 2039.
18. Boileau, A. C., Merchen, N. R., Wasson, K., Atkinson, C. A. and Erdman, Jr, J. W., *J. Nutr.*, 1999, **129**, 1176.
19. Castenmiller, J. J. M. and West, C. E., *Annu. Rev. Nutr.*, 1998, **18**, 19.
20. Schalch, W., Dayhaw-Barker, P. and Barker II, F. M., 'The carotenoids of the human retina', in *Nutritional and Environmental Influences on Vision*, A. Taylor (Ed.), CRC Press, Boca Raton, FL, 1999, Ch. 20, pp. 215–250.
21. Lacker, T., Strohschein, S. and Albert, K., *J. Chromatogr., A*, 1999, **854**, 37.
22. Hagiwara, T., Yasuno, T., Funayama, K. and Suzuki, S., *J. Chromatogr., B*, 1998, **708**, 67.
23. Dachtler, M., Glaser, T., Kohler, K. and Albert, K., *Anal. Chem.*, 2001, **73**, 667.
24. van Breemen, R. B., *Pure Appl. Chem.*, 1997, **69**, 2061.
25. Strohschein, S., Pursch, M. and Albert, K., *J. Pharm. Biomed. Anal.*, 1999, **21**, 669.
26. Barker, S. A., *J. Chromatogr., A*, 2000, **885**, 115.
27. Barker, S. A., *J. Chromatogr., A*, 2000, **880**, 63.
28. Glaser, T., Dachtler, M. and Albert, K., *GIT Fachz. Lab.*, 1999, **9**, 904.
29. Dachtler, M., Kohler, K. and Albert, K., *J. Chromatogr., B*, 1998, **720**, 211.

30. *USDA-NCC Carotenoid Database for US Foods* (1998): http://www.nal.usda.gov/fnic/foodcomp/Data/car98/car98.html.
31. Sommerburg, O., Keunen, J. E. E., Bird, A. C. and van Kuijk, F. J. G. M., *Br. J. Ophthalmol.*, 1998, **82**, 907.
32. Clinton, S. K., *Nutr. Rev.*, 1998, **56**, 35.
33. Albert, K., *Trends Anal. Chem.*, 1998, **17**, 648.
34. Sander, L. C., Pursch, M., Märker, B. and Wise, S. A., *Anal. Chem.*, 1999, **71**, 3477.
35. Dachtler, M., Glaser, T., Händel, H., Lacker, T., Tseng, L.-H. and Albert, K., in *Encyclopedia of Separation Science*, 2nd Edn, I. D. Wilson, E. R. Adlard, M. Cooke and C. F. Poole (Eds), Academic Press, London, 2000, pp. 747–760.
36. Albert, K., Dachtler, M., Glaser, T., Händel, H., Lacker, T., Schlotterbeck, G., Strohschein, S. and Tseng, L.-H., *J. High Resol. Chromatogr.*, 1999, **22**, 135.
37. Albert, K., *J. Chromatogr., A*, 1999, **856**, 199.
38. Paetau, I., Khachik, F., Brown, E. D., Beecher, G. R., Kramer, T. R., Chittams, J. and Clevidence, B. A., *Am. J. Clin. Nutr.*, 1998, **68**, 1187.
39. Johnson, E. J., Qin, J., Krinsky, N. I. and Russell, R. M., *J. Nutr.*, 1997, **127**, 1993.
40. Yeum, K.-J., Booth, S. L., Sadowski, J. A., Liu, C., Tang, G., Krinsky, N. I. and Russell, R. M., *Am. J. Clin. Nutr.*, 1996, **64**, 594.

6 LC–NMR in Environmental Analysis

ALFRED PREISS
Fraunhofer Institut für Toxikologie und Aerosolforschung, Hannover, Germany

and

MARKUS GODEJOHANN
Bruker BioSpin GmbH, Rheinstetten, Germany

6.1 INTRODUCTION

The introduction of chemicals into the environment is considerable. Large amounts of organic compounds are released into the environment every year by industrial and agricultural processes, traffic, urban waste disposal and ecological disasters. Once present in the environment, they are subjected on the one hand to transport processes in air, water and soil and, on the other hand, they are subjected to the influence of the 'reactor environment', i.e. transformation products may be formed by chemical, photochemical and microbiological transformation processes. Chemical reactions with other pollutants present in the environment can also take place. As a result of these processes, a variety of new and unexpected compounds can be formed from the originally released pollutants and, as a rule, they are more polar than the parent compounds.

It is interesting that, in analytical chemistry, besides the efforts to increase the sample throughput and to decrease the detection limits, another trend can be observed which is directed to the analysis of more and more complex mixtures without laborious sample preparation and separation steps. This development was triggered by the requirements of bio- and environmental analysis and is closely connected to the development of multidimensional analytical methods, as well as hyphenated techniques which provide much more selectivity than one-dimensional analytical methods. Among the range of hyphenated techniques, those which combine a high separation efficiency with a maximum of structural information are of particular importance. These are hyphenated techniques such as GC–MS, LC–MS, LC–NMR and LC–NMR–MS.

On-Line LC–NMR and Related Techniques
Edited by K. Albert.

6.2 TARGET AND NON-TARGET ANALYSIS

In the past, the first to be analysed were the parent pollutants that were released into the environment originally. Later on, in some cases, important known degradation products were included in the characterization of environmental samples. Since the chemical and physical properties of these compounds were known in most cases, specific analytical methods for their determination in air, water and soil could be developed. Chromatographic methods such as gas chromatography (GC) and high performance liquid chromatography (HPLC), in connection with specific detection systems, such as electron capture (EC), nitrogen–phosphorous (NP), photodiode array (PDA), mass spectrometry (MS) and fluorescence (FI) detectors, played an important role in this. The permitted contents for these target compounds, for example, pesticides, were strictly regulated by national and European legislation and consequently monitored in rigid analytical schemes (target analysis). All of these detectors are much more sensitive than the NMR detector. Therefore, in target analysis the use of the LC–NMR coupling is not meaningful.

In the case of degradation products and metabolites, the situation is different. Generally, these compounds were not analysed because in most cases they are not regulated and no effective analytical methods exist for their determination. This means that a correct diagnosis of the environmental situation cannot be made and, as a consequence, no appropriate action can be taken. Therefore, in order to improve the risk assessment of a hazardous waste site for example, as many compounds as possible should be analysed at the beginning of the investigations (non-target analysis).

This is especially true in cases where the pollutants are non-persistent, and degradation products, as well as metabolites, can be expected. After the identification of the pollutants, suitable leading components should then be selected for further monitoring.

Unfortunately, non-target analysis is much more difficult to perform than target analysis. In the former, sample extraction and separation must take into account the very different (and *a priori* unknown) physical and chemical properties of the individual organic compounds, while the detection methods must provide a maximum of structural information. Furthermore, the separation of complex mixtures is often incomplete, especially when liquid chromatography is used. For compounds amenable to GC (volatile and semi-volatile thermally stable compounds), the situation is less critical since the separation efficiency of gas chromatography is high and the mass detector under electron-impact conditions provides valuable structural information. Furthermore, large MS libraries are available for compound identification. In principle, many polar compounds can also be analysed by gas chromatography after derivation but this approach is time-consuming and not suitable for non-target analysis since the chemical properties of the degradation products and metabolites are unknown. Polar compounds can be separated better by HPLC but, for a long time, only the UV or PDA detector were available and the structural information of a UV spectrum is completely insufficient for non-target analysis. Over the last ten years, however, the LC–MS coupling has been used as a routine method for the target analysis of polar compounds. Due to the extended possibilities offered by the MS–MS and MS^n techniques, it has also been increasingly used for the identification of unknown compounds. However, the LC–MS technique has some disadvantages which limit its use for non-target analysis, as follows:

1. The possibilities for structural elucidation of unknowns are restricted; in particular, the distinction between isomers is only possible in exceptional cases.

2. There are no universal ionization conditions under which any unknown compound can be expected to be ionized, and the ionization efficiency depends strongly on the chemical properties of the analytes.

3. For the quantification of the identified compounds, 'genuine' reference compounds are needed, which are often not available for degradation products and metabolites.

The most important method for structural analysis is nuclear magnetic resonance (NMR) spectroscopy and its information is often complementary to that of mass spectrometry. It therefore seems very meaningful to also use LC–NMR coupling alone, or in combination with LC–MS for the non-target analysis of complex environmental samples [1]. In the past, this failed because of the low sensitivity of the NMR spectrometer and the limited dynamic range of the digitizer and receiver which did not allow very weak analyte signals to be detected in the presence of strong solvent signals. Today, the sensitivity of a high-field NMR spectrometer (> 500 MHz) is much higher and there are a number of very efficient methods for the suppression of the solvent signals.

6.3 LC–NMR COUPLING IN NON-TARGET ANALYSIS

6.3.1 HOW DO WE OBTAIN A REALISTIC PICTURE OF THE SAMPLE?

For non-target analyses, LC–NMR coupling has the advantage that the NMR detector (here, only ^1H NMR is regarded) is a *universal* detector, i.e. all proton-carrying compounds (these are nearly all organic compounds) are detected. It is a further advantage that the NMR detector more directly reflects the concentration of a component in the sample. (For a given concentration, the signal intensity only depends on the molecular mass and the number of equivalent protons.) As a rule, this leads to a more realistic picture of the composition of the sample than is provided by the UV or MS detectors.

6.3.2 IMPROVEMENT OF SELECTIVITY

The coupling of two different analytical methods (here, liquid chromatography and NMR spectroscopy) can lead to a significant improvement of the selectivity compared to a single analytical method. In the case of LC–NMR coupling, the extent that can be reached depends strongly on the nature of the sample. The highest increase can be expected if the mixtures consist of small molecules that have simple NMR spectra. Good examples of this are aromatic compounds whose NMR spectra consist of a limited number of signals with simple coupling patterns. Therefore, in complex environmental mixtures of aromatic compounds, components which overlap in the chromatographic dimension can be resolved in most cases in the spectroscopic dimension.

6.3.3 WHICH CLASSES OF COMPOUNDS?

From the selectivity point of view, LC–NMR coupling is especially suited to the analysis of compound classes such as nitroaromatics, phenols, aromatic amines, aromatic carboxylic acids, polyaromatic hydrocarbons (PAHs), polychlorinated biphenyls (PCBs), and azo- and anthraquinone dyes. Another advantage of LC–NMR coupling for the investigation of aromatic compounds in environmental samples is that the position of substituents on the aromatic ring, e.g. in unknown metabolites or degradation products, can best be determined by NMR spectroscopy.

For aliphatic compounds with longer alkyl chains, such as surfactants, the NMR detector can contribute little to an increased selectivity of the LC–NMR coupling since, in the range of aliphatic protons, the spectra are often complex. Moreover, analyte signals around 2 ppm can be suppressed or influenced by the solvent suppression when acetonitrile is used as the organic component of the eluent. Since surfactants are present in many environmental samples, they pose problems for non-target analysis, not only because of their complex spectra but also because they can influence the separating properties of the analytical column by their surface activity [2].

6.3.4 QUANTIFICATION

While chromatographic methods with commonly used detectors (UV, MS, etc.) for quantification always require calibration data of genuine reference compounds, the situation for the NMR detector is different, because the NMR method is almost an absolute method. Here, it is only necessary that the compound is unambiguously identified and that a suitable signal is available for quantification. Then, after addition of a standard with a known concentration, the compound can be quantified. This is advantageous in those cases in which no reference compounds are commercially available for metabolites or degradation products for which the structures have been elucidated. However, in LC–NMR, quantification up until now can only be carried out in the on-flow mode [3] and the precision depends on the signal-to-noise ratio which, in most cases, is low in on-flow runs.

6.3.5 CONDITIONS FOR THE ON-FLOW MODE

The general concepts of LC–NMR coupling have been described in previous chapters of this volume. Here, only the on-flow mode is discussed with respect to the non-target analysis of environmental samples. This approach is particularly suited to obtaining a good overview of all organic compounds (including UV-inactive) present in the sample. In the on-flow mode the mobile phase with the eluent flows through the NMR cell while the spectrometer continuously acquires NMR data. This simple and rugged technique has the advantage that the chromatographic separation will not be affected in any way. However, the detection limits achieved in the on-flow mode are relatively high (ppb range), as the NMR data of a peak can only be acquired while the peak passes the NMR flow cell. Therefore, the possibility of improving the signal-to-noise (S/N) ratio by accumulation is limited, and chromatographic, as well as spectroscopic, parameters have to be optimized in order to obtain better detection limits. One possibility is to optimize the chromatographic separation in such a way that the peak volume (2 $\times$ peak width at half-height $\times$ flow rate) is as small as possible, i.e. the analyte concentration in the chromatographic peak is as high as possible [4]. The best conditions are achieved when the peak volume is equal or smaller than the volume of the NMR flow cell (for example, the active volume of the Bruker 4 mm flow probe is 120 μl). Another simple approach is to reduce the flow rate. As investigations have shown [5], the reduction of the flow rate by a factor of 20 (e.g. from 1.0 to 0.05 ml/min) leads only to a slight increase in the peak volume or a slight decrease in the analyte concentration in the peak volume. As the S/N ratio increases with the square root of the number of scans, an improvement in sensitivity by a factor of 2–3 can be achieved. Another point is that the chromatographic resolution in the flow probe is obviously maintained to a higher degree at low flow rates, as can be seen from Figure 6.1. Thus, in the on-flow mode under optimized conditions for the aromatic protons in a mixture of the four *p*-hydroxybenzoic acid esters (methyl, ethyl, propyl and butyl), absolute detection limits of 2 μg were determined with a 4 mm inverse probe head (Bruker). For comparison, detection

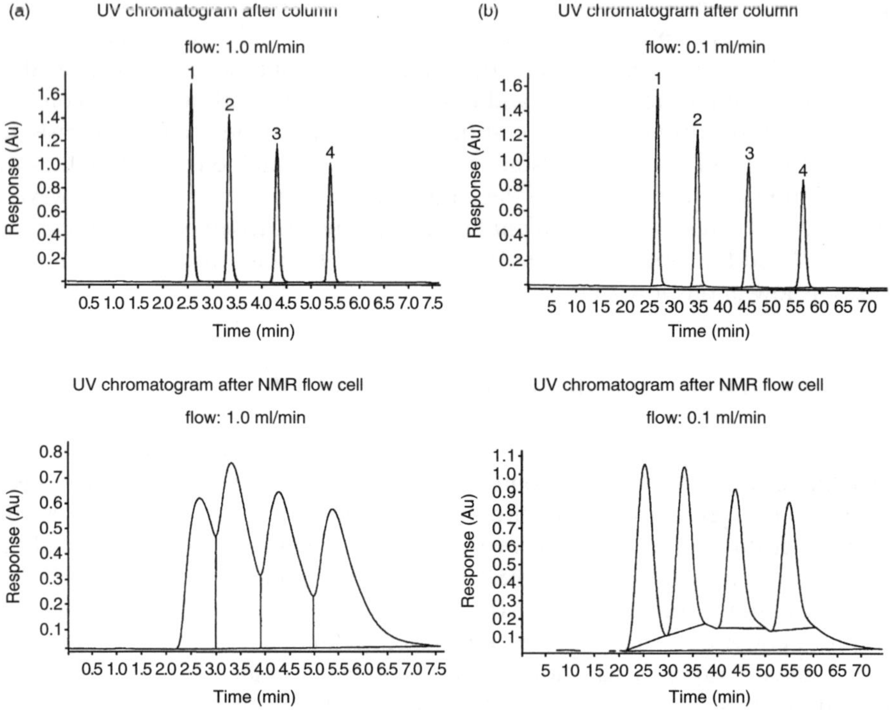

Figure 6.1 UV chromatograms of the test mixture of four *p*-hydroxybenzoic acid esters (1, methyl; 2, ethyl; 3, propyl; 4, butyl) after the column and after the NMR flow cell at flow rates of (a) 1.0 and (b) 0.1 ml/min conditions: column, LiChrospher RP select B, 125×4 mm id, 5 μm, spectrometer, Bruker DRX 600; probe head, 4 mm *z*-gradient LC probe, active volume 120 μl; eluents, acetonitrile (A) and D_2O (B); gradient, $t = 0$ min A/B (40/60), $t = 8$ min A/B (70/30) at a flow rate of 1.0 ml/min and t = 80 min A/B (70/30) at a flow rate of 0.1 ml/min

limits in the nanogram range are achieved in the stop-flow mode but in this detection mode the chromatographic process is disturbed.

6.4 APPLICATION OF LC–NMR AND LC–NMR IN COMBINATION WITH LC–MS TO ENVIRONMENTAL SAMPLES

6.4.1 AMMUNITION HAZARDOUS WASTE SITES

As a result of the extensive production of ammunition before and during World War II, a large number of hazardous waste sites still exist in Germany, where both soil and water are polluted by explosives and their transformation products. The plants were often located near large ground water reservoirs which today, in some instances, represent important drinking water reservoirs. For the risk assessment of these sites, water and soil samples were analysed, where the analysis was usually restricted to target compounds such as explosives expected or found on the site, but also to several by-products and neutral transformation products formed by photo-or biodegradation of the parent compounds. Recent studies [6–9] have revealed that aqueous samples from such ammunition hazardous waste sites may contain a variety of polar nitroaromatic compounds which are in some cases major components. In previous investigations during target analysis, these compounds were overlooked.

Figure 6.2 demonstrates this in the case of a ground water sample from the former ammunition site in Elsnig (Germany). Many unknown compounds could be identified in the non-target analysis, e.g. 2,4-dinitrobenzoic acid (2,4-DNBA) and 3,5-dinitrophenol (3,5-DNP) which are major components (see Figure 6.2(a) and Table 6.1).

At present, physical and biological methods are applied in Germany for the disposal of ammunition waste. Figure 6.3 shows the LC–NMR chromatogram of an extract of a TNT-contaminated soil sample from Tanne (Lower Saxony, Germany) before biological treatment.

In the non-target analysis of this sample, carried out by LC–NMR and LC–MS, TNT-typical biodegradation products such as 2-amino-4,6-dinitrotoluene and 4-amino-2,6-dinitrotoluene were detected, although smaller quantities of 2-amino-4,6-dinitrobenzoic acid, trinitrobenzene and 2,2′6,6′-tetranitro-4,4′-azoxytoluene could also be identified. The NMR chromatogram further reveals that the soil was also contaminated by PAHs. Several late-eluting compounds could be identified as PAHs.

6.4.2 INDUSTRIAL EFFLUENTS AND LEACHATE FROM INDUSTRIAL LANDFILLS

The emissions of industrial chemicals in liquid effluents are regulated in the EU by several directives and guidelines listing compounds which should not exceed a given concentration (European Union EU Directive). On the other hand, industrial effluents may contain a high number of chemicals which are not considered by such a directive and which are, in many cases, unknown. Moreover, the composition of the chemicals typically present at the end of the production cycle is usually changed by effluent treatment. The same is true in the case of leachate from industrial landfills. Besides the parent pollutants, the leachate may contain additional transformation products which are produced during the ageing of the waste.

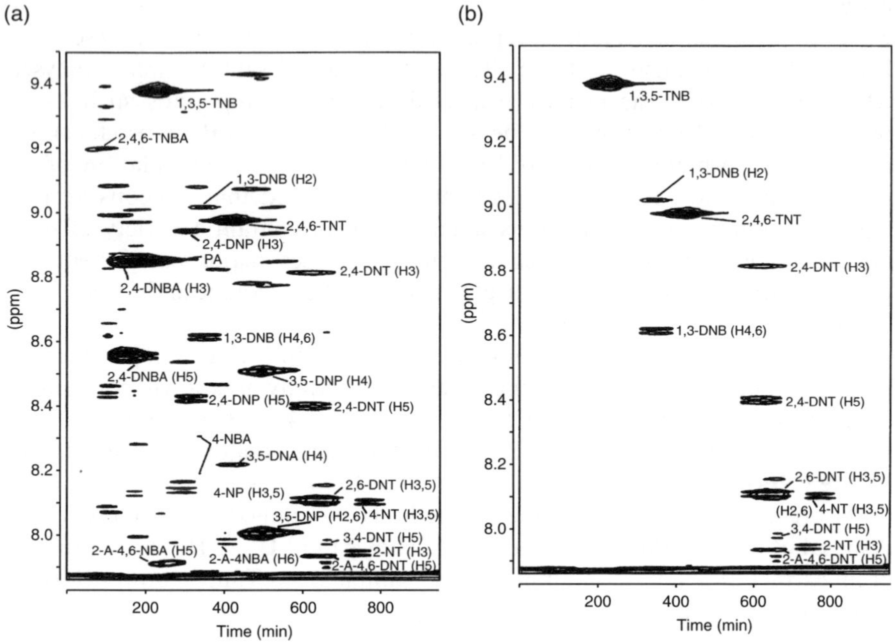

Figure 6.2 (a) Low-field part of the on-flow NMR chromatogram of a ground water extract from the former ammunition site in Elsnig (Germany). (see Table 6.1 for abbreviations). The numbers in parentheses assign the NMR signals to the protons generating this signal. Reprinted with permission from Godejohann, M., Astratou, M., Preiss, A., Levsen, K. and Mügge, C., *Anal. Chem.*, **70**, 4104–4110 (1998). Copyright (1998) American Chemical Society. (b) The same NMR chromatogram but showing only signals from those compounds which were determined in the target analysis by HPLC. All other signals were deleted

Table 6.1 Abbreviations used for compounds identified in the analyses of samples from former ammunition sites in Germany (c.f. Figures 6.2, 6.3 and 6.14).

Abbreviation	Full name
2-NT	2-Nitrotoluene
3-NT	4-Nitrotoluene
2,4-DNT	2,4-Dinitrotoluene
2,6-DNT	2,6-Dinitrotoluene
3,4-DNT	3,4-Dinitrotoluene
2,4,6-TNT	2,4,6-Trinitrotoluene
2-A-4,6-DNT	2-Amino-4,6-dinitrotoluene
4-A-2,6-DNT	4-Amino-2,6-dinitrotoluene
1,3-DNB	1,3-Dinitrobenzene
1,3,5-TNB	1,3,5-Trinitrobenzene
2,4-DNP	2,4-Dinitrophenol

Abbreviation	Full name
3,5-DNP	3,5-Dinitrophenol
PA	Picric acid (2,4,6-Trinitrophenol)
3,5-DNA	3,5-Dinitroaniline
4-NBA	4-Nitrobenzoic acid
2,4-DNBA	2,4-Dinitrobenzoic acid
2,4,6-TNBA	2,4,6-Trinitrobenzoic acid
2-A-4-NBA	2-Amino-4-nitrobenzoic acid
2-A-4,6-DNBA	2-Amino-4,6-dinitrobenzoic acid

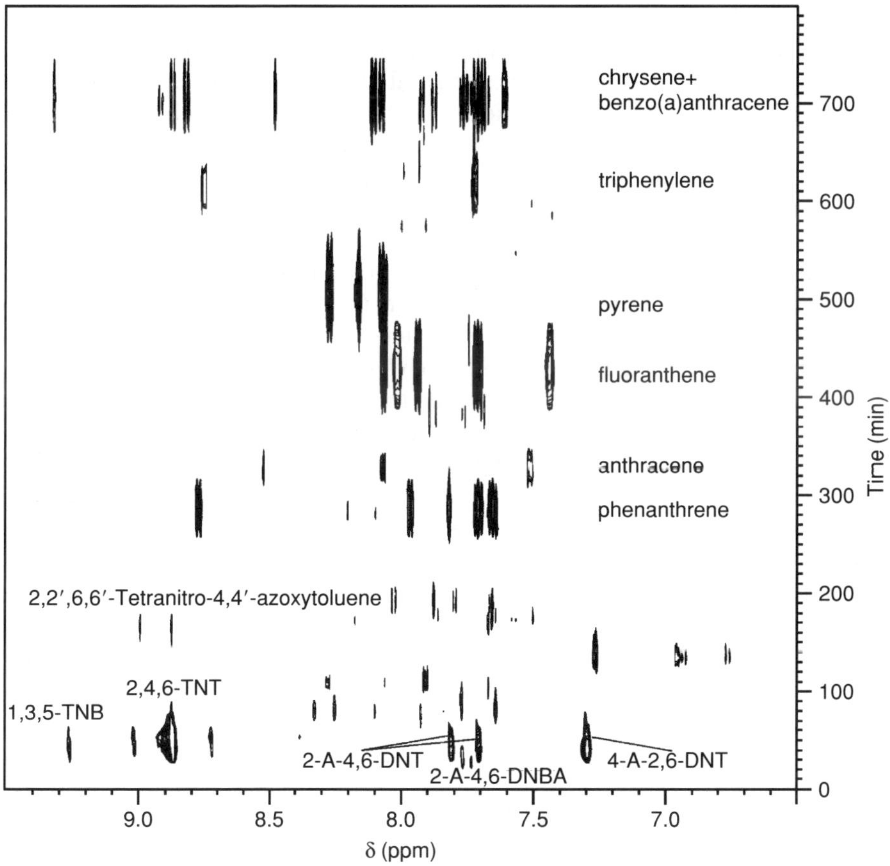

Figure 6.3 Low-field part of the on-flow NMR chromatogram of a soil extract from the former ammunition site in Tanne (Germany) (see Table 6.1 for abbreviations). Conditions: column, LiChrospher 100 RP18, 125 × 4 mm id, 5 μm; spectrometer, Bruker AMX 600; probe head, 4 mm *z*-gradient LC probe; eluent, acetonitrile/D_2O (70/30), flow 0.05 ml/min; 48 scans per row were required

6.4.3 EFFLUENT FROM A TEXTILE COMPANY

Effluents, released from textile companies, may contain dyes and auxiliaries used in the textile industry. The dyes themselves often form complex mixtures that contain considerable quantities of manufacturing precursors and by-products. However, for non-target analysis not only the large variety of compounds but also the large differences in the volatility, solubility and polarity of individual components pose problems. Most of the dyes are non-volatile or thermally unstable. Thus, in recent years predominantly LC–MS techniques have been used for the analysis of dyes [10]. However, the combined use of LC–NMR and LC–MS offers extended possibilities which are illustrated by the analysis of an untreated waste water sample from a textile company [11].

For LC–NMR analysis, the sample concentration should be as high as possible. However, it is limited by the solubility of the analytes in the mobile phase at the starting composition of the eluent. The chromatographic separation of this sample started at 60 % acetonitrile. At this percentage, the solubility of the late-eluting compounds was not high enough for NMR measurements. Therefore, the extract was fractionated before LC–NMR analysis into two fractions (A and B), of which fraction B contained the late-eluting components. The separation of this fraction started with a considerably higher content of acetonitrile (85 %). The LC–UV chromatogram shown in Figure 6.4 gives an overview of the early-eluting components (fraction A), while Figures 6.5 and 6.6 show relevant regions of the stop-flow NMR spectra of selected peaks in fractions A and B. Although these spectra provided valuable structural information, e.g. the anthraquinone skeleton of the dyes could be recognized, additional LC–MS investigations were necessary for the complete identification of the compounds and were performed using atmospheric pressure chemical ionization (APCI). The NMR and MS data, as well as the structure assignments, are summarized in Table 6.2.

Several anthraquinone-type dyes and their by-products, a fluorescent brightener, and a by-product of polyester production, as well as anionic and non-ionic surfactants and their degradation products, could be identified in this sample.

6.4.4 ORGANIC ACIDS IN LEACHATE FROM INDUSTRIAL LANDFILL

Leachate samples from industrial landfills were analysed by LC–NMR and LC–thermospray (TSP)-MS with emphasis on organic acids [12]. After removal of the neutral and basic components by pre-extraction with methylene chloride,

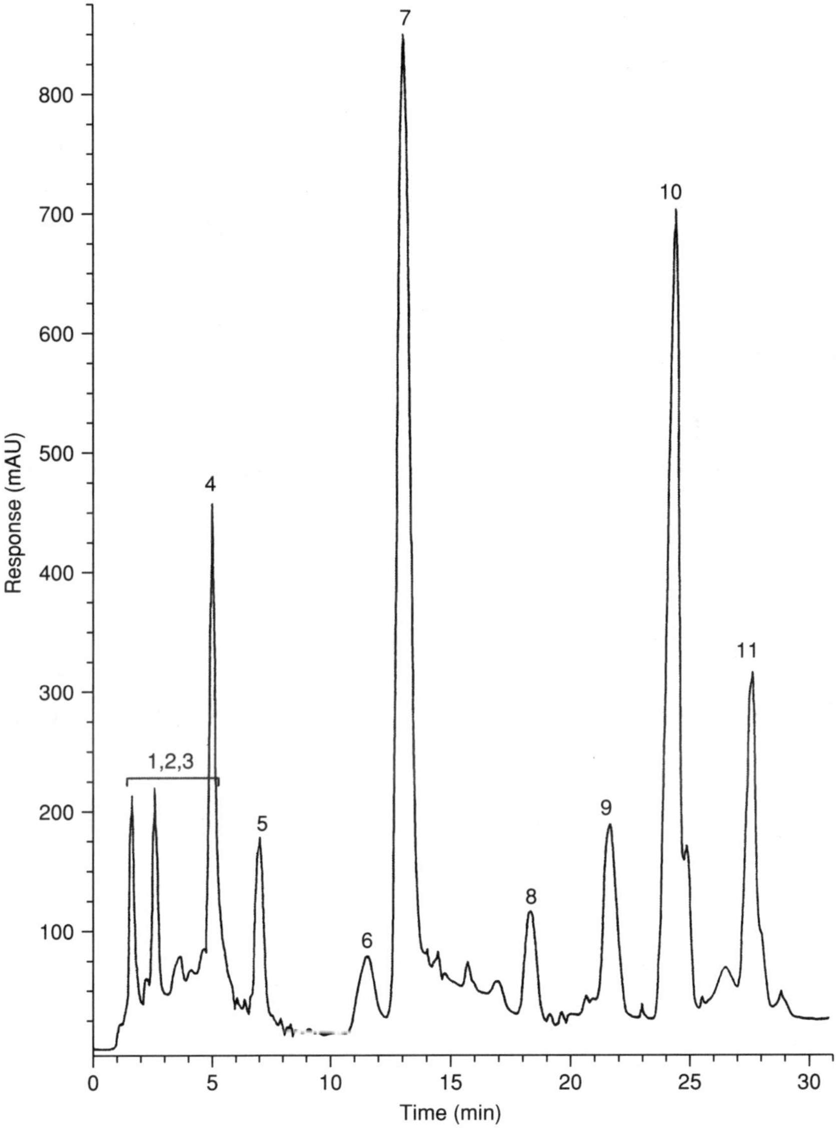

Figure 6.4 UV chromatogram as obtained in the LC–NMR run of fraction A. The peak numbers correspond to the compound numbers given in Table 6.2. Reprinted with permission from Preiss, A., Sänger, U., Karfich, N., Levsen, K. and Mügge, C., *Anal. Chem.*, **72**, 992–998 (2000). Copyright (2000) American Chemical Society

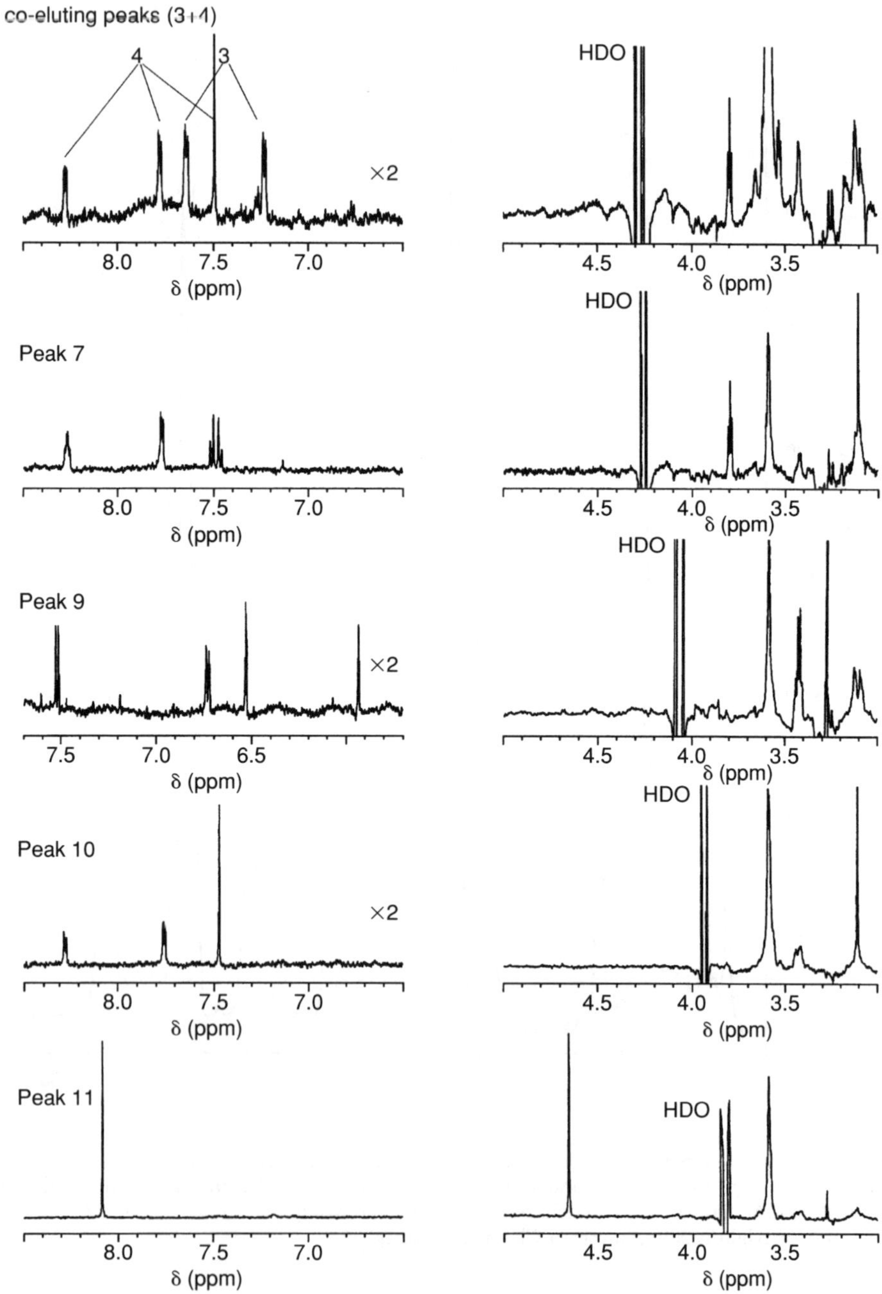

Figure 6.5 Stop-flow NMR spectra of selected peaks in fraction A. Reprinted with permission from Preiss, A., Sänger, U., Karfich, N., Levsen, K. and Mügge, C., *Anal. Chem.*, **72**, 992–998 (2000). Copyright (2000) American Chemical Society

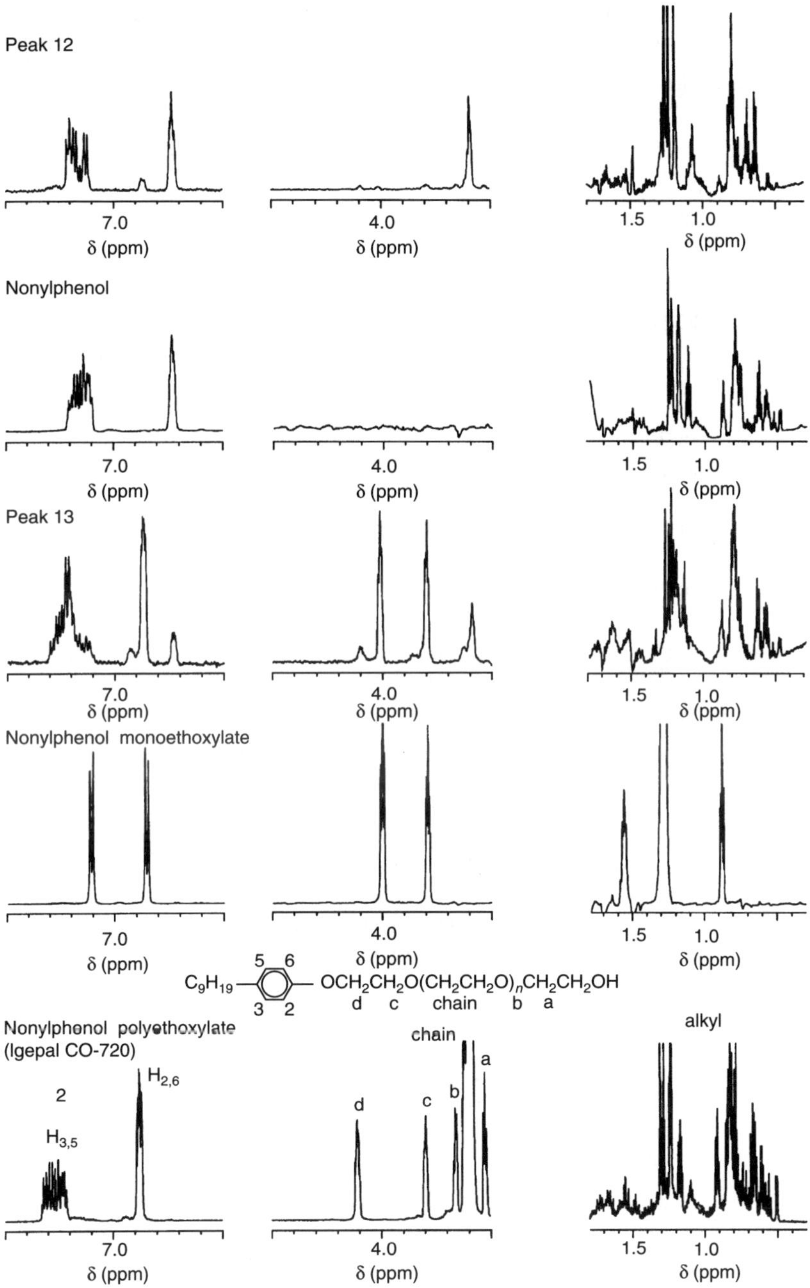

Figure 6.6 Stop-flow NMR spectra of selected peaks in fraction B. Reprinted with permission from Preiss, A., Sänger, U., Karfich, N., Levsen, K. and Mügge, C., *Anal. Chem.*, **72**, 992–998 (2000). Copyright (2000) American Chemical Society

Table 6.2 LC/NMR and LC/MS data of compounds identified or tentatively identified in the waste water sample from a textile company. Reprinted with permission from Preiss, A. Sänger, U., Karfich, N., Levsen, K. and Mügge, C., *Anal. Chem.*, **72**, 992–998 (2000). Copyright (2000) American Chemical Society.

Compound	Structure	NMR data[a]	MS data[b]
1	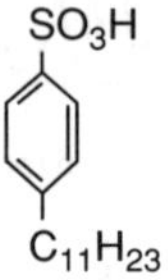	—	311 $[M-H]^-$ ↓ 183 ($-C_9H_{20}$)
2	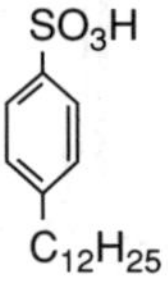	—	325 $[M-H]^-$ ↓ 183 ($-C_{10}H_{22}$)
3	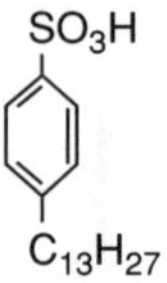	H_2, H_6: 7.64 (pd) H_3, H_5: 7.23 (pd) $-C_{13}H_{27}$: 0.50–1.90	339 $[M-H]^-$ ↓ 183 ($-C_{11}H_{24}$)
4	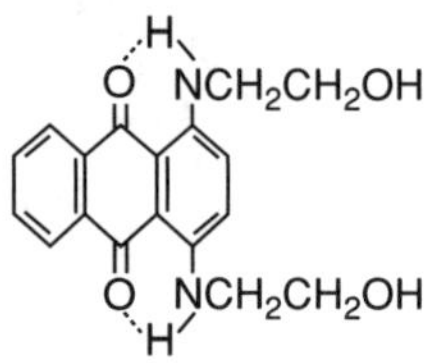	H_2, H_3: 7.49 (s) H_5, H_8: 8.27 (m) H_6, H_7: 7.78 (m) $-NH\underline{CH_2}CH_2OH$: 3.53 (t) $-NHCH_2\underline{CH_2}OH$: 3.80 (t)	326 $[M]^-$/327$[M+H]^+$ ↓ 282 ($-C_2H_4OH$) ↓ 251 ($-CH_2OH$)
5	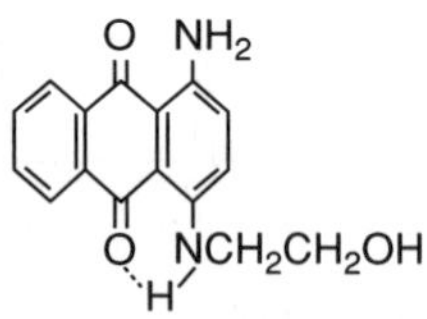	H_2, H_3: 7.25 (d), 7.37 (d) H_5, H_8: 8.26 (m) H_6, H_7: 7.78 (m) $-NH\underline{CH_2}CH_2OH$: 3.56 (t) $-NHCH_2\underline{CH_2}OH$: 3.79 (t)	282 $[M]^-$/283$[M+H]^+$ ↓ 238 ($-C_2H_4OH$)
6	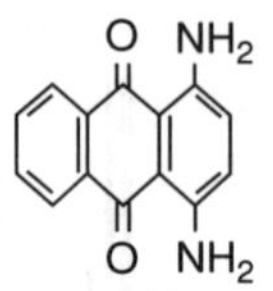	H_2, H_3: 7.12 (s) H_5, H_8: 8.32 (m) H_6, H_7: 7.79 (m)	239 $[M+H]^+$ ↓ 221 ($-H_2O$)
7	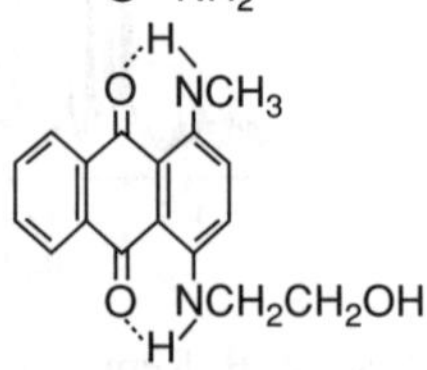	H_2, H_3: 7.47 (d), 7.51 (d) H_5, H_8: 8.27 (m) H_6, H_7: 7.77 (m) $-NHCH_3$: 3.11 (s) $-NH\underline{CH_2}CH_2OH$: 3.59 (t) $-NHCH_2\underline{CH_2}OH$: 3.80 (t)	296 $[M]^-$/297$[M+H]^+$ ↓ 282 ($-CH_3$) ↓ 251 ($-CH_2OH$) ↓ 223 ($-H_2CN$)

8	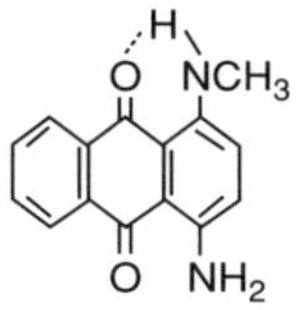	H_2, H_3: 7.26 (d), 7.34 (d) H_5, H_8: 8.27 (m) H_6, H_7: 7.77 (m) $-NH\underline{CH_3}$: 3.09 (s)	253 $[M + H]^+$ ↓ 238 ($-$ CH_3)
9	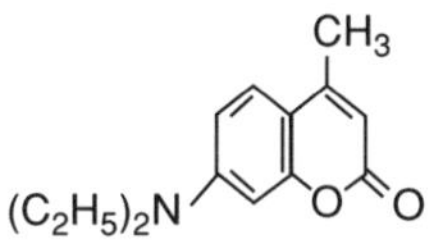	H_2: 5.93 (q), H_6: 6.52 (d) H_8: 6.73 (dd), H_9: 7.52 (d) $-CH_3$: 2.35 (d) $-N(\underline{CH_2}CH_3)_2$: 3.43 (q), $-N(CH_2\underline{CH_3})_2$: 1.15 (t)	232 $[M + H]^+$ ↓ 204 ($-$ C_2H_4),203($-$ C_2H_5) ↓ 176 ($-C_2H_4$ or CO),160($-CO_2$)
10	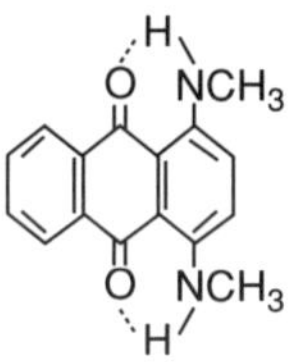	H_2, H_3: 7.47 (s) H_5, H_8: 8.28 (m) H_6, H_7: 7.76 (m) $-NH\underline{CH_3}$: 3.10 (s)	266 $[M]^-$/267 $[M + H]^+$ ↓ 252 ($-$ CH_3) ↓ 235 ($-$OH),237($-CH_3$)
11	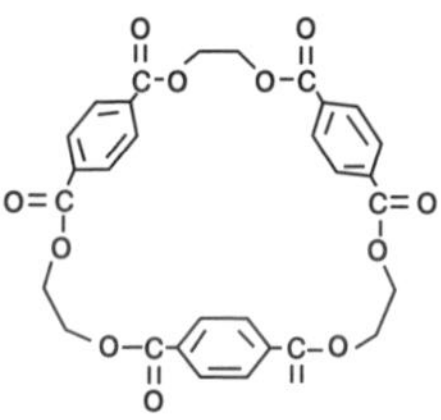	Aromatic H: 8.07 (s) Aliphatic H: 4.66 (s)	576 $[M]^-$/577 $[M + H]^+$ ↓ 385 $(-C_2H_4CO_2C_6H_4CO_2)^+$ 193 $(-2C_2H_4CO_2C_6H_4CO_2)^+$ ↓ 149 ($-CO_2$)
12	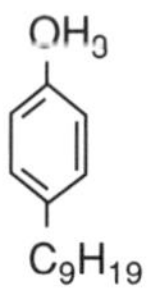	H_2, H_6: 6.70–6.76 (m) H_3, H_5: 7.10–7.22 (m) C_9H_{19}: 0.50–1.80	219 $[M–H]^-$
13	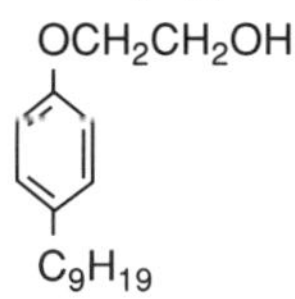	H_2, H_6: 6.82–6.90 (m) H_3, H_5: 7.20–7.32 (m) C_9H_{19}: 0.50–1.80 $-O\underline{CH_2}CH_2OH$: 4.01 (t) $-OCH_2\underline{CH_2}OH$: 3.81 (t)	—
14	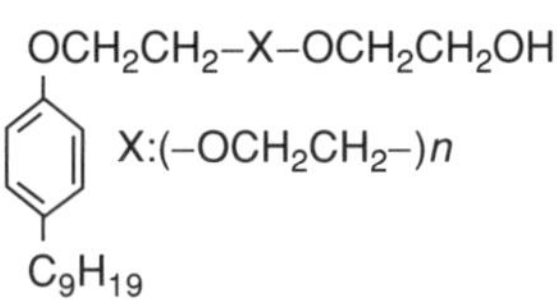	H_2, H_6: 6.88–6.89 (m) H_3, H_5: 7.22–7.32 (m) C_9H_{19}: 0.50–1.80 a: 3.52 (s) b: 3.63 (s) X: 3.60 (s) c: 3.80 (s) d: 4.10 (s)	353–969 $[M + H]^+$ X: $n = 1$–15

[a] δ-values in ppm. Signal multiplicities: s, singlet; d, doublet; pd, pseudo-doublet; t, triplet; m, multiplet.
[b] Obtained using atmospheric pressure chemical ionization (APCI), in negative / positive modes.

the aqueous phase was acidified and the organic acids were extracted by solid-phase extraction (SPE).

Figure 6.7 shows the chromatograms of an extract of a leachate sample at two different detection wavelengths which reveal that the separation efficiency

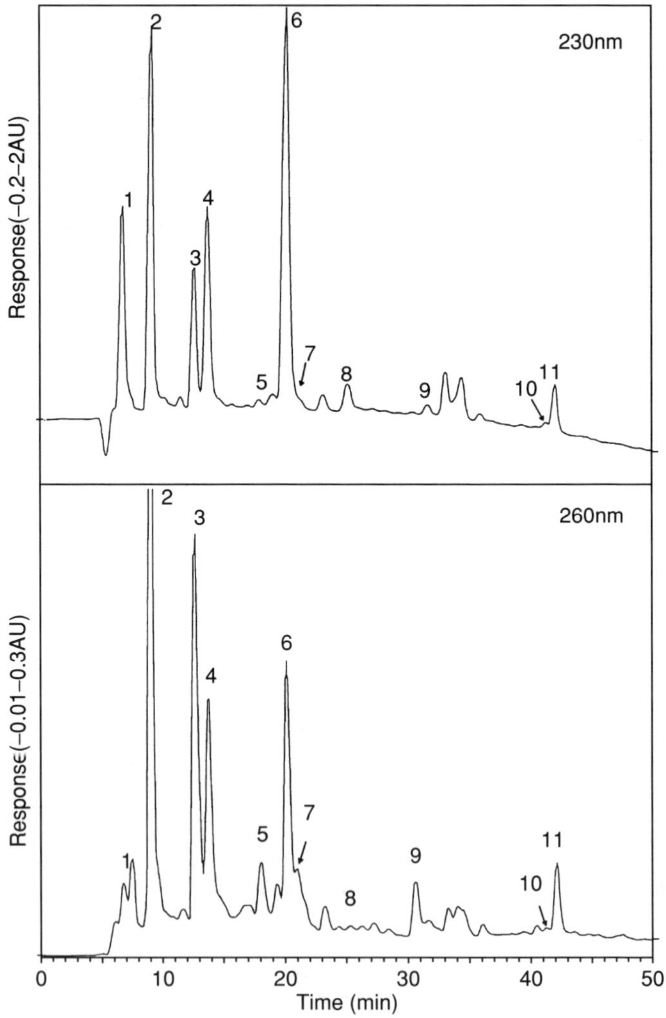

Figure 6.7 LC chromatograms with photodiode array detection at two different wavelengths of the leachate sample (for peak assignment, see Table 6.3). Reprinted from Benfenati, E., Pierucci, P., Fanelli, R., Preiss, A., Godejohann, M., Astratov, M., Levsen, K., and Barcelo, D., *J. Chromatogr., A*, **831**, 243–256, copyright (1999), with permission of Elsevier Science

of the LC method was insufficient. Compound identification by the UV spectrum alone could not be achieved. Furthermore, UV-inactive compounds, such as aliphatic carboxylic acids or sulfates, cannot be detected.

Figure 6.8 shows the contour plot of the on-flow NMR chromatogram of the same sample. In the range of aliphatic protons, the resolution of the NMR information is relatively poor because the protons of many acidic aliphatic compounds, e.g. carboxylic acids or alkyl sulfates (which have of course different retention times), resonate at nearly the same chemical shift values (0.9, 1.1, 1.3, 1.5 and 2.3 ppm). As a consequence, broad traces are observed along the chromatographic axis. Furthermore, the residual signals of the solvent acetonitrile and its ^{13}C satellites, as well as the signals of propionitrile (an impurity of acetonitrile), make the analysis of this part of the NMR chromatogram rather difficult.

On the other hand, the aromatic part of the on-flow NMR chromatogram is better resolved, both in the NMR chemical shift and along the chromatographic axis. As can be seen from the extracted rows of the NMR chromatogram

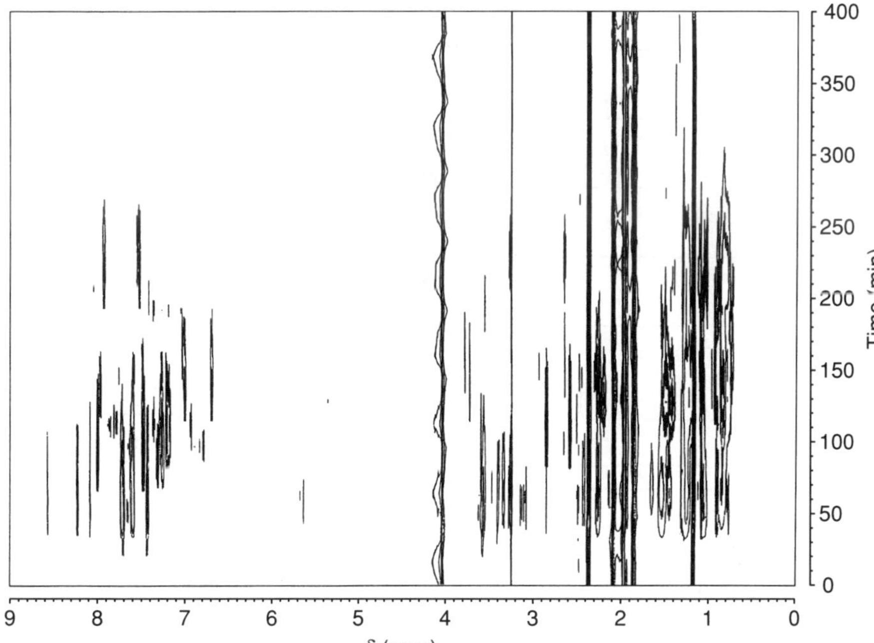

Figure 6.8 Contour plot of the on-flow NMR chromatogram of the leachate sample. Reprinted from Benfenati, E., Pierucci, P., Fanelli, R., Preiss, A., Godejohann, M., Astratov, M., Levsen, K. and Barcelo, D. *J. Chromatogr., A*, **831**, 243–256, copyright (1999), with permission of Elsevier Science

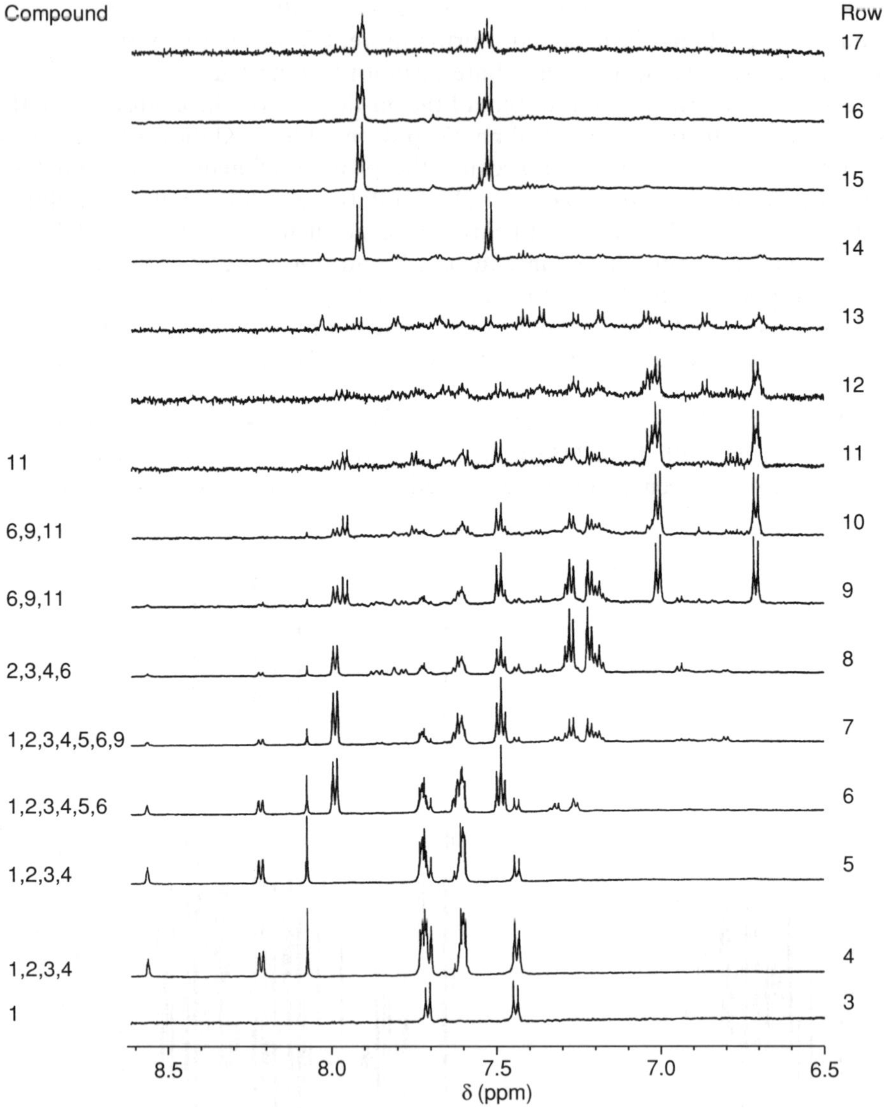

Figure 6.9 Selected rows of the on-flow NMR chromatogram of the leachate sample (chemical shift range of the aromatic protons) (for peak assignment, see Table 6.3). Reprinted from Benfenati, E., Pierucci, P., Fanelli, R., Preiss, A., Godejohann, M., Astratou, M., Levsen, K. and Barcelo, D., *J. Chromatogr., A*, **831**, 243–256, copyright (1999), with permission of Elsevier Science

(Figure 6.9), the spectra are rather simple and could be analysed without the use of further NMR techniques. Eight organic acids could be identified on the basis of TSP-MS and NMR data (NMR data of various reference compounds are given in Table 6.3,), while another three acids (minor components) could only be detected by TSP-MS. Later on, the very early-eluting compounds of this sample were re-examined under modified chromatographic conditions in the stop-flow mode in which further additional organic acids (*p*-aminobenzoic acid, *m*-chlorobenzenesulfonic acid and 1,2,4-benzenetricarboxylic acid) could be identified (Figure 6.10). The TSP-MS data are given in Table 6.4. The more volatile compounds of this sample were analysed by GC–MS, as reported in the same publication [12].

6.4.5 ORGANOPHOSPHORUS COMPOUNDS IN A SOIL SAMPLE

An interesting example deals with the identification of organophosphorus compounds in soil samples [13]. Such compounds can occur as degradation products of 'chemical weapons' (CWs) in soil and other matrices. The convention for the prohibition of chemical weapons sets up procedures for the

Table 6.3 ^{1}H NMR chemical shift values[a] of reference compounds A–C, and compounds 1–11 identified in a leachate from an industrial landfill site. Reprinted from Benfenati, E., Pierucci, P., Fanelli, R., Preiss, A., Godejohann, M., Astratov, M., Levson, K. and Barcelo, D.,[c] *J. Chromatogr., A*, **831**, 243–256, copyright (1999) with permission of Elsevier Science.

Identifies Compound	H2	H3	H4	H5	H6	$-CH_2-$
A *p*-Aminobenzoic acid	8.07 (pd)	7.38 (pd)	—	7.38 (pd)	8.07 (pd)	—
B *m*-Chlorobenzene-sulfonic acid	7.76 (d)	—	7.49 (dd)	7.41 (t)	7.68 (dd)	—
C 1,2,4-Benzenetri-carboxylic acid	—	8.35 (d)	—	8.19 (dd)	7.79 (d)	—
1 *p*-Chlorobenzene-sulfonic acid	7.45 (pd)	7.72 (pd)	—	7.72 (pd)	7.45 (pd)	—
2 Phthalic acid	—	7.74 (m)	7.61 (m)	7.61 (m)	7.74 (m)	—
3 Terephthalic acid	8.06 (s)	8.06 (s)	—	8.06 (s)	8.06 (s)	—
4 Isophthalic acid	8.54 (d)	—	8.19 (dd)	8.60 (t)	8.19 (dd)	—
5 Phenylacetic acid	7.22 (pd)	7.27 (pt)	7.20 (pt)	7.27 (pt)	7.22 (pd)	3.60 (s)
6 Benzoic acid	7.97 (pd)	7.47 (pt)	7.60 (pt)	7.47 (pt)	7.98 (pd)	—
7 *o*-Chlorbenzoic acid	—	7.52 (d)	7.41 (t)[b]	7.40 (t)[b]	7.82 (d)	—
8 *o*-Hydroxybenzoic acid	—	6.94 (dd)	7.50 (dt)	6.92 (dt)	7.84 (dd)	—
9 3-Phenylpropionic acid	7.22 (d)	7.27 (t)	7.20 (t)	7.27 (t)	7.22 (d)	2.60 (t), 287 (t)
10 *m*-Chlorbenzoic acid	7.95 (s)	—	7.60 (d)	7.46 (t)	7.90 (d)	—
11 *p*-Chlorbenzoic acid	7.97 (pd)	7.50 (pd)	—	7.50 (pd)	9.97 (pd)	—

[a] t: Multiplicities of the signals: s, singlet; d, doublet; t, triplet, dd, double doublet; m, multiplet; pd, pseudo-double δ(ppm), refs reduced to the solvent peak of aceto nitride (δ =2.00 ppm) pt, pseudo-triplet.
[b] Assignment may be interchanged.

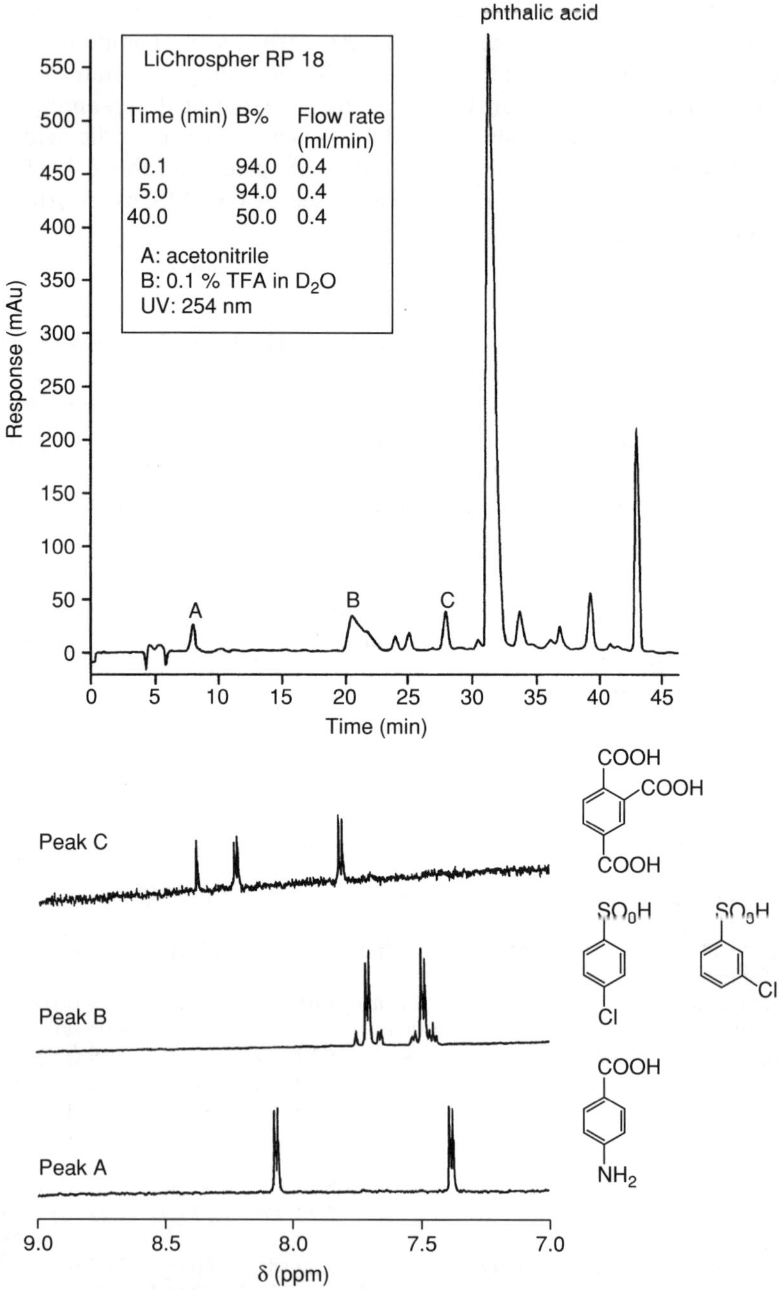

Figure 6.10 Stop-flow NMR spectra of the very early-eluting compounds of the leachate sample; the chromatographic conditions are given on the figure. Spectrometer, Bruker DRX 600; probe head, 4 mm *z*-gradient LC probe; between 256 and 1024 scans were acquired

Table 6.4 Aromatic carboxylic and sulfonic acids identified in a leachate sample by LC – thermospray-MS. Reprinted from Benfenati, E., Pierucci, P., Fanelli, R., Preiss, A., Godejohann, M., Astratov M., Levsen, K. and Barcelo, D., *J. Chromatogr., A*, **831**, 243–256, copyright (1999) with permission of Elsevier Science.

Number	Compound	M_r	t_r (min)	Relative abundance (%)					
				Negative ions			Positive ions		
				100	20 < | < 100		100	20 < | < 100	
1	*p*-Chlorobenzenesulfonic acid	192	5.8	156 [M–Cl–H]$^-$	191 [M–H]$^-$	(50)	—	—	
2	Phthalic acid	166	8.7	165 [M–H]$^-$	148 [M–H_2O]$^-$	(20)	184 [M + NH_4]$^+$	167 [M + H]$^+$	(80)
3	Terephthalic acid	166	12.5	165 [M–H]$^-$	211 [M + COOH]$^-$	(80)	184 [M + NH_4]$^+$	—	
4	Isophthalic acid	166	14.0	211 [M + COOH]$^-$	165 [M–H]$^-$	(30)	184 [M + NH_4]$^+$	—	
5	Phenylacetic acid	136	17.3	181 [M + COOH]$^-$	—		154 [M + NH_4]$^+$	—	
6	Benzoic acid	122	20.3	167 [M + COOH]$^-$	—		140 [M + NH_4]$^+$	—	
7	*o*-Chlorobenzoic acid	156	21.6	201 [M + COOH]$^-$	155 [M–H]$^-$	(30)	—	—	
8	o-Hydroxybenzoic acid	138	25.0	183 [M + COOH]$^-$	—		—	—	
9	3-Phenylpropionic acid	150	30.7	195 [M + COOH]$^-$	—		168 [M + NH_4]$^+$	—	
10	*m*-Chlorobenzoic acid	155	41.0	201 [M + COOH]$^-$	155 [M–H]$^-$	(30)	—	—	
11	*p*-Chlorobenzoic acid	155	41.5	201 [M + COOH]$^-$	155 [M–H]$^-$	(30)	—	—	

control of CW use or disposal, which involve expert laboratories 'designated for handling and analysing samples possibly containing CW-related material'. To award this designation, i.e. to assay the skill of candidate laboratories, international proficiency tests have been organized under the aegis of the *Organization for the Prohibition of Chemical Weapons* (OPCW). GC–MS or LC–MS are the methods of choice for the identification but some of the compounds related to the convention cannot be directly detected by GC–MS. Since proficiency-test criteria demand at least two analytical methods for the identification of a compound, the NMR approach can be used in addition to confirm the results obtained by the LC–MS method [14]. In principle, ^{31}P NMR spectroscopy is useful to check for the presence of organophosphorus compounds in a sample, although the sensitivity of this spectroscopy is low and the different classes of organophosphorus compounds related to the convention resonate in relatively narrow spectral regions. On the other hand, ^{1}H NMR spectroscopy is more sensitive but these spectra often present strong overlapping signals (resonance signals of matrix components often interfere with the signals of interest). One possibility for solving this problem is the application of two-dimensional ^{1}H–^{31}P inverse NMR spectroscopy [15], while a second is the use of the hyphenated LC–NMR technique.

Figure 6.11 shows the pseudo-on-flow NMR chromatograms of a soil sample extracted with water. The soil sample was prepared by a Polish laboratory as part of the sixth proficiency test. It was spiked with five compounds, with two of them (compounds 3 and 4) being related to chemical weapons.

The ^{1}H NMR spectra of the soil extract were run in the 'time-slice' mode, i.e. the chromatographic run was stopped every 20 s and the ^{1}H NMR spectrum measured twice, with (a) and without (b) ^{31}P decoupling. Processing of the data files led to the pseudo-on-flow NMR chromatograms.

Comparison of the two pseudo-on-flow NMR chromatograms reveals that compounds 1,2 and 3 are organophosphorus compounds (Table 6.5), but only compounds 2 and 3 are phosphonates and related to chemical weapons. In fact, compound 2 is a degradation product of compound 3

An additional 2D TOCSY spectrum (Figure 6.12) measured for row 11 (stop at 3.3 min) shows that one of the two co-eluting compounds (2 and 5) is a non-phosphorus material (compound 5).

6.5 SIMULATION OF ENVIRONMENTAL PROCESSES

The determination of known organic pollutants in the environment (target analysis) and the identification of all possible organic pollutants, i.e. in hazardous waste sites (non-target analysis), are important tasks in environmental analysis. However, the chemical, photochemical and microbiological

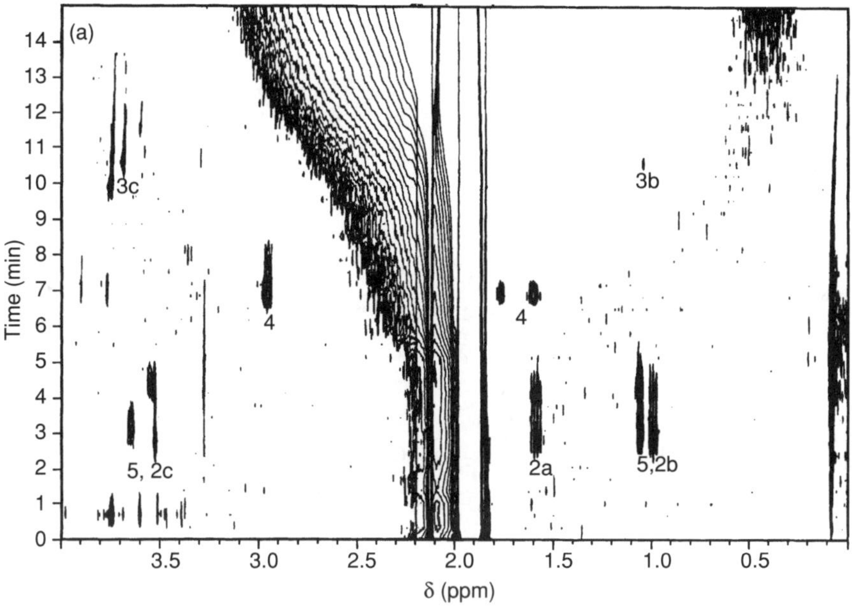

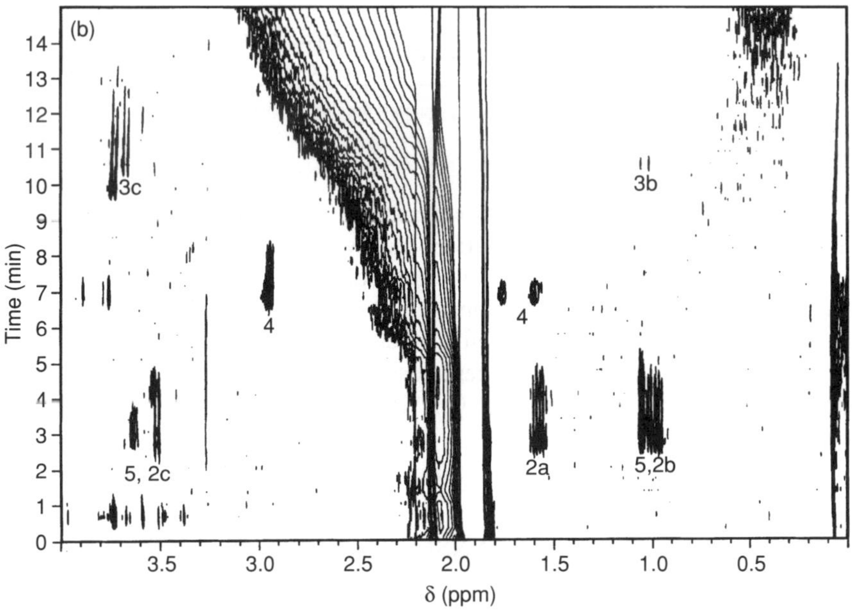

Figure 6.11 Pseudo-on-flow ^{1}H NMR chromatograms of a spiked soil sample (a) with and (b) without ^{31}P decoupling. Conditions: column, Purospher RP18, 250 × 4 mm, 5 μm; eluents, A – acetonitrile, B – 1 % formic acid in D_2O; gradient, $t = 0$ min A/B (1/99), $t = 5$ min A/B (10/90), $t = 10$ min A/B (90/10); flow, 1.0 ml/min; spectrometer, Bruker Avance 500 MHz; probe head, TXI ^{1}H/^{13}C/^{31}P 4 mm LC probe; 64 scans were acquired per row

Table 6.5 ^{1}H NMR data of organophosphorus identified in a spiked soil sample.

Compound	Structure	NMR data		
		δ (ppm)	Signal multiplicities[a]	
			^{1}H	^{1}H{^{31}P}[b]
trimethyl phosphate (1)	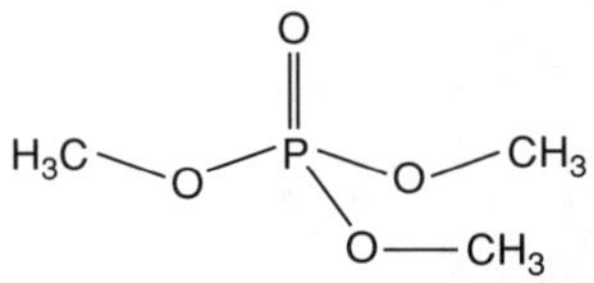	CH_3–, 3.73	(d)	(s)
methylethyl phosphonate (2)	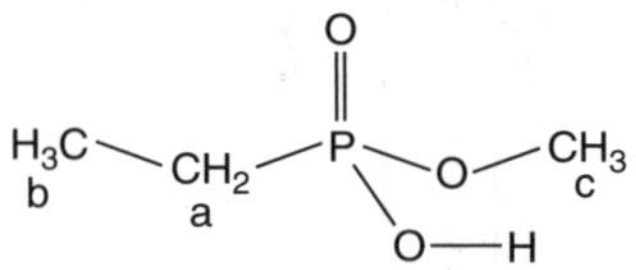	–CH_2–, 1.59 a CH_3–, 0.98 b CH_3–, 3.52 c	(dq) (dt) (d)	(q) (t) (s)
dimethylethyl phosphonate (3)	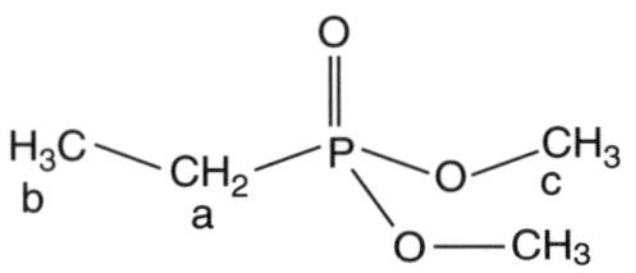	–CH_2–, n.d[c] a CH_3–, 1.05 b CH_3–, 3.68 c	 (dt) (d)	 (t) (s)

[a] s, singlet; d, doublet; t, triplet; dt, double triplet; q, quartet; dq, double quartet.
[b] Multiplicities in the ^{1}H{^{31}P} spectrum.
[c] Not detectable because of solvent suppression.

processes that occur in the environment are also investigated intensively in order to understand the degradation pathways or to be able to influence them. In the first step of such investigations, environmental processes are often simulated by model reactions. In this field, the LC–NMR technique can also significantly complement other analytical methods and its use is favoured by the fact that, in these cases, sufficient amounts of the investigated compounds are available. This is illustrated by the following three examples:

6.5.1 PHOTOCHEMICAL TRANSFORMATION OF TNT

Trinitrotoluence (TNT) released into the environment may also undergo photochemical transformation processes which lead to highly polar, water-soluble nitroaromatic compounds. This fast reaction can be observed by irradiation of an aqueous solution of TNT with UV or natural sunlight: the formerly colourless solution immediately turns to pink.

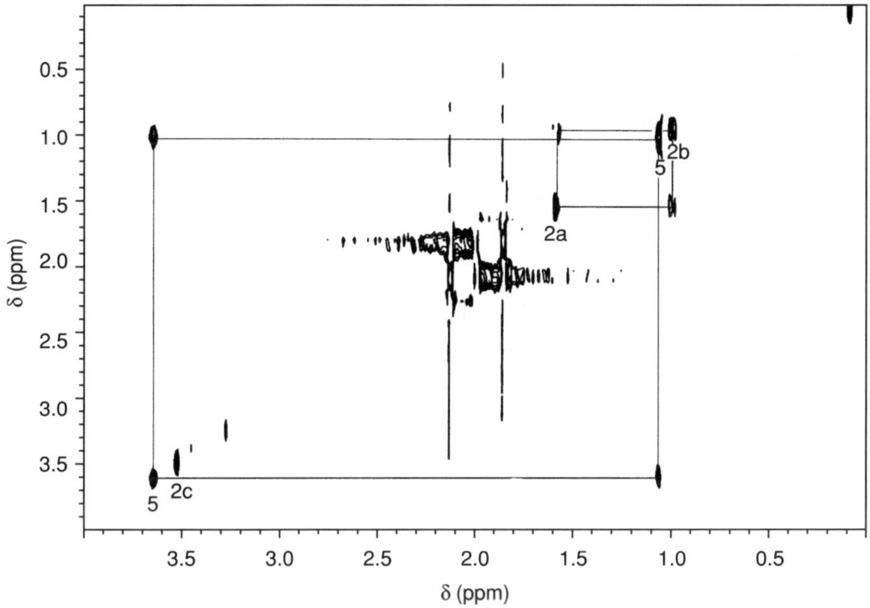

Figure 6.12 2D TOCSY spectrum with ^{31}P decoupling acquired at row 11 (stop at 3.3 min): spectral size, 4k × 160 data points; 8 scans per increment; 65 ms TOCSY mixing time

The photochemical transformation of TNT had already been investigated by the early 1980s and some of the unknown transformation products had been isolated and identified. While Spanggord *et al.* [16] examined primarily the compounds which arise in waste water under the influence of natural sunlight, Kaplan *et al.* investigated the photoproducts which are formed by irradiation of an aqueous TNT solution with a mercury lamp [17]. During these studies, the isolated unknown compounds were identified off-line by spectroscopic methods, followed by syntheses of various reference compounds. Nowadays, this can be done much more effectively by using the LC–NMR and LC–MS hyphenated techniques. One example is the on line identification of the transformation products in an aqueous TNT solution which was exposed to sunlight (microbial degradation of TNT was prevented by the addition of sodium azide) by the combined use of LC–NMR and LC–TSP-MS [9]. Here, only the identification of the three components (a, d and f) is discussed as an example for the combined use of LC–NMR and LC–MS. The total ion current chromatogram (TIC) of the extract is shown in Figure 6.13(a), while the mass spectra of peaks a,d and f, as extracted from this chromatogram, are presented in Figure 6.13(b). Peak d could readily be identified as trinitrobenzene, but the very early-eluting compound has almost the same mass spectrum as 1,3,5-trinitrobenzene, although this compound should be much more polar. Therefore, it was assumed that the base peak at *m/z* 213 was not the molecular ion peak but a fragment

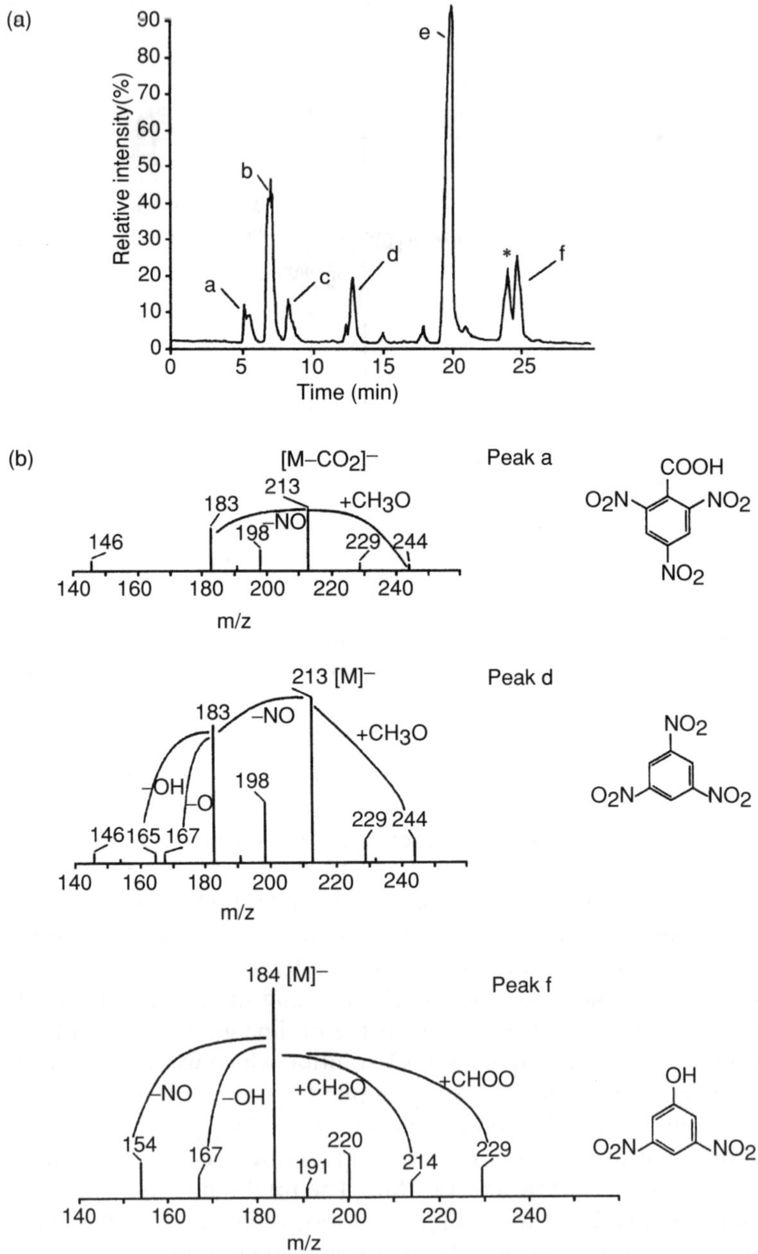

Figure 6.13 (a) Total ion current chromatogram of the extract of the irradiated TNT solution. (b) Mass spectra of peaks a, d and f, as extracted from this chromatogram. Reprinted with permission from Godejohann, M., Astratov, M., Preiss, A., Levson, K. and Mügge, C., *Anal. Chem.*, **70**, 4104–4110 (1998). Copyright (1998) American Chemical Society

peak after the loss of carbon dioxide and thus peak a in the TIC should originate from 2,4,6-trinitrobenzoic acid, which is known to decarboxylate very easily. From the interpretation of the mass spectral data of peak f, a dinitrophenol was suggested but no information could be obtained on the substitution pattern of the aromatic ring. In both cases, further information was necessary and this was obtained from the on-flow LC–NMR chromatogram shown in Figure 6.14(a). On the chromatographic axis, the retention times of peaks a-f are marked. The retention times in the TIC and in the NMR chromatogram cannot be compared directly, since different chromatographic conditions were used. Nevertheless, the elution order is the same and the relative retention time and peak intensities are very similar for both separation methods. Therefore, a correlation between the total ion current chromatogram and the peaks in the NMR chromatogram is possible. The signal at 9.21 ppm (marked a) in the NMR chromatogram thus belongs to a very early-eluting compound, corresponding to peak a in the TIC. The observation of a singlet and the chemical shift value are consistent with 2,4,6-trinitrobenzoic acid. On the other hand, the two signals at 8.52 and 8.01 ppm, marked f, belong to the last eluting peak, that is peak f in the TIC. The observation of the two signals and their fine structures (doublet and triplet) is only consistent with 3,5-dinitrophenol.

Comparison with the NMR chromatogram Figure 6.14(b) of a ground water extract from the former ammunition site Elsnig (Saxony, Germany) shows that these two compounds are also present in this sample.

6.5.2 HYDROLYSIS OF SULFONATED AZO DYES

Sulfonated azo dyes (reactive dyes) are widely used in the textile industry. Due to the simultaneous hydrolysis in the dyeing process, 15–60 % of the reactive dyes reach the waste water system. In addition, their use strongly increases with the colouring of natural fibres, and so a total of 60 % of all dyes emitted to the waste water are reactive materials.

Whereas non-azo dyes are almost resistant to bacterial decolorization, azo dyes can be decolorized by bacteria under anaerobic conditions to form aromatic amines which show both toxic and carcinogenic potential [18]. Further aerobic mineralization can therefore only be initiated by additional chemical oxidation, for example, by partial ozonation, as reported by Krull *et al.* [19].

The most commonly used waste water treatment procedures are not sufficient to remove these dyes. Considerable efforts are therefore made to improve the efficiency of the physical, chemical and biological treatment procedures. However, in order to achieve this aim, reliable analytical methods for the identification and quantification of the azo dyes and their by- and degradation products are required. Since most of the dyes are non-volatile and thermally labile, LC–NMR and LC–MS are the methods of choice.

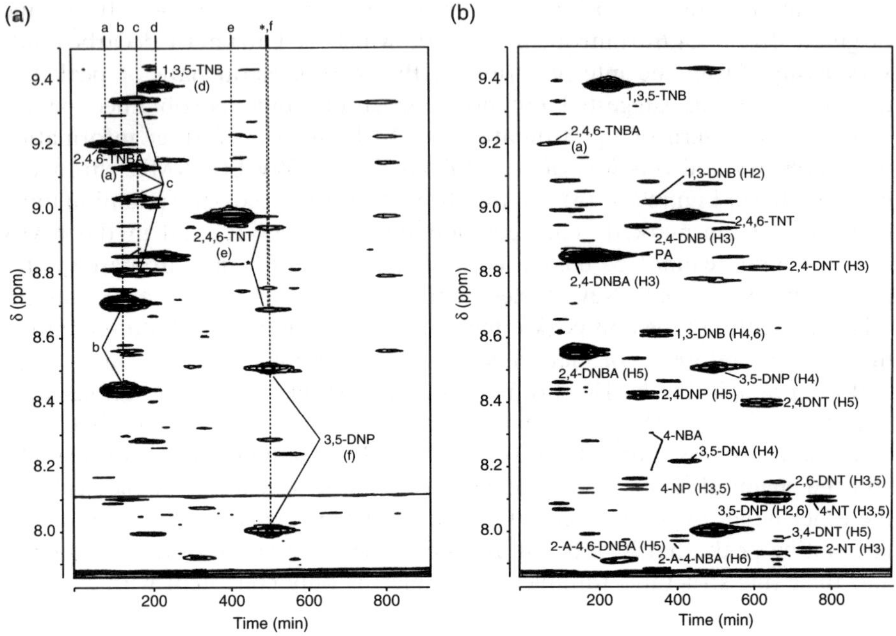

Figure 6.14 (a) On-flow NMR chromatogram (low-field part) of the extract of the irradiated TNT solution (see Table 6.1 for abbreviations). The numbers in parentheses assign the NMR signals to the protons generating this signal. (b) Low-field part of the on-flow NMR chromatogram of a ground water extract from the former ammunition site in Elsnig (Germany) (see Table 6.1 for abbreviations). The numbers in parentheses assign the NMR signals to the protons generating this signal. Both parts reprinted with permission from Godejohann, M., Astratov, M., Preiss, A., Levsen, K. and Mügge, C., *Anal. Chem.*, **70**, 4104–4110 (1998). Copyright (1998) American Chemical Society

Figure 6.15 presents the LC–UV chromatogram of the hydrolysis products from the first step of simulated waste water treatment of Remazol Black 5 (RB5), a commercially important textile dye, while Figure 6.16 shows a series of stop-flow LC–NMR spectra acquired in an LC–NMR–MS run. The NMR and MS data of the tentatively identified compounds are shown in Tables 6.6 and 6.7, respectively. These are only by- or degradation products which elute earlier than the hydrolysed Remazol Black. Peaks which elute later consist of co-eluting dye components which have not yet been identified.

The MS data (negative ESI mode) provide information on the molecular mass of the partially deuterated molecules due to the use of D_2O in the eluent. The MS^n experiments, however, were carried out via flow injection analysis (FIA-MS) on peaks that were cut from a separate chromatographic run in which H_2O was used instead of D_2O. In this way, in addition to the fragmentation, the

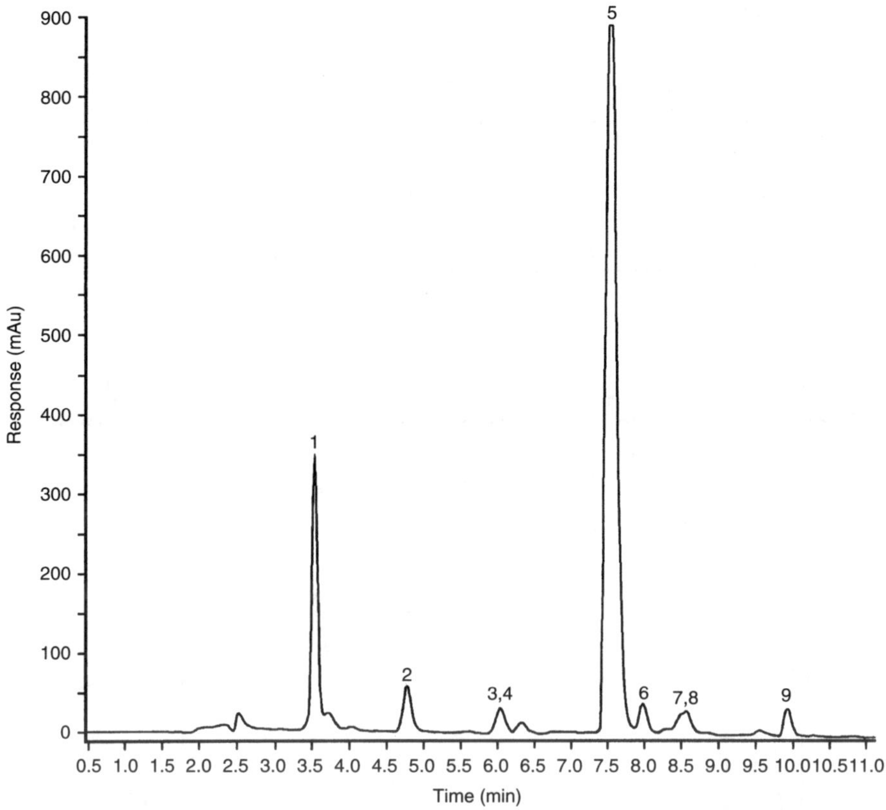

Figure 6.15 LC chromatogram of the hydrolysis products from the first step of simulated waste water treatment of Remazol Black 5. Conditions: column, LiChrospher RP 18e, 250×4.0 mm, 5 μm; eluent, A – 20 mM NH_4COOH in CH_3CN/D_2O (70/30), B – 20 mM NH_4COOH in CH_3CN/D_2O (10/90); gradient, $t = 0$ min A/B (10/90), $t = 15$ min A/B (70/30); flow, 0.8 ml/min

number of exchangeable protons could also be determined. This was sometimes helpful for distinction between alternative structures.

Furthermore, components could already be detected in this mixture (peaks 2 and 3) which are assumed to be degradation products formed by the reductive cleavage of the azo linkage.

It is interesting to note that compound 4, which was found only in small quantities in the first step of the simulated waste water treatment, is one of the main components in the second step, that is, the decolorization of the hydrolysed Remazol Black by treatment with the white rot fungus *Trametes versicolor* in sequencing batch reactors [20].

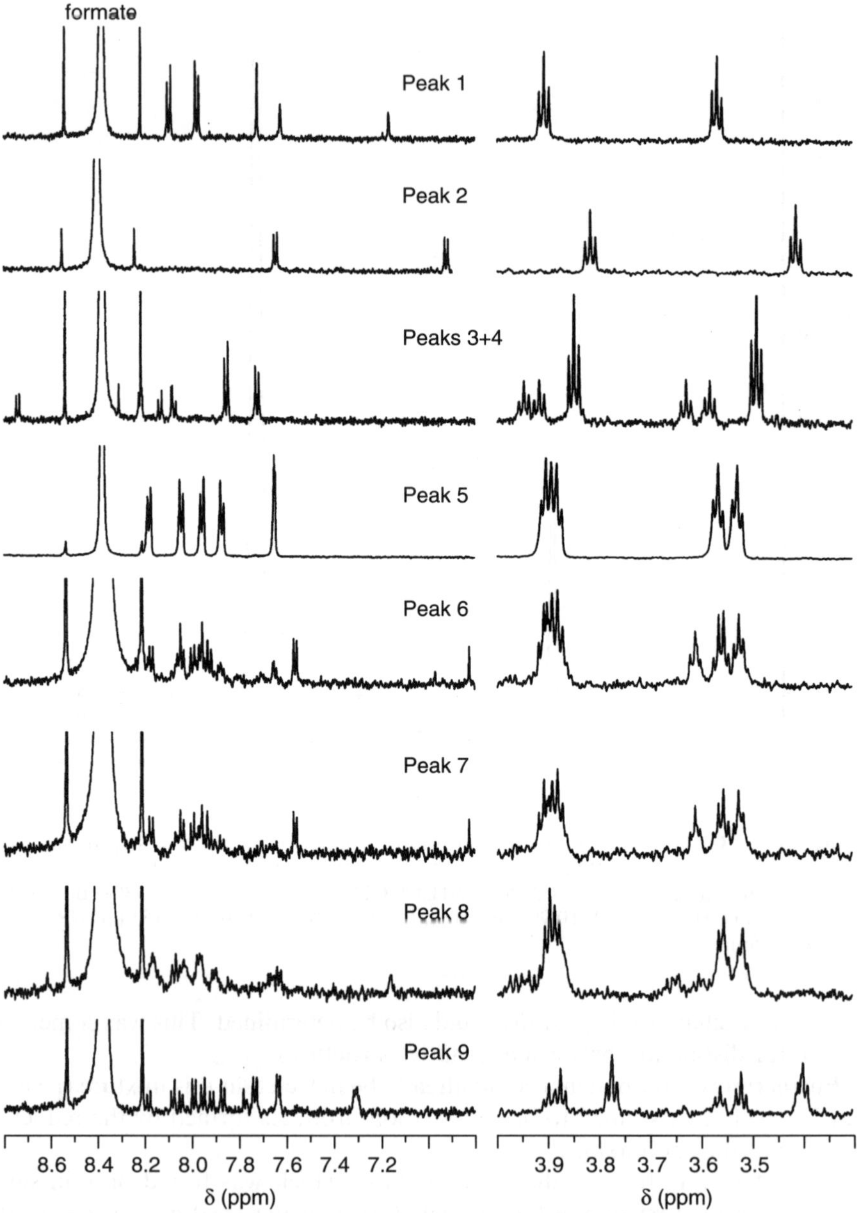

Figure 6.16 Stop-flow NMR spectra of the hydrolysis products of Remazol Black 5 acquired in an LC–NMR–MS run; 95% of the eluent was transfered to the NMR spectrometer and 5% to the mass spectrometer. Equipment: NMR spectrometer, Bruker DRX 600; probe head, 4 mm *z*-gradient LC probe; between 256 and 1024 scans were acquired; mass spectrometer, Bruker Esquire-LC ion-trap machine, equipped with an ESI ionization source

Table 6.6 ^{1}H NMR data obtained for the identified or tentatively identified hydrolysis products of Remazol Black 5.

Peak	t_r (min)	Structure	NMR data: δ (ppm)	NMR data: Signal multiplicities[a]
1	3.52	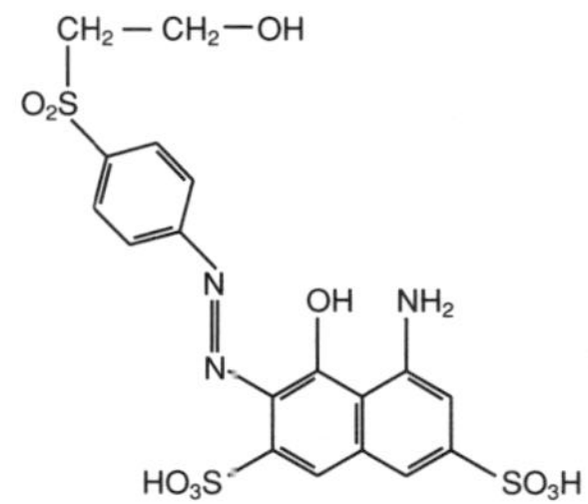 or 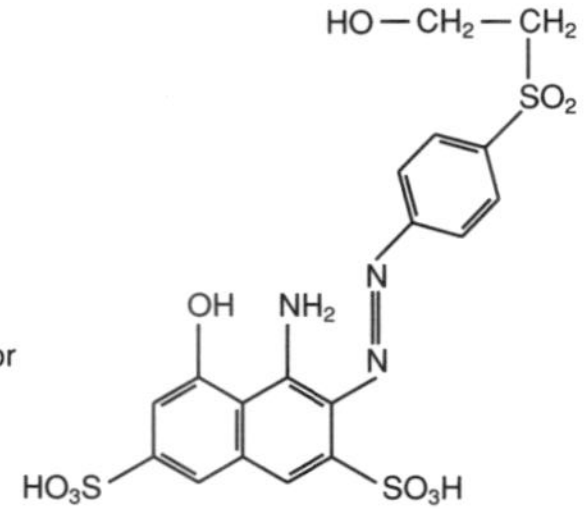	$H_{aromatic}$, 8.11 8.00 7.73 7.65 7.18 $-SO_2CH_2\underline{CH_2}OH$, 3.91 $-SO_2\underline{CH_2}-CH_2OH$, 3.57	2H(pd) 2H(pd) 1H(s) 1H(d) 1H(d) 2H(t) 2H(t)
2	4.77	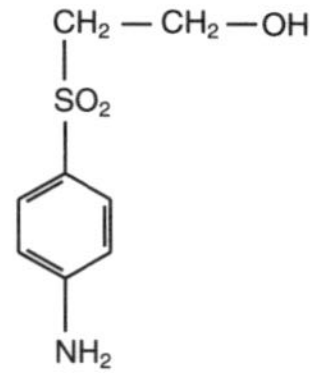	$H_{2,6}$, 7.59 $H_{3,5}$, 6.83 $-SO_2CH_2\underline{CH_2}OH$, 3.82 $-SO_2\underline{CH_2}CH_2OH$, 3.42	(pd) (pd) 2H(t) 2H(t)
3	6.04	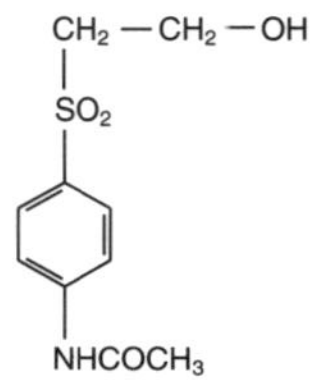	$H_{2,6}$, 7.73 $H_{3,5}$, 7.86 $-SO_2CH_2\underline{CH_2}OH$, 3.85 $-SO_2\underline{CH_2}CH_2OH$, 3.50 $-COCH_3$	(pd) (pd) 2H(t) 2H(t)

(continues)

Table 6.6 (*continued*)

Peak	t_r (min)	Structure	NMR data		
			δ ppm		Signal multiplicities[a]
4	6.04	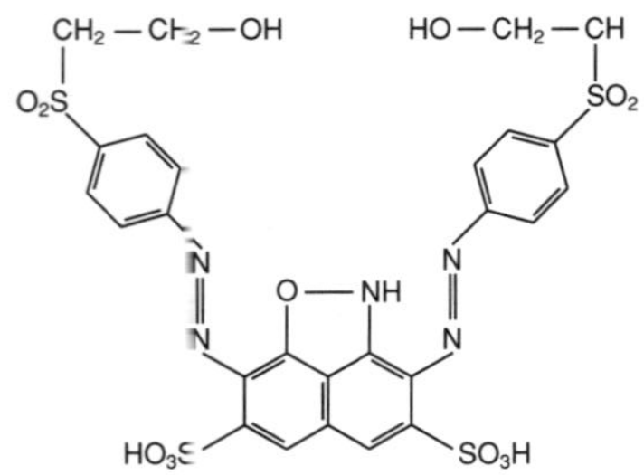	$H_{aromatic}$,	8.73	2H(pd)
				8.21	2H(pd)
				8.13	2H(pd)
				8.07	2H(pd)
				8.30	1H(s)
				8.08	1H(s)
			$-SO_2CH_2\underline{CH_2}OH$,	3.95	2H(t)
			$-SO_2\underline{CH_2}CH_2OH$,	3.63	2H(t)
			$-SO_2CH_2\underline{CH_2}OH$,	3.92	2H(t)
			$-SO_2\underline{CH_2}CH_2OH$,	3.59	2H(t)
5	7.51	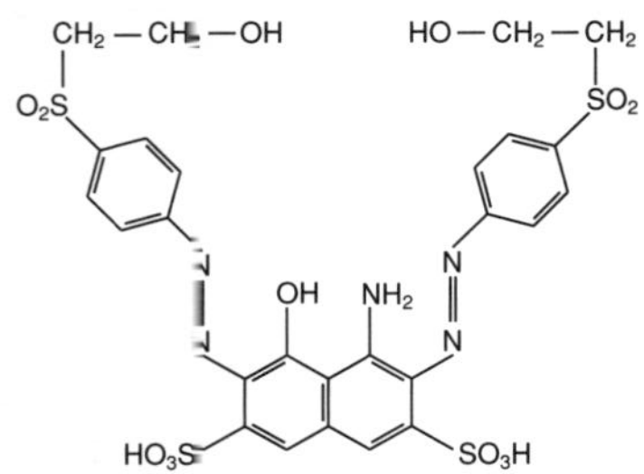	$H_{aromatic}$,	8.18	2H(pd)
				8.04	2H(pd)
				7.96	2H(pd)
				7.87	2H(pd)
				7.66	1H(s)
				7.65	1H(s)
			$-SO_2CH_2\underline{CH_2}OH$,	3.91	2H(t)
			$-SO_2\underline{CH_2}CH_2OH$,	3.57	2H(t)
			$-SO_2CH_2\underline{CH_2}OH$,	3.88	2H(t)
			$-SO_2\underline{CH_2}CH_2OH$,	3.53	2H(t)

[a] s, singlet; d, doublet; pd, pseudo-double; t, triplet.
[b] Not detectable because of solvent suppression.

Table 6.7 MS data obtained for the identified or tentatively identified hydrolysis products of Remazol Black 5.

Peak	Structure	H[a]	MS and MS^n Data[b]	
1	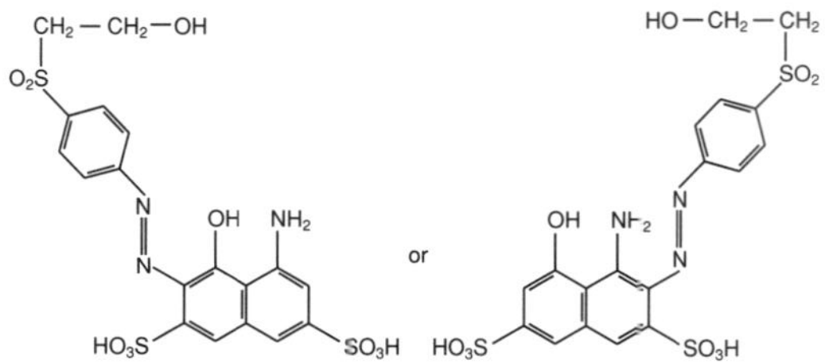	6	$[M-H]^-$ 530 466 317	—
2	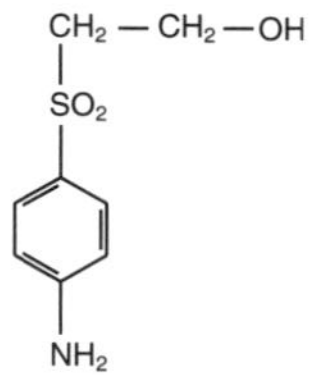	—	—	$[M+H]^+$ 202 138 184 93 120 156 108 92
3	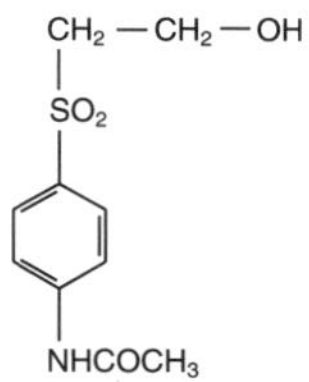	—	—	$[M+H]^+$ 244 202 136 138 184 93 94 120 156 108 92

(continues)

Table 6.7 *(continued)*

Peak	Structure	H[a]	MS and MS^n Data[b]	
4	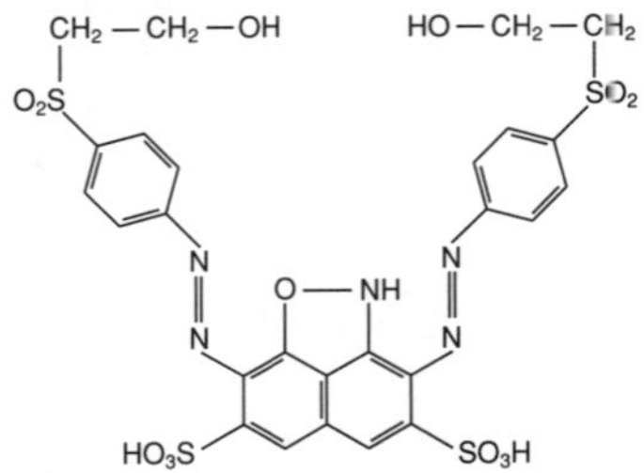	5	[M−H]$^-$ 740 676 527 463 463 354 399 290 290 186	—
5	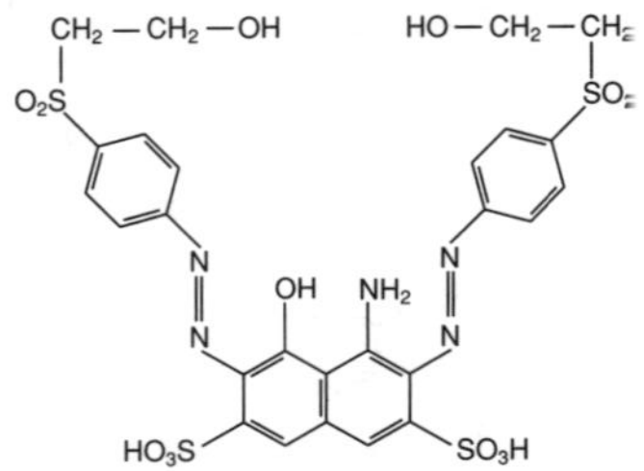	7	[M−H]$^-$ 742 678 465 570 354 540 461 290 431	—

[a] Number of exchangeable protons.
[b] Using electrospray ionization (ESI) in negative/positive modes.

6.5.3 CHLORINATED HYDROXYBIPHENYLS

Polychlorinated biphenyls (PCBs) belong to those pollutants that are widely dispersed and accumulated in the environment. During the biodegradation of PCBs, dechlorination and hydroxylation are the initial steps. For risk

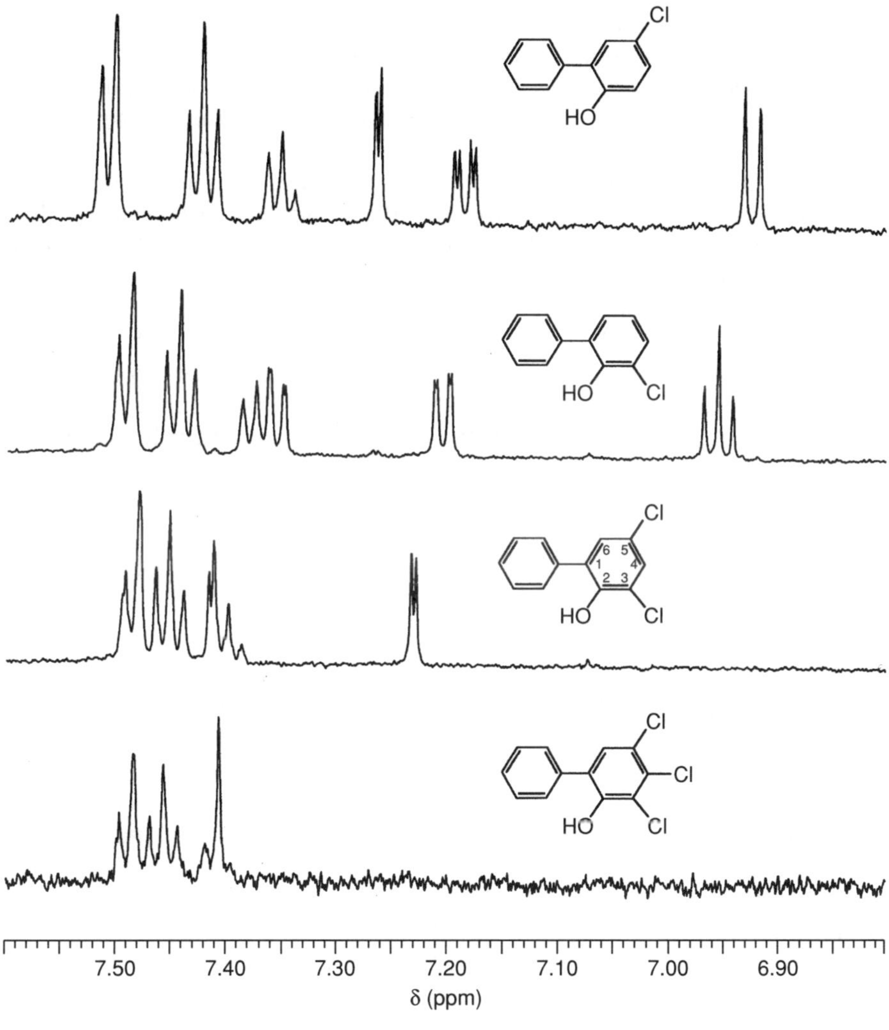

Figure 6.17 Stop-flow NMR spectra of the chlorination products of 2-hydroxybiphenyl. Conditions: column, LiChrospher RP18e, 125 × 4 mm, 5 μm; eluent, acetonitrile (A), D_2O (B); gradient, $t = 0$ min A/B (55/45), $t = 35$ min A/B (75/25); flow, 1.0 ml/min; spectrometer, Bruker DRX 600; probe head, 4 mm z-gradient LC probe; between 24 and 256 scans were acquired

assessment and mechanistical interpretation of the degradation pathways, the identification of individual metabolites is very important [21]. Unfortunately, the group of hydroxylated PCBs includes 839 congeners with different toxic potentials. An assignment of individual congeners requires genuine reference compounds that are commercially not available at this time. Moreover, a distinction between structural isomers by mass spectroscopy is only possible in exceptional cases. As a consequence, many of the metabolites cannot be unambiguously identified. A promising approach to determine congeneric compounds such as PCB-ols is the use of gas chromatographic retention indices (RIs) [22]. Because of 'missing' standards, mathematical models can be used to pre-calculate RI values for the identification of individual congeners [23]. However, procedures for such pre-calculations should be validated as much as possible by the use of experimental retention values of selected congeners. Such compounds can be isolated from synthetic mixtures of congeners which are synthethized by catalytic chlorination of hydroxybiphenyls [24]. Separation and identification of the congeners can be carried out very simply and rapidly by LC–NMR, as demonstrated here for the chlorination products of 2-hydroxybiphenyl. Figure 6.17 shows the stop-flow ^{1}H NMR spectra of the four main compounds, namely the congeners 3-, and 5-chloro-2-hydroxybiphenyl, 3,5-dichloro-2-hydroxybiphenyl and 3,4,5-trichloro-2-hydroxy-biphenyl. In more complicated cases, identification of the compounds can be further confirmed by stop-flow 2D TOCSY spectra, as shown in Figure 6.18 for 3,5-dichloro-2-hydroxybiphenyl.

6.6 CONCLUSIONS

In general, LC–NMR coupling will not be meaningful for target analyses of environmental samples. Rather, after preconcentration, the method gives a complete overview of the organic compounds present in the sample (in the ppb range) and is therefore suitable for non-target analyses. In combination with LC–MS, the on-line identification of unknowns can be achieved in many cases; in other cases, valuable structural information on the pollutants can be obtained. On the basis of LC–NMR and LC–MS investigations, simpler and less expensive methods may be developed in a second step for the target analysis of important identified pollutants or metabolites.

At present, applications of LC–NMR in environmental analysis are still limited by the relatively low sensitivity of the NMR detector. Further progress is expected when cryoprobes and new capillary flow cells hopefully become available in the near future.

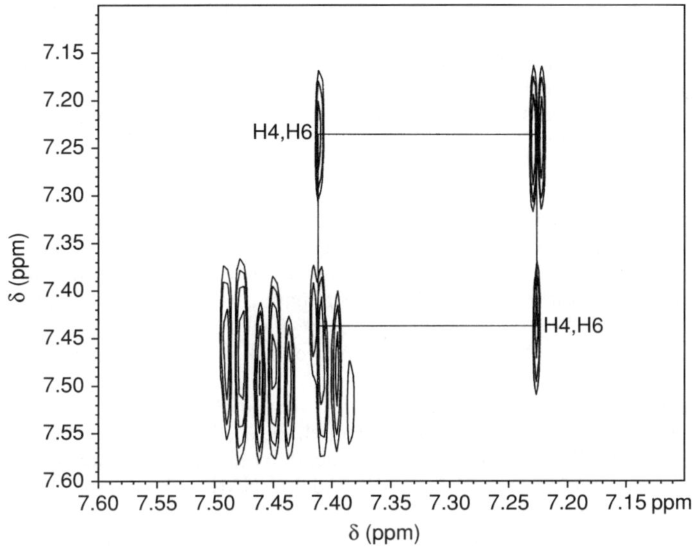

Figure 6.18 Stop-flow 2D WET-TOCSY spectrum of 3,5-dichloro-2-hydroxybiphenyl. Protons (H4,H6) of the chlorinated ring can be identified due to the magnetization transfer between the signals at 7.23 (H4) and 7.41 (H6) ppm. Conditions: spectrometer, Bruker DRX 600; probe head, 4 mm *z*-gradient LC probe. ^{13}C decoupling was applied during the WET-pulse train and acquisition. The spectral size was 8k × 512 data points, with 16 scans per increment, and a 64 ms TOCSY mixing time

REFERENCES

1. Levsen, K., Preiss, A. and Godejohann, M., *Trends Anal. Chem.*, 2000, **19**, 27.
2. Schröder, H. Fr., *J. Chromatogr.*, 1993, **643**, 145.
3. Godejohann, M., Preiss, A. and Mügge, C., *Anal. Chem.*, 1998, **70**, 590.
4. Griffiths, L., *Anal. Chem.*, 1995, **67**, 409.
5. Godejohann, M., Preiss, A., Mügge, C. and Wünsch, G., *Anal. Chem.*, 1997, **69**, 3832.
6. Preiss, A., Lewin, U., Wennrich, L., Findeisen, M. and Efer, J., *Fresenius' J. Anal. Chem.*, 1997, **357**, 676.
7. Godejohann, M., Preiss, A., Levsen, K., Wollin, K. M. and Mügge, C., *Acta Hydrochim. Hydrobiol.*, 1998, **26**, 330.
8. Schmidt, T. C., Steinbach, K., van Löw, E. and Stork, G., *Chemosphere*, 1998, **37**, 1079.
9. Godejohann, M., Astratov, M., Preiss, A., Levsen, K. and Mügge, C., *Anal. Chem.*, 1998, **70**, 4104.
10. Yinon, J., Betowski, L. D. and Voyksner, R. D., 'Applications of LC–MS in environmental chemistry', *Journal of Chromatography Library*, 1996, **59**, pp. 187–261.

11. Preiss, A., Sänger, U., Karfich, N., Levsen, K. and Mügge, C., *Anal. Chem.*, 2000, **72**, 992.
12. Benfenati, E., Pierucci, P., Fanelli, R., Preiss, A., Godejohann, M., Astratou, M., Levsen, K. and Barcelo, D., *J. Chromatogr., A*, 1999, **831**, 243.
13. Albaret, Ch., Froment, D. and Godejohann, M., unpublished results.
14. Mesilaakso, M. and Tolppa, E.-L., *Anal. Chem.*, 1996, **68**, 2313.
15. Albaret, Ch., Loeillet, D., Augé, D. and Fortier, P.-L., *Anal. Chem.*, 1997, **69**, 2694.
16. Spanggord, R. J., Mabey, W. R., Mill, T., Chou, T.-W., Smith, J. H., Lee, S. and Roberts, D., 'Environmental fate studies on certain ammunition wastewater constituents: Phase IV. Lagoon Model Studies', Report AD-A138550, US Army Medical Research and Development Command, Fort Detrick, Frederick, MD, 1984.
17. Kaplan, L. A., Burlinson, N. E. and Sitzmann, M. E., 'Photochemistry of TNT; Investigation of the "Pink Water" problem, Part II, Report NSWC/WOL/TR 75–152, Naval Surface Weapons Center White Oak Laboratory, Silver Spring, MD, 1975.
18. Chung, K. T. and Cerniglia, C. E., *Mutat. Res.*, 1992, **77**, 201.
19. Krull, R., Hemmi, M., Otto P. and Hempel, D. C., *Water Sci. Tech.*, 1998, **38**, 339.
20. Pham, T. H. L., Rotard, W., Preiss, A., Elend, M., 'Identifizierung von Reaktivfarbstoffabbauproduckten mittels LC-MS und LC-NMR', Anakon 2001, 04.-07. 2001 Konstanz, Tagungsband S. 145.
21. Köllner, G., Möder, M. and Czihal, K., *Chemosphere*, 2000, **41**, 1827.
22. Kurz, J. and Ballschmiter, K., *Fresenius' J. Anal. Chem.*, 1994, **349**, 533.
23. Zenkevich, I. G. and Zibulskaya, I. P., *J. Phys. Chem.*, 1997, **71**, 341 (in Russian).
24. Möder, M., Zenkevich, I. G., Köllner, G., and Popp, P. 'New Approach in the precalculation of GC-Retention Indices of Hydroxylated Polychlorinated Biphenyls on HP-5 Liquid Phase', Proceedings of the 20th International Symposium on Chromatography, 25-29.05. 1998, Riva, A 15, in Fundamentals, CD produced by Naxos Software Soultions, Schriesheim, Germany.

7 Related Techniques – Introduction

Despite the fact that High Performance Liquid Chromatography (HPLC) using analytical columns (250 × 4.60 mm) is currently the most widely employed technique in separation science, several very important related techniques such as Gel Permeation Chromatography (GPC), Supercritical Fluid Chromatography (SFC), and the newly emerging capillary HPLC, together with electro-driven separation techniques such as capillary electrophoresis and capillary electrochromatography, have to be considered. The separation mechanisms of the individual related techniques are totally different and can be verified in chromatographic textbooks, because the common feature of this section of this present volume is the application of the closed-loop separation–detection system, as in LC–NMR, which shows major benefits in comparison with the conventional off-line system. The unsurpassed advantage of on-line GPC–NMR is the severe saving in overall analysis time, as outlined below in Chapter 7.1.

Supercritical Fluid Chromatography, as well as Supercritical Fluid Extraction (SFE), are very often considered as niche detection techniques. However, it is obvious from Chapter 7.2 that SFC–NMR and SFE–NMR show the inherent advantage that no solvent suppression technique is needed and the whole proton chemical shift range can be used without any distortions of solvent signals and impurities.

Nanolitre NMR techniques employ capillaries and will be used in the near future to an increasing extent. The two major current approaches for NMR detection in the nanogram range are outlined side-by-side. The coupling of electro-driven capillary separation techniques, together with ^{1}H NMR spectroscopy, has been pioneered by the group of Ernst Bayer in Tübingen. Special connections for capillaries had been developed and interesting applications in the fields of natural products and drugs have been performed, thus preparing the basis for major real-world applications.

In summary, the variety of applications outlined in this section provides evidence for the concept of on-line hyphenation and demonstrates the enormous potential for the solution of structure elucidation problems in all major research areas.

On-Line LC–NMR and Related Techniques
Edited by K. Albert.

7.1 GPC–NMR Coupling

HEIDRUN HÄNDEL and KLAUS ALBERT

Institut für Organische Chemie, Universität Tübingen, Tübingen, Germany

One of the most impressive advantages of continuous-flow ^{1}H NMR spectroscopy is the direct monitoring of the change in stereochemical and chemical composition of polymers and copolymers during Gel Permeation Chromatography (GPC)/Size Exclusion Chromatography (SEC) [1–13]. The on-line GPC–NMR coupling method opens a new opportunity for the characterization of polymers in general and is particularly helpful in the various application examples discussed in the following.

All of the related articles demonstrate the feasibility of the NMR spectrometer as a detector in GPC. The experimental set-up for GPC–NMR does not differ from the LC–NMR technique in principle, and has previously been described above in detail in Chapter 1.

Some examples for demonstrating the method of polymer characterization have been selected, and will be described in the following sections.

7.1.1 DIRECT DETERMINATION OF MOLECULAR WEIGHT DISTRIBUTION WITHOUT CALIBRATION

Polymers with well-defined chemical structures are a prerequisite for direct determinations of molecular weight distributions (MWDs) [1–3]. Ute and coworkers [6] have used on-line SEC–NMR coupling to determine the molecular weight distribution in isotactic poly(methyl methacrylate) (PMMA) samples.

The on-line SEC–NMR measurements were performed on a 750 MHz spectrometer with an inverse-geometry LC–NMR probe with pulsed-field gradient coils. The detection volume of the flow cell was approximately 60 μl. A Shodex 'linear' K-805L SEC column (8 mm × 300 mm) with a maximum porosity of 4 $\times 10^6$ was used for the chromatography.

Figure 7.1.1 shows the stacked trace plot of the serial spectra for isotactic PMMA.

The NMR information can be seen, on the x-axis, to be dependent on the elution time (in min) (y-axis). The chemical structure of isotactic PMMA is defined, as a consequence of the synthesis, as possessing a t-C_4H_9 group from the initiator at the chain end:

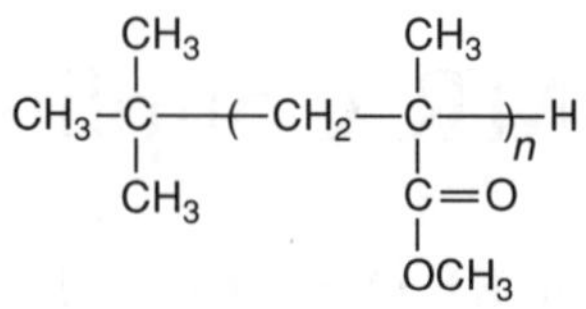

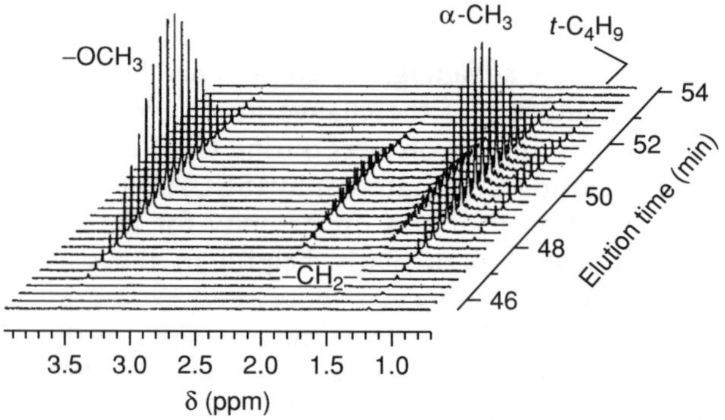

Figure 7.1.1 750 MHz on-line SEC–NMR data obtained for isotactic PMMA

Thus it is possible to determine the degree of polymerization from the integrals of the NMR signals of the t-C_4H_9 group (0.85 ppm) and the α-CH_3 or OCH_3 group resonances (1.18 and 3.6 ppm, respectively) in the chain.

These measurements were performed on a 750 MHz NMR spectrometer. A comparison of the spectra obtained at the elution maximum using the 750 MHz spectrometer, and previously a 500 MHz spectrometer, measured under similar conditions [2], shows a signal-to-noise ratio (S/N) which was 9.2 times higher in the 750 MHz spectrum than that measured at 500 MHz. In this way, it is possible to determine both the M_n and M_w/M_n over a molecular weight region greater than 2×10^4.

7.1.2 MOLECULAR WEIGHT DEPENDENCE OF TACTICITY

The properties of polymers depend on the stereoregularity of the (polymer) chains [1,4,5]. For instance, the glass transition temperature (T_g) increases according to the order isotactic < heterotactic < syndiotactic PMMA, while the melting points differ according to the order isotactic < syndiotactic < heterotactic PMMA. However the stereoregularity itself is, in addition, dependent on the chain length. Hatada and co-workers [7] investigated PMMA obtained by anionic polymerization with t-C_4H_9/MgCl (Mg^{2+}/t-C_4H_9Mg = 0.87) in toluene at −78°C by using GPC–NMR. On-line GPC–NMR measurements were performed with a 500 MHz spectrometer and a chromatographic system equipped with a GPC column with a maximum porosity of 3×10^7. Chloroform was used as the eluent, and the flow rate was 0.2 ml/min, with an injected sample of 1 mg. Figure 7.1.2 demonstrates the results that were obtained.

In this figure, the top chromatogram, obtained by monitoring the OCH_3 signal at 3.59 ppm, shows a trimodal molecular weight distribution with the marked ranges F1, F2, and F3 (Figure 7.1.2(a)). The splitting of the α-CH_3 proton resonances at 0.86 and 1.2 ppm illustrates the *rr*- and *mm*-triads, respectively. The concentration of these triads as a function of the elution time is shown in Figure 7.1.2(b). The α-CH_3 signal eluted in the periods F1 (Figure 7.1.2(c)), F2 (Figure 7.1.2(d)), and F3 (Figure 7.1.2(e)) shows the higher concentration of *rr*-triads (syndiotactic) in the higher-molecular-weight part F1 and the higher concentration of *mm*-triads (isotactic) in the lower-molecular-weight part F3.

Figure 7.1.3, in comparison to Figure 7.1.2, demonstrates the influence of the initiator, in this case 1.1-diphenylhexyllithium (DPHLi), on the microstructure.

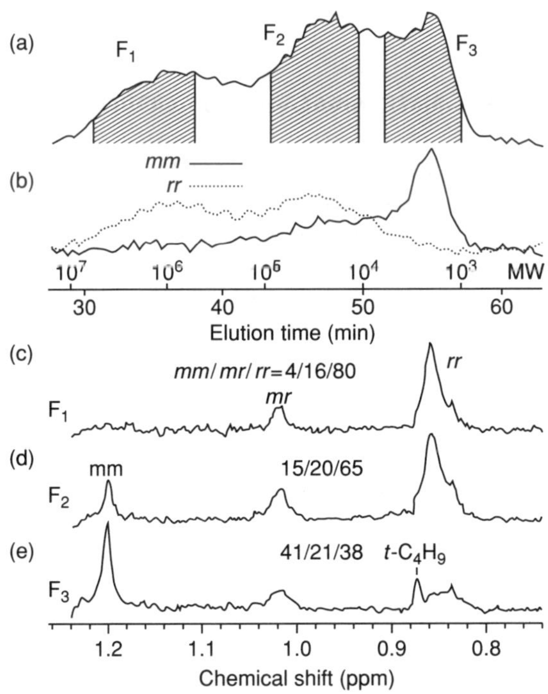

Figure 7.1.2 (a) The ^{1}H NMR-detected GPC traces obtained by monitoring the methoxy proton resonances at 3.59 ppm, and (b) the α-methyl proton resonances at 0.86 ppm (· · · ·) and 1.2 ppm (—) due to *rr*- and *mm*-triads, respectively. The NMR signals due to the α-methyl protons of PMMA eluted in the elution periods F1 (c), F2 (d), and F3 (e) are also shown

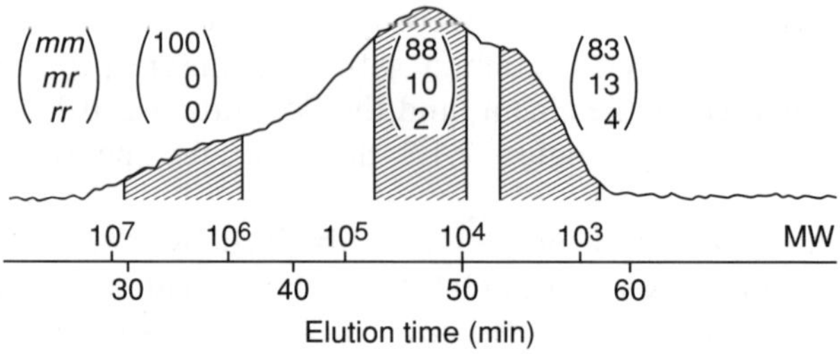

Figure 7.1.3 ^{1}H NMR-detected GPC trace of a sample of PMMA prepared with 1,1-diphenylhexyllithium in toluene at −78 °C (monitoring the methoxy proton resonance at 3.59 ppm), plus the triad tacticities of the eluting fractions

The molecular weight distribution is very broad, and the high-molecular-weight part only contains *mm*-(isotactic) triads. Further studies with other initiators are described in the same paper [7]. All of the results show a dependence of the microstructure on the initiator being used and allow an insight into the mechanism of the polymerization process. The nature of the active species in these special anionic polymerizations is responsible for the actual course of the polymerization.

7.1.3 ON-LINE GPC–NMR ANALYSIS OF COPOLYMERS

The properties of copolymers primarily depend upon their chemical composition but also on the microstructure of the polymer chains. However, the chemical composition of copolymers can often depend on their molecular weight fractions [1,8–10].

Pioneering work from Hatada and co-workers [8] has illustrated the molecular weight dependence of the chemical composition of block and random (alkyl) methacrylate copolymers, for example:

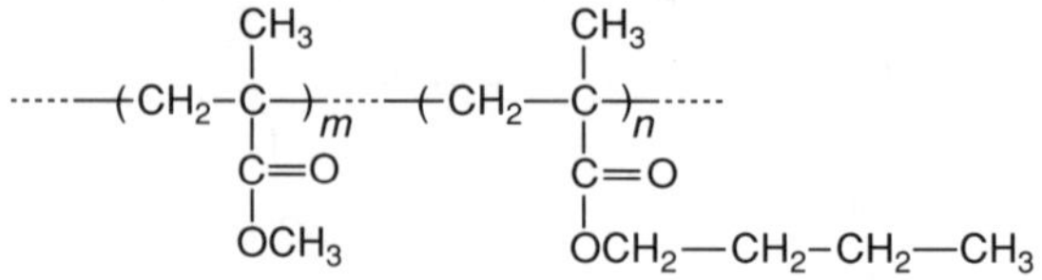

methyl methacrylate(MMA) *n*-butyl methacraylate(*n*-BuMA)

In order to derive information about the molecular weight dependence of the chemical composition, a GPC separation of 1 g of the above mentioned copolymer was performed with a 30 × 0.8 cm GPC column, using chloroform-d$_1$ as the eluent at a flow rate of 0.2 ml min^{-1}. NMR spectra were recorded with a 60 μl flow cell on a 500 MHz instrument (see Figure 7.1.4).

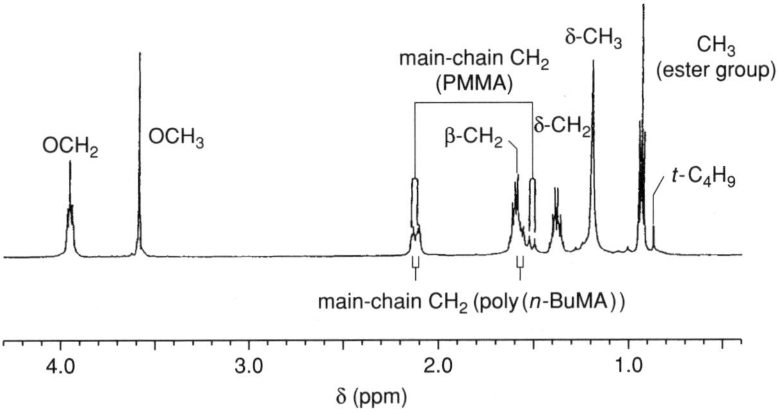

Figure 7.1.4 ^1H NMR spectrum (500 MHz) of PMMA-*block*-poly(*n*-butyl methacrylate)

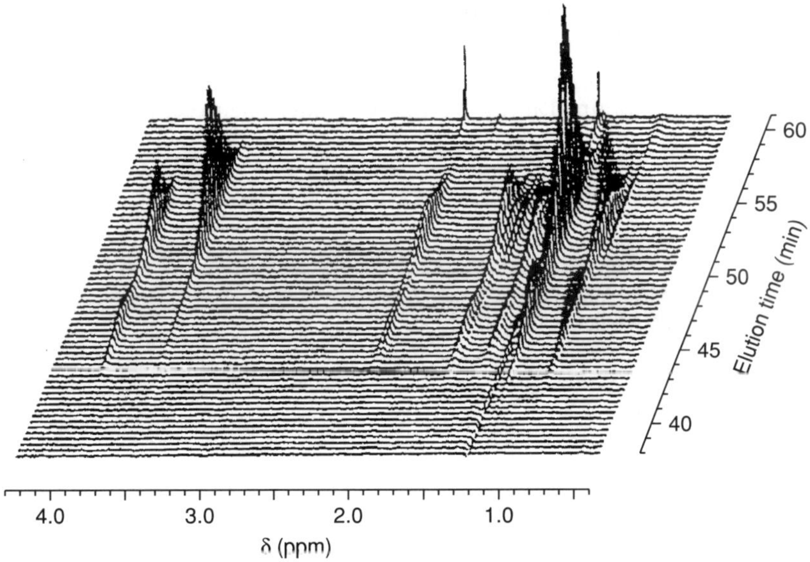

Figure 7.1.5 ^1H NMR chromatogram (stacked plot, 500 MHz) of a GPC separation of PMMA-block-poly(*n*-butylmethacrylate)

The Fourier-transformed spectrum results in a 'row' in the two-dimensional plot of the ^1H chemical shifts versus the retention times. Within one separation run, up to 150 rows are accumulated, resulting in an overall acquisition time of 60 min. Figure 7.1.5 shows the stacked plot of the GPC–NMR separation, where the varying intensities of the OCH_3 signals of PMMA at 3.58 ppm, in contrast to the OCH_2 signal of poly(*n*-BuMA) at 3.95 ppm, can be readily seen.

Thus, the copolymer composition can be directly determined from the elution curves of both signals at any row of the chromatogram. Figure 7.1.5

clearly shows that the first eluting high-molecular-weight fraction is rich in *n*-BuMA units (intense signal of the OCH_2 groups at 3.95 ppm), whereas the low-molecular-weight fraction is rich in MMA units (intense signal of OCH_3 groups at 3.56 ppm). This example demonstrates the great time-saving nature of the hyphenation of a chromatographic separation method with NMR spectroscopy. In order to yield the same information as in the on-line GPC–NMR run, 150 fractions of the GPC separation have to be collected and then 150 routine ^{1}H NMR spectra have to be recorded. Whereas the (hyphenated) GPC–NMR data were obtained within 1 h, off-line separation and NMR registration would take at least take 3 h. A comparison of the GPC–NMR data for block and random copolymers of MMA and *n*-BuMA is shown in Figure 7.1.6 [1]. This representation of the data demonstrates that the comonomer composition of the random copolymer is independent of the molecular weight.

The monomer-selective 'living' copolymerization of *t*-butyl acrylate (*t*-BuA) and ethyl methacrylate (EMA) was studied on a 750 MHz spectrometer with an ^{1}H inverse-geometry LC–NMR probe with pulsed-field gradient coils [10]. The detection volume of the flow cell was ca. 60 μl. The measurements were performed in chloroform-d$_1$, with a flow rate of 0.2 ml/min, at 296 K. The copolymers were obtained using bis (2,6-di-*t*-butylphenoxy) methylaluminium

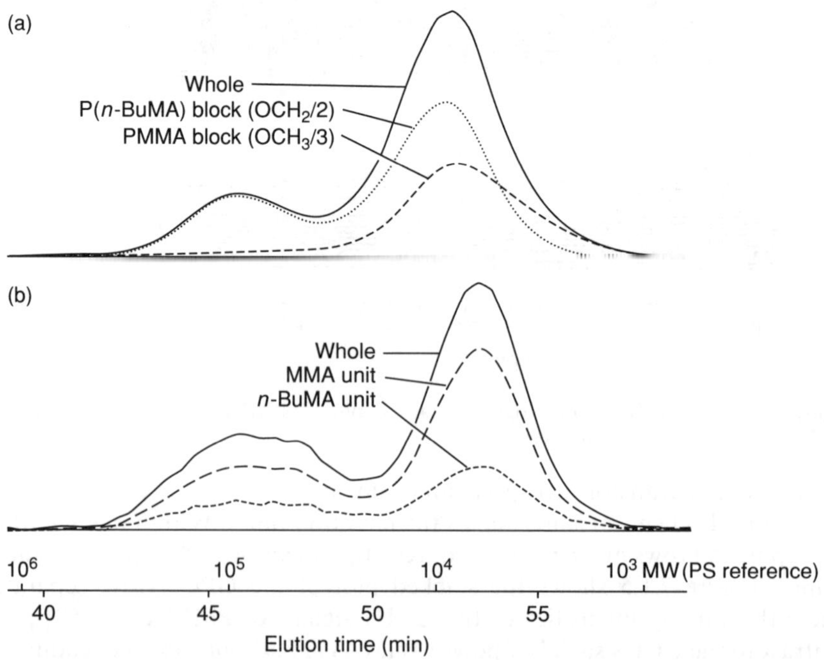

Figure 7.1.6 ^{1}H NMR-detected GPC curves of the (a) block and (b) random copolymers of MMA and *n*-BuMA

(t-BuLi/MeAl(ODBP)$_2$), with different polymerization times and over several different temperatures ranges (−60 to − 20°C).

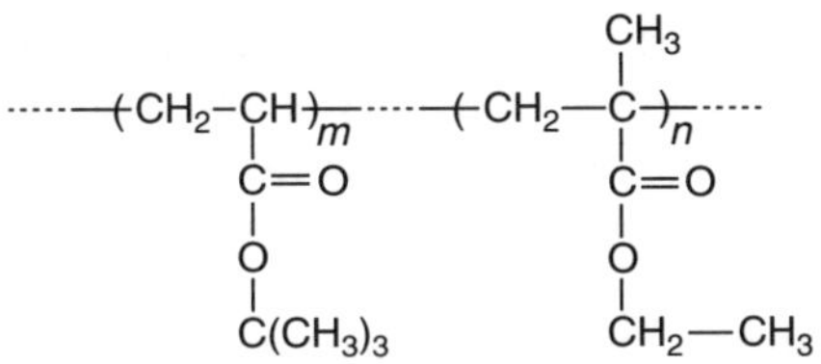

Figure 7.1.7 shows the results obtained for different polymerization times. These chromatograms were reconstructed using the intensities of the t-butyl signal (OC(CH$_3$)$_3$) from the t-BuA units at 1.26 ppm and the OCH$_2$ signal at 4.04 ppm from the EMA units. The copolymer composition was determined at any specified region in the chromatogram. The copolymer obtained after 6 h (Figure 7.1.7(a)) has a unimodal distribution. In this time interval, the t-BuA is

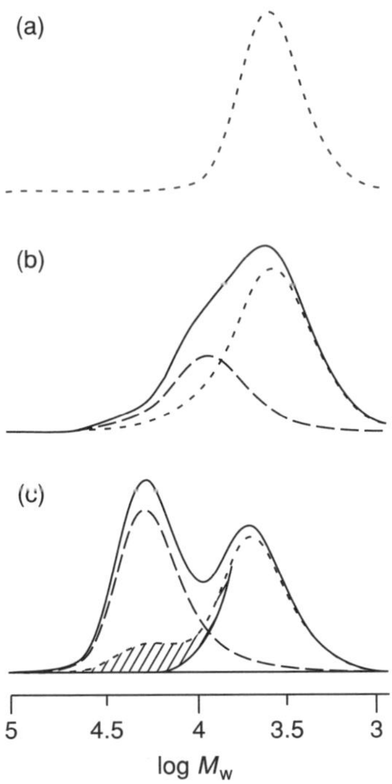

Figure 7.1.7 On-line SEC–NMR analysis of the copolymers of t-BuA and EMA prepared with t-BuLi/MeAl(ODBP)$_2$ (1/5) in toluene at −60 °C for (a) 6 h, (b) 9 h, and (c) 12 h: (· · ·) t-BuA; (- - -) EMA; (—) whole unit

polymerized preferentially. After 9 h (Figure 7.1.7(b)), the EMA also polymerizes, and the chromatogram has a shoulder on the higher-molecular-weight side. At 100% conversion, after 12 h, a bimodal chromatogram was obtained (Figure 7.1.7 (c)). The higher-molecular-weight part of the chromatogram mainly contains EMA units with only a few *t*-BuA units. The time-course of the copolymerization shows that the investigated final copolymer is a block copolymer with a long poly (EMA) block and a short poly(*t*-BuA) block.

Figure 7.1.8 demonstrates the influence of the polymerization temperatures on the molecular weight distribution. It can be seen from this figure that the block efficiency increases with increasing polymerization temperature.

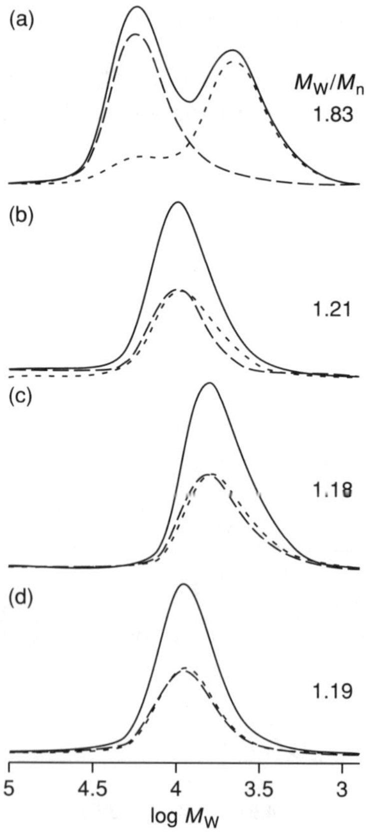

Figure 7.1.8 On-line SEC–NMR analysis of the copolymers of *t*-BuA and EMA prepared with *t*-BuLi/MeAl(ODBP)$_2$ at (a) −60, (b) −40, (c) −30, and (d) −20 °C: (· · ·) t-BuA; (- - -) EMA; (—) whole unit

Händel and co-workers [11] have investigated the influence of the polymerization conditions on the molecular weight and chemical composition of styrene–butyl acrylate copolymers:

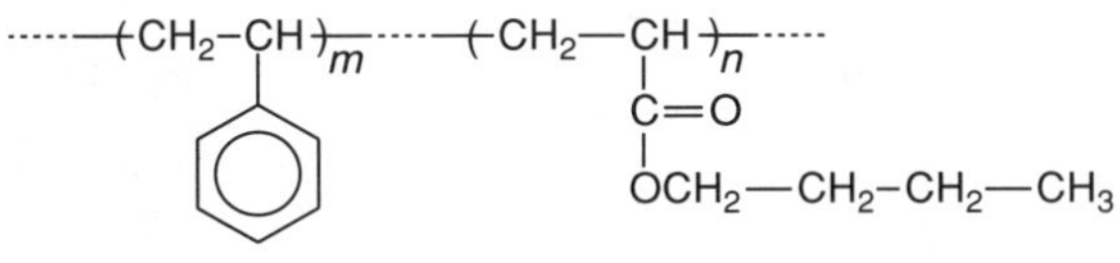

The NMR spectra were recorded on a 400 MHz spectrometer equipped with an LC probe selective for protons. The NMR flow-through cell had a volume of 120 μl. A PLGel mixed B 10 μm (2 × 300 mm × 7.5 mm) column was used for the GPC separation. The eluent was dichloromethane (CH_2Cl_2), with 5% deuterated dichloromethane used for the field-frequency lock. The injection volume was 50 μl of a 10% solution of the copolymer in CH_2Cl_2. For all of the experiments, a flow rate of 0.4 ml/min was used, and the separation was monitored by a UV detector (at 254 nm). The NMR spectra were recorded at 300 K. Suppression of the solvent resonance was achieved by application of a NOESY-type pulse train. The studied copolymer samples were synthesized by 'semi-continuous' radical emulsion polymerization. During the first stage, styrene and butyl acrylate are added continuously to the reactor, while the second stage feed consists of styrene alone. All samples were made under the same polymerization conditions in relation to the initiator, acid used, and polymerization stage. The only difference in conditions relates to the concentration of the charge-transfer agent (CTA) used in the second stage of the (co)polymerization process.

A contour plot obtained for a typical GPC–NMR on-line coupling experiment is shown in Figure 7.1.9. In this figure, the GPC chromatogram depicted along the vertical dimension was reconstructed by summation over the ^1H NMR signal intensities. The chromatography stage was completed after 1 h. Rows 40 and 65 were selected as examples for evaluation of the dependence of the chemical composition on the molecular weight (higher-numbered rows represent low-molecular-weight ranges of the copolymer). All of the three investigated copolymer samples contained 42 mol% butylacrylate, although their measured physical and chemical properties were completely different.

The results of the on-line GPC–NMR analysis are summarized in Figure 7.1.10. The chemical compositions – above all, the fractions with higher-molecular-weight ranges – of the three copolymers are dependent on both the concentration and the stage at which the charge-transfer (CTA) is added to the reaction system. Sample 1 was made by using a constant CTA concentration in both stages, and here the butyl acrylate content is constant over a wide range. Sample 2 was produced by using a high concentration of CTA in the second

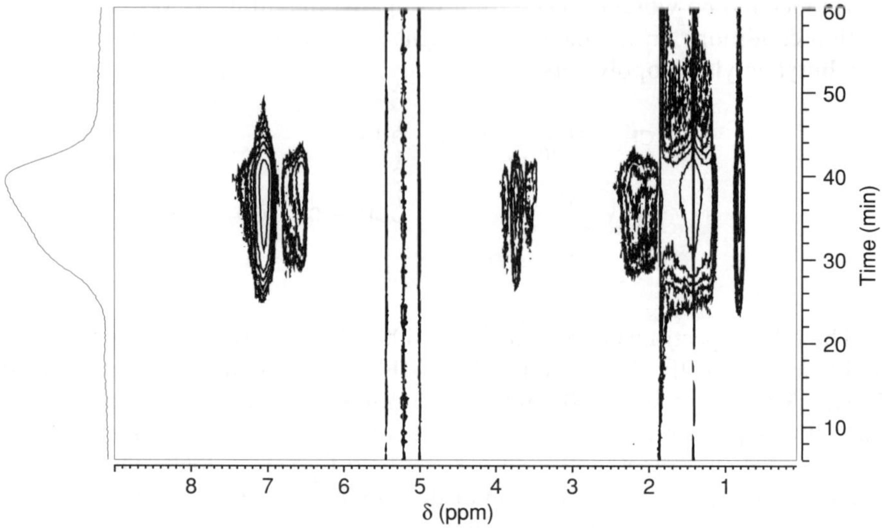

Figure 7.1.9 Contour plot obtained for a typical GPC–NMR on-line coupling analysis of a styrene–butyl acrylate copolymer

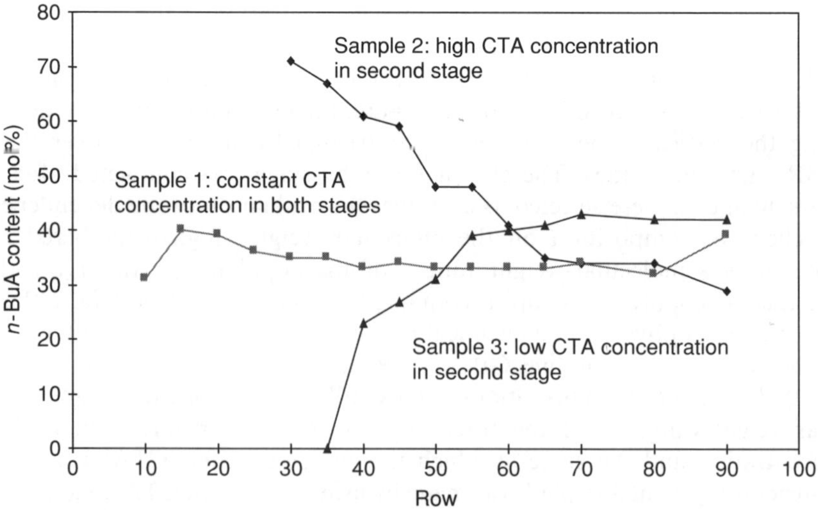

Figure 7.1.10 Dependence of the butyl acrylate content on the molecular weight of the copolymer, showing the influence of the charge-transfar agent (CTA): (–◆–) high CTA content; (–■–) medium CTA content; (–▲–) low CTA content

stage. In this system, the butyl acrylate content shows a strong variation, i.e. from more than 70 mol% in the longer chains to less than 30 mol% in the shorter chains. A low concentration of the CTA in the second stage leads to the opposite behaviour (see sample 3 in Figure 7.1.10). Here, the butyl acrylate content is higher in the shorter chains. In further studies, samples were taken at different times during the reaction – from 15–390 min from the beginning of polymerization – and the chemical compositions of the copolymers obtained were then determined (constant addition of CTA). On-line GPC–NMR analysis was carried out for each time-interval of the polymer steps. Figure 7.1.11 presents some of the results obtained for the chemical composition as a function of the molecular weight of the final copolymer.

It can be seen from this figure that compared to the first time-interval, in the second interval the copolymer chains that are made contain more butyl acrylate. In further steps (60 and 90 min), the butyl acrylate content continues to increase, as shown in Figure 7.1.12.

At the end of the first stage (190 min), a constant average composition exists. During the second stage (involving only the addition of styrene) from 220–390 min, the butyl acrylate content decreases, although the overall distribution of the chemical composition remains constant.

This study of the chemical composition of styrene–ethyl acrylate copolymers as a function of their molecular weight fractions show a broader molecular weight distribution for ethyl acrylate-rich copolymers than for styrene-rich copolymers.

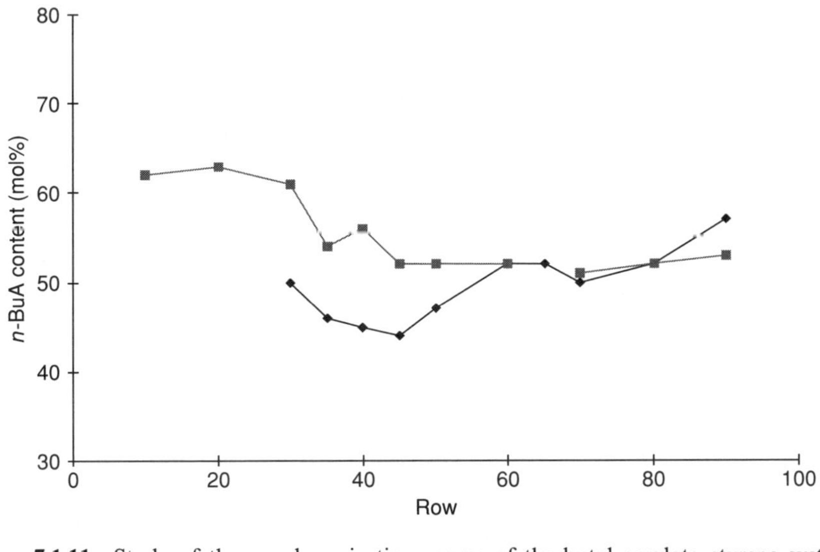

Figure 7.1.11 Study of the copolymerization course of the butyl acrylate–styrene system, using GPC–NMR coupling, in the first two time-intervals: (–◆–) 15 min; (–■–): 30 min

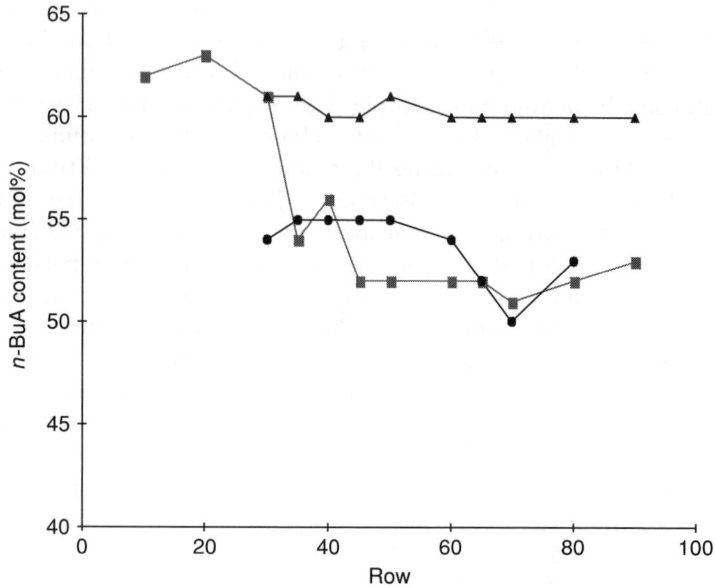

Figure 7.1.12 Study of the copolymerization course of the butyl acuylate–styrene system, using GPC–NMR coupling, in the middle time-intervals: (-■-) 30 min; (-●-) 60 min; (-▲-) 190 min

7.1.4 ON-LINE GPC–NMR ANALYSIS OF OLIGOMERS

The possibilities for analysis of oligomeric systems can be demonstrated by considering chlorinated materials [13]. On-line GPC–NMR has been applied to mixtures of the following chlorinated oligomers:

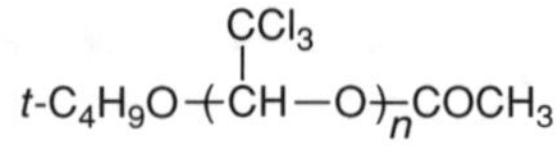

The stacked plot of the on-line GPC–NMR coupling of such oligomers is shown in Figure 7.1.13. In this system, the chemical shift of the single acetal protons depends on their different chemical environments in the various oligomers. The single spectra for the separated oligomers with $n = 3$–7, obtained from the on-line GPC–NMR coupling with a suitable GPC column of low porosity, are shown in Figure 7.1.14.

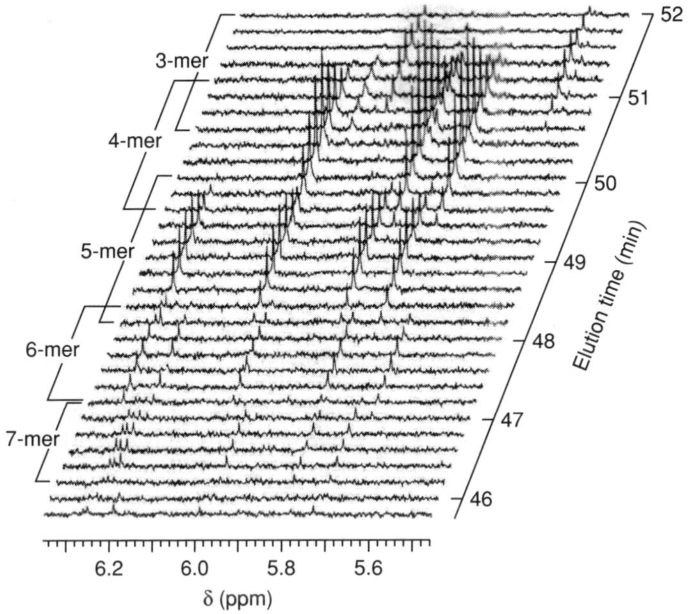

Figure 7.1.13 On-line GPC–NMR data obtained for a mixture of chlorinated oligomers

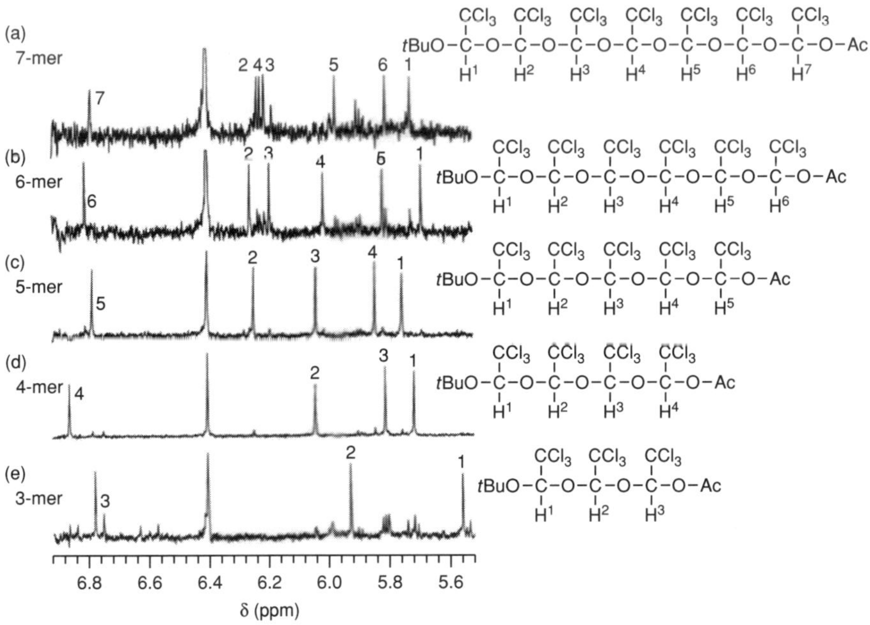

Figure 7.1.14 ^{1}H NMR spectra of the separated chlorinated oligomers obtained from the on-line GPC–NMR data

In conclusion, on-line GPC–NMR coupling provides a fast and efficient method for the characterization of polymers in general and, in particular, for copolymers of different chemical compositions and stereoregularity.

REFERENCES

1. Ute, K. and Hatada, K., *Anal. Sci.*, 1991, **7**, 1629.
2. Hatada, K., Ute, K., Okamoto, Y., Imanari, M. and Fujii N., *Polym. Bull.*, 1988, **20**, 317.
3. Hatada, K., Ute, K., Kashiyama, M. and Imanari, M., *Polym. J.*, 1990, **22**, 218.
4. Hatada, K., *J. Polym. Sci., Polym. Chem. Edn*, 1999, **37**, 245.
5. Hatada, K., Kitayama, T., Ute, K. and Nishimura, T., *Macromol. Symp.*, 1999, **143**, 111.
6. Ute, K., Niimi, R., Hongo, S. and Hatada, K., *Polym. J.*, 1998, **30**, 439.
7. Hatada, K., Ute, K., Kitayama, T., Nishimura, T., Kashiyama, M. and Fujimoto, N., *Polym. Bull.*, 1990, **22**, 549.
8. Hatada, K., Ute, K., Kitayama, T., Yamamoto, M., Nishimura, T. and Kashiyama, M., *Polym. Bull.*, 1989, **21**, 489.
9. Albert, K. and Bayer, E., *Anal. Methods Instrum.*, 1995, **2**, 302.
10. Kitayama, T., Tabuchi, M. and Hatada, K., *Polym. J.*, 2000, **32**, 796.
11. Händel, H., Lazarus, E., Emilie, B., Schlotterbeck, G., Streck, R. and Albert K., to be published.
12. Krämer, I., Pasch, H., Händel, H. and Albert, K., *Makromol. Chem. Phys.*, 1999, **200**, 1734.
13. Ute, K., Kashiyama, M., Oka, K., Hatada, K. and Vogl, O., *Makromol. Chem., Rapid Commun.*, 1990, **11**, 31.

7.2 SFC–NMR and SFE–NMR

HOLGER FISCHER and KLAUS ALBERT
Institut für Organische Chemie, Universität Tübingen, Tübingen, Germany

7.2.1 INTRODUCTION

This article treats the benefits, possibilities and drawbacks of supercritical fluid chromatography (SFC) and supercritical fluid extraction (SFE) coupled to nuclear magnetic resonance spectroscopy. After a general overview and consideration of the motivation for such techniques, the design of high-pressure flow probes, as well as the principle experimental set-ups, are described. By means of several applications and comparison to HPLC–NMR, the utility of these hyphenated techniques is demonstrated.

7.2.2 OVERVIEW AND MOTIVATION

The work horses of separation techniques are, and maybe always will be, high performance liquid chromatography (HPLC) and capillary high performance liquid chromatography (CHPLC), respectively, followed by gas chromatography (GC), which is applicable for volatile and temperature-resistant compounds. Thus, it is only logical, that HPLC was the first separation technique which was coupled to NMR spectroscopy, thus leading nowadays to a powerful tool in structure elucidation.

The fact that separated compounds in HPLC are diluted in a multiple surplus of one and more predominant proton-carrying solvents results in NMR spectra, where usually only the solvent signals can be seen. This is caused by the reduced sensitivity of the receiver, which has to be adjusted to avoid signal overflow.

This main difficulty in coupling HPLC to NMR spectroscopy is faced by methods known as solvent suppression techniques, where the large solvent signals are reduced by special pulse sequences, switched prior to the information-selecting and acquisition pulses. Therefore, many efforts have been made to develop effective and minor-disturbing pulse sequences, such as presaturation, zero excitation and PFG-pulse sequences (WET) (see Chapter 1 and the following chapters). Despite the possibility of also suppressing several of the

On-Line LC–NMR and Related Techniques
Edited by K. Albert.

solvent signals, at the positions of the latter signals and close to them, information from the signals of the analyte is also affected by the suppression and can be erased.

In principle, it would be possible to use fully deuterated solvents for HPLC–NMR coupling, and then solvent suppression techniques would be unnecessary; however, due to the high costs of these solvents, only the use of D_2O is economically acceptable. Solutions to this problem can be seen in the development of hyphenating capillary HPLC to NMR spectroscopy, and there, because of the reduced solvent consumption, the use of fully deuterated solvents would be reasonable. However, this approach is difficult and the development is still an on-going process (see Chapter 7.3 below). From the viewpoint of NMR spectroscopy, there could be a completely new situation, when separation techniques could be found which use proton-free solvents as the mobile phases.

This requirement is fulfilled in gas chromatography (GC), where nitrogen or helium is used as the carrier gas. The low density of gases correlates with their high diffusion coefficients, and therefore in GC fast separations at high flow rates can be achieved when compared with LC. In Figure 7.2.1, the diffusion coefficient versus density diagram shows the areas occupied by the mobile phases in HPLC and GC.

The first attempt to combine GC and NMR was carried out in 1965 by Brame [1]. This author used specific NMR tubes for direct sample trapping.

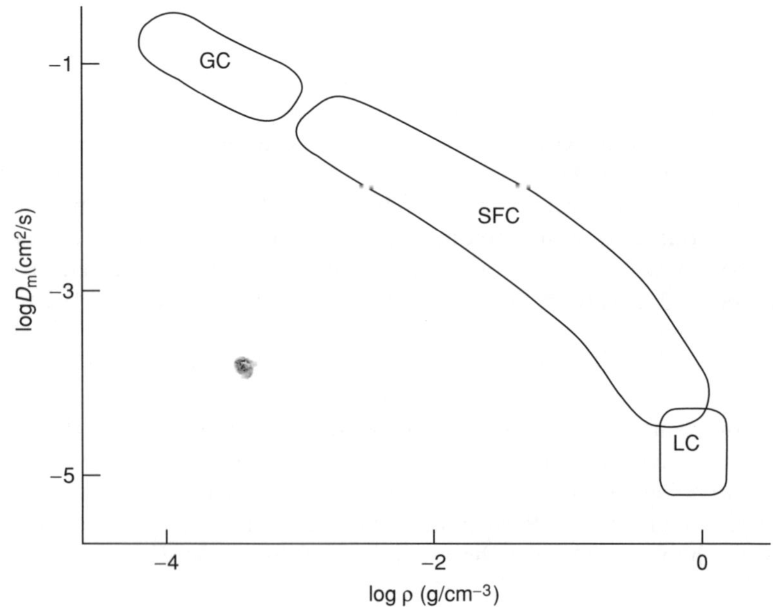

Figure 7.2.1 Diffusion coefficient versus density diagram

Further improvements were made by Tsuda *et al.* [2]. In the early 1980s, Buddrus and co-workers [3,4] developed an on-line GC-NMR apparatus to separate and identify hydrocarbon and terpene mixtures in the gaseous state. Spectra were recorded on a 100 MHz Fourier-transform spectrometer (2.3 T iron magnet) with an NMR signal line width on the half-signal height of around 2 Hz.

One of the mayor drawbacks is that only volatile and temperature-resistant compounds can be investigated. Gases are magnetized faster than liquids, because they have shorter spin–lattice relaxation times (T_1), due to an effective spin rotation mechanism. Therefore, pulse repetition times in flow experiments can be in the range of 1 s and some dozen transients can be accumulated per separated peak. Nevertheless, the sample amounts used nowadays in capillary GC are far from the detection limit of NMR spectroscopy, and therefore the sensitivity is low or insufficient, due to the small number of gas molecules per volume at atmospheric pressure in the NMR flow cell. In addition, high-boiling components ($> 100\,^{\circ}$C) are not easy to handle in NMR flow probes and can condense on colder parts of the apparatus, thus reducing their sensitivity in NMR spectroscopy.

Supercritical fluid chromatography (SFC) [5–7] is located between the GC and HPLC separation methods and bridges these two separation techniques (see Figure 7.2.1). Supercritical fluid separations are performed by using a pressure gradient to control the solvating power and this can be increased by adding a polar modifier (e.g. methanol) to the supercritical eluent. SFC is used to separate petroleum distillates [8] and by-products, synthetic polymers, pesticides [9] and herbicides, food and biologically active components, natural products [10], and more recently for the separation of chiral compounds, such as pharmaceutical drugs [11,12]. SFC is most commonly camed out by using a normal-phase retention mechanism and is for this reason an orthogonal separation methods to LC. The advantages of SFC over LC include higher efficiency separations and faster speed of analysis. The advantage of SFC over GC is that separation of thermally labile compounds is possible. The limited polarity of the mobile phase is disadvantageous, however, even when polar modifiers are added. At low pressures (< 100 bar), supercritical CO_2 has the solvating power of aliphatic hydrocarbons, while at higher pressures (about 400 bar) its solvating power is similar to dichloromethane. Whereas many analysts are sure that all existing separation problems can be solved by using either HPLC or GC, some of these chromatographers still believe in the benefits of SFC, and so it exists more or less as a niche technique.

However, regarding its coupling to NMR spectroscopy, SFC occupies an exceptional position, due to its physico-chemical properties and the chemical characteristics of its mobile phases.

The most common solvent used in SFC is carbon dioxide (CO_2), which is a proton-free molecule, so it represents an ideal solvent for ^{1}H NMR spectroscopy. The necessity of difficult solvent suppression techniques can be discarded and consequently the whole spectral range can be used for structure elucidation purposes.

While supercritical fluid chromatography seems to have a controversial benefit, supercritical fluid extraction (SFE) is a technique used world-wide and plays an important role in many industrial processes, such as the decaffeination of coffee. Due to the high information content of NMR spectroscopy, the hyphenation of SFE with NMR spectroscopy promises a deep insight into the extraction process itself, thus leading to a better understanding and development of more efficient extractions.

7.2.3 A SHORT HISTORY OF SFC–NMR COUPLING

NMR spectroscopy under high-pressure conditions is as old as NMR itself. Benedick and Purcell [13] first utilized a pressure vessel surrounded by an RF coil in 1954, while Jonas and co-workers [14,15] continually developed this approach over some three decades.

Studies of high-pressure NMR spectra have been carried out in by one of two ways, known, respectively, as the high-pressure probe method and the high-pressure cell method. High-pressure probes usually show a better sensitivity, due to the better filling factor.

The first flow probe suitable for on-line monitoring of chromatographic separations under supercritical fluid conditions was developed by Dorn and co-workers [16] in 1988. In a separation of a hydrocarbon mixture, these researchers could achieve a signal resolution under static conditions of 1.4 Hz and 4–5 Hz with a 2 ml/min flow. Years later, Braumann was able to acquire highly resolved proton NMR spectra suitable for structure elucidation with a flow probe developed and build by Bruker (Rheinstetten, Germany). In the following years, several applications of SFC–NMR coupling in polymer [17,18] and biomedical science [19], plus SFE–NMR coupling in polymer and nutrition science [20], were published as well as one review on SFC–NMR coupling [21]. Recent developments endeavour to solve specific NMR-related problems caused by the increased spin–lattice relaxation time T_1 of compounds in supercritical fluids [22].

7.2.4 HIGH-PRESSURE FLOW PROBES

The main, and most crucial, part of the experimental set-up is the probe, which has to be pressure-and temperature-stable at the same time and must have a flow cell with an inlet and an outlet to supply and take away the samples. For the development of SFC–NMR probes, one has to keep in mind the aim of the application, i.e. the structure elucidation of unknown compounds. For the investigation of new, unknown substances, the resolution of the probe has to be of the same order as that of conventional flow probes. Therefore, the available materials to construct such a probe are restricted. There exist a great variety of probes which are suitable for either high-temperature, high-pressure or flow approaches, but a combination of all three features, together with a very good resolution, is the main difficulty.

Several designs of high-pressure NMR flow probes suitable for the recording of spectra of supercritical fluids are known [23,24]. The first approach used for continuous-flow detection was developed by Dorn and co-workers [16]. This pioneering design used a combination of premagnetization volume for the equilibrium of the Boltzmann magnetization, together with a pressure-stable flow cell, capable of operating at pressures up to 100 bar and temperatures up to 373 K (Figure 7.2.2). In this design, a 0.25 in OD alumina ceramic tube (g) used for the observation cell was connected to stainless steel tubing of the

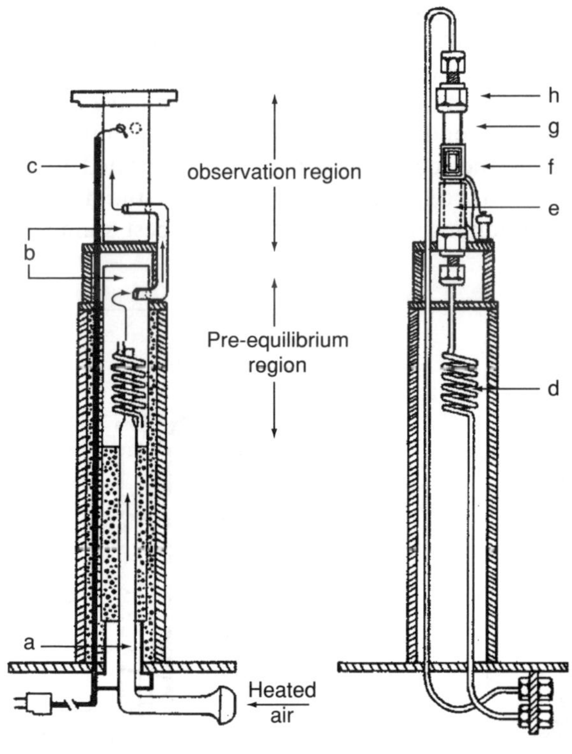

Figure 7.2.2 Schematic diagram of the flow probe developed by Dorn and co-workers and used for the direct coupling of SFC to NMR: (a) insulated glass transfer line; (b) glass insert; (c) Cu/constantin thermocouple; (d) stainless steel equilibrium coil; (e) brass shield; (f) Helmholtz coil; (g) ceramic flow cell; (h) brass 'Swagelok' fitting. Reprinted with permission from Allen, L. A., Glass, T. E. and Dorn, H. C., *Anal. Chem.*, **60**, 390–394 (1988). Copyright (1988) American Chemical Society

equilibration coil (d) with brass 'Swagelok' reducing fittings (h), with the brass ferrules being replaced by graphite/Vespel-blended ferrule. Thus, an easy interchange of ceramic cells with different internal diameter tubes could be performed. Detection volumes range between 20 and 120 μl, with a ceramic tube with internal diameter of 1.6 mm, resulting in a detection volume of 20 μl, being used for most SFC–NMR experiments. With this experimental design, a static NMR resolution of 1.4 Hz could be obtained, whereas under continuous-flow separation conditions of 2.0 ml/min, a signal line width of the order of 4–5 Hz was obtained.

Other experimental designs are derived from continuous-flow probes used for liquid-state detection. In Figure 7.2.3(a), a probe is shown where the inner glass tube of the originating HPLC–NMR probe is substituted with a titanium–sapphire flow cell (od 5 mm, id 3 mm, detection volume 120 μl), and where the PEEK capillaries used in the HPLC–NMR probe are replaced by titanium tubing. A double-turned proton/deuterium coil is directly fixed to the sapphire flow cell The whole arrangement is centred in the glassware of a conventional probe body, in which a thermocouple is inserted, thus allowing the execution of well-defined temperature-dependent measurements.

Figure 7.2.3 (b) depicts a probe which is also suitable for ^{13}C detection, due to the inclusion of a second coil. The main advantage of this probe compared to all of the others is the lock capillary, which is located in the centre of the titanium–sapphire flow cell. This can be used for the deuterated solvent, i.e. for locking and shimming the magnet. For the best shimming, however, the

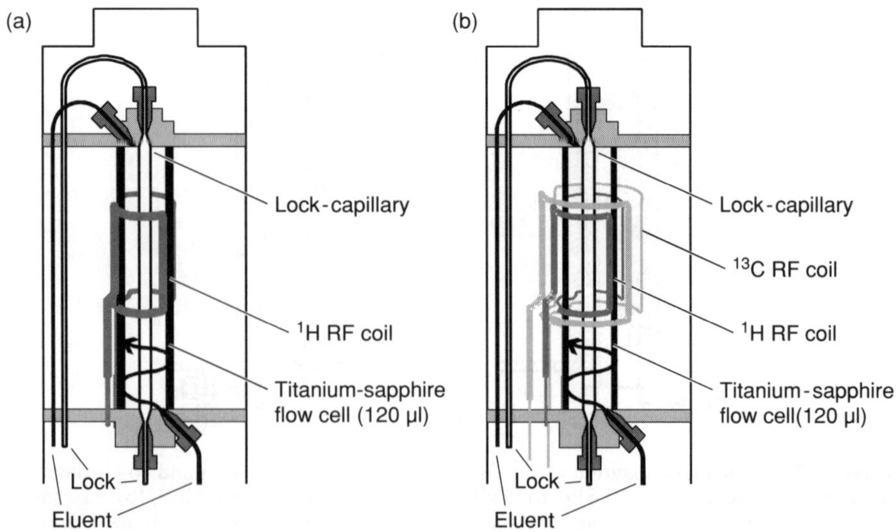

Figure 7.2.3 Schematic diagrams of two designs for SFC–NMR probes which have been derived from LC–NMR probes

outer, sapphire flow cell should be used prior to the filling of the cell with the supercritical fluid.

The non-rotation of the sapphire flow cell, together with the detection volume and the employed flow rate, determine the resulting NMR line width. Whereas with rotation of the NMR tube in a conventional NMR probe, the signal line width of chloroform at the height of the ^{13}C satellites in degassed acetone-d_6 is about 3–4 Hz, SFC continuous-flow probes show values of the order of 15–18 Hz in the liquid and in the supercritical state. This 'hump test' also indicates that there is no change in the signal line width in the ^1H NMR spectra in the liquid and in the supercritical state.

7.2.5 EXPERIMENTAL SET-UP

As an example for coupling an NMR spectrometer to a SFC separation system, a feasible experimental set-up is shown in Figure 7.2.4.

The SFC system is located at a distance of 2–2.5 m from the conventional superconducting cryomagnet; however, new developments in active shielded cryomagnet design have allowed the reduction of this distance to a minimum of 50 cm at a field strength of 9.4 T, where the 5 G line is inside the magnet. Here, the chromatographic system is assembled from commercially available GC and HPLC components. The pump heads of the CO_2 pump are cooled to approximately 275 K with a cryostat; the CO_2 is then supplied from a gas cylinder with a dip-tube, being cooled through 2 m of steel capillary, which is also positioned in the cryostat. Column heating and establishment of the supercritical state is performed within a GC oven. In order to maintain supercritical conditions throughout the system, including the NMR detection cell, a back-pressure regulator is connected to the outlet of the NMR flow cell. Alternatively, an adjustable restrictor can be used for this purpose. The back-pressure regulator, together with the whole chromatographic system, is controlled with a personal computer, using a special software program for analytical purposes.

7.2.6 STOP- AND CONTINUOUS-FLOW SFC–NMR MEASUREMENTS

7.2.6.1 SFC–NMR VERSUS LC–NMR

Despite the more elaborate set-up of the SFC–NMR experiment, Figure 7.2.5 illustrates very impressively why SFC–NMR can be so advantageous when compared to the LC–NMR technique.

In the upper spectrum taken from a LC–NMR run (Figure 7.2.5(a)), despite solvent suppression there are large remaining solvent signals, i.e. water at 3.7 ppm and acetonitrile with ^{13}C satellites at 2.1 ppm. In cases where the sample peaks lie under the solvent signals, structure elucidation is difficult or

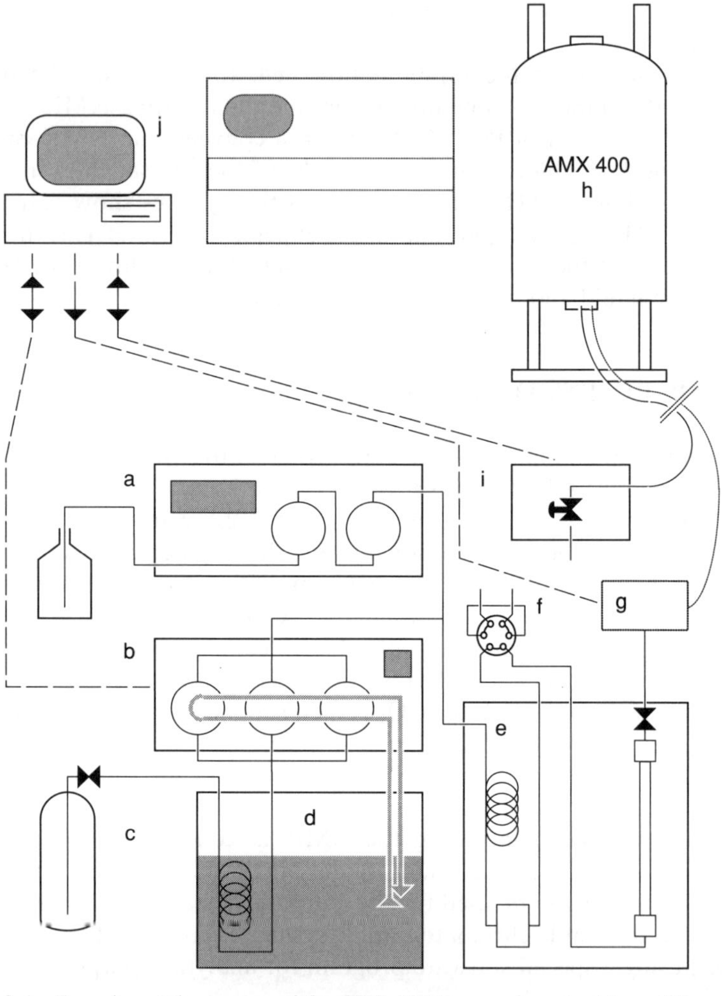

Figure 7.2.4 Experimental set-up used for SFC–NMR experiments: (a) modifier pump; (b) SFC pump; (c) CO_2 cylinder with dip-tube; (d) cryostat; (e) GC oven with mixing chamber and separation column; (f) injection valve; (g) UV detector; (h) NMR magnet; (i) back-pressure regulator or restrictor; (j) hardware control unit

even impossible, due to lost information. In Figure 7.2.5 (b), the clean spectrum of an SFC–NMR run is depicted. The whole range of the spectrum is free from large disturbing solvent signals, shows a flat base line and can thus be completely used for structure elucidation purposes.

By using the F1 direction as the time axis, the separation process, combined with the structure information in the F2 direction, can be investigated. Figure 7.2.6 shows the contour plot of the separation of five plasticizers, recorded by using a 400 MHz spectrometer with a 120 μl flow cell. The separation was

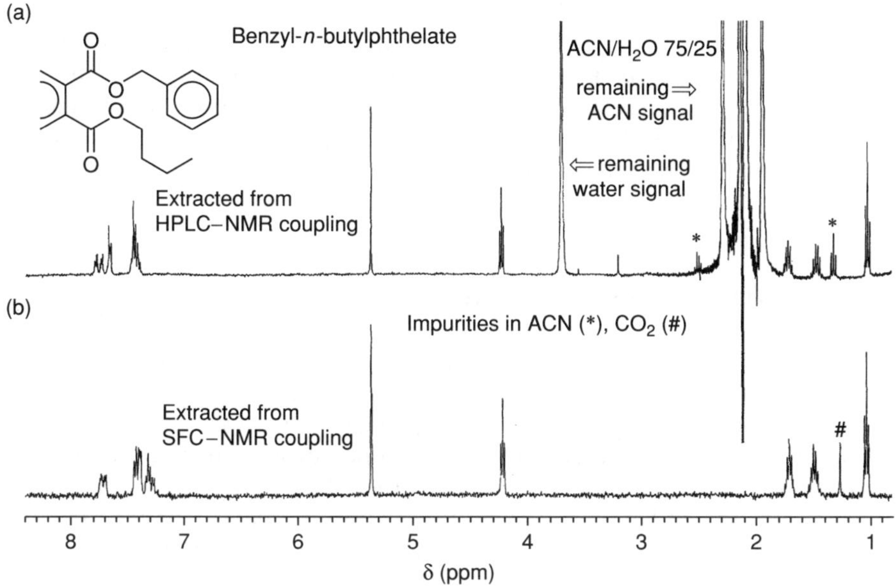

Figure 7.2.5 Phthalate spectra recorded under (a) LC-and (b) SFC–NMR conditions

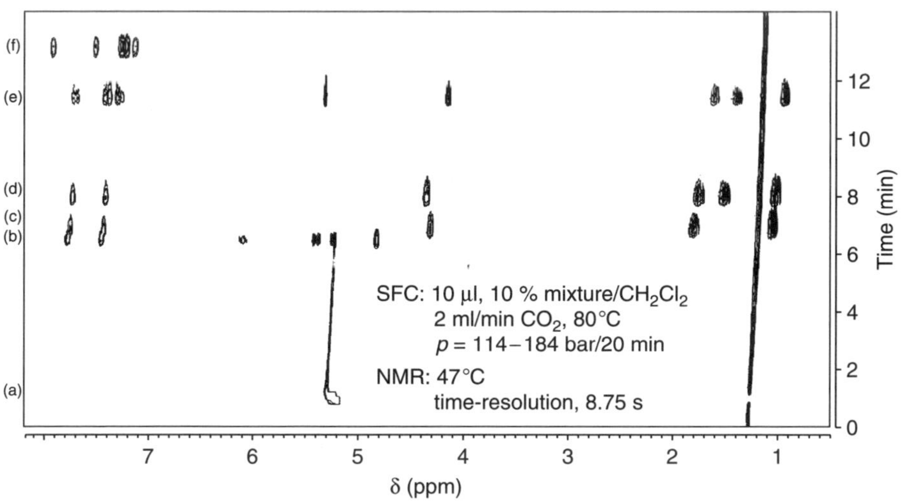

Figure 7.2.6 1H NMR chromatogram (contour plot, 400 MHz) of an SFC separation of five plasticizers in supercritical CO_2: (a) CH_2Cl_2; (b) diallylphthalate; (c) di-*n*-propylphthalate; (d) di-*n*-butylphthalate; (e) benzyl-*n*-butylphthalate; (f) diphenylphthalate

carried out in supercritical CO_2, with a bonded C_8 stationary phase at a flow rate of 2.0 ml/min, employing a linear pressure gradient from 115 to 180 bar over 20 min at a temperature of 353 K. Besides an impurity resonance at 1.25 ppm, the whole spectral range is not obscured by any solvent signals. Figure 7.2.7 displays the expanded ^1H NMR spectrum of diallylphthalate. Here, even small long-range coupling constants of the order of 1.4 Hz can be determined.

7.2.6.2 TWO-DIMENSIONAL NMR SPECTROSCOPY

Signal assignment in routine NMR spectroscopy is usually facilitated by the application of two-dimensional assignment techniques such as homonuclear proton–proton shift correlated spectroscopy ($^1H/^1H$-COSY).

Figure 7.2.8 shows the contour plot of one constituent of the phthalate separation. Here the dead volume between the UV detector and the SFC flow cell was determined before the separation. After an adequate delay after the occurrence of the UV signal of benzyl-*n*-butylphthalate in the UV detector, the SFC separation was stopped and the two-dimensional acquisition was started. The pressure proved to be stable for several hours, which was sufficient for the acquisition of the two-dimensional COSY spectrum. Despite the intense

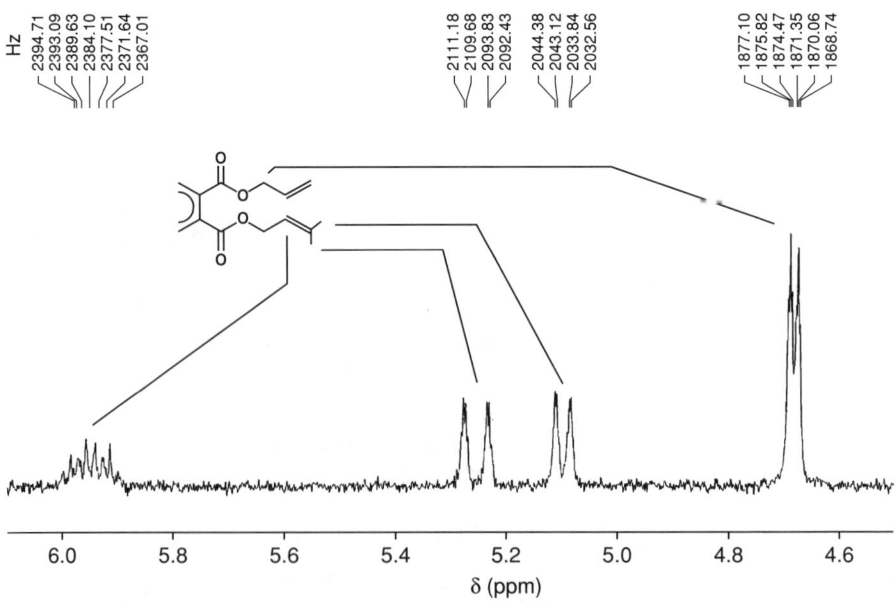

Figure 7.2.7 Continuous-flow ^1H NMR spectrum (400 MHz) of diallylphthalate in supercritical CO_2, taken from the SFC–NMR run shown in Figure 7.2.6

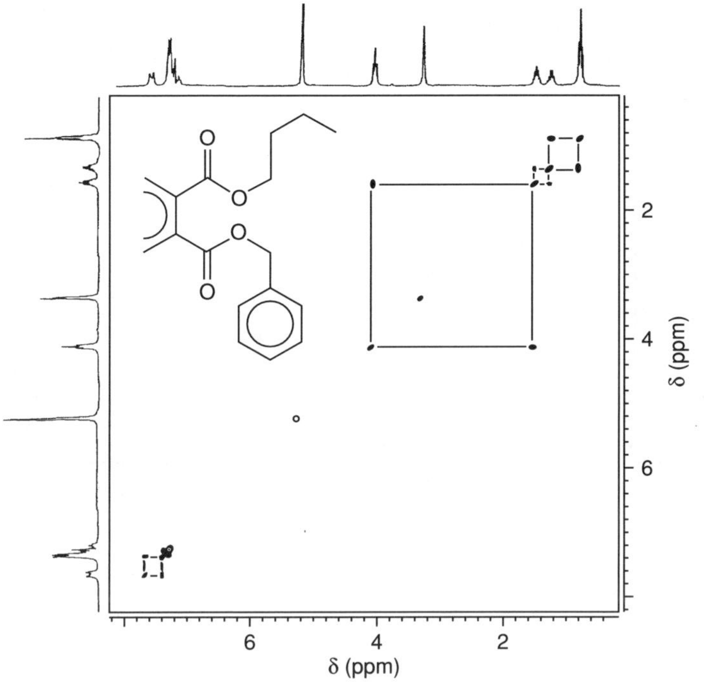

Figure 7.2.8 Two-dimensional NMR spectroscopy in supercritical fluids: ^{1}H, ^{1}H-COSY of benzyl-*n*-butylphthalate, acquired in the stop-flow mode

response of the undeuterated methanol, which was used as a modifier in the SFC separation, the ^{1}H NMR signal connectivities are clearly visible. The corresponding cross-peaks in the aromatic system and the aliphatic part of the butyl chain are indicated on the contour plot. This example proves that even under the 'exotic' supercritical fluid conditions, elaborate two-dimensional assignment techniques can be used.

7.2.6.3 RESOLUTION

Comparison of the stop-flow spectra from ethylbenzene in the liquid and in the supercritical state, recorded at a temperature of 323 K and a pressure of 165 bar, shows that there is no degradation in resolution in going from the liquid to the supercritical state (Figure 7.2.9). Because the NMR signal line widths have a reciprocal relationship to the spin–spin relaxation time T_2, it is evident that T_2 for protons does not change dramatically in the supercritical state. However, this is not the case with respect to the spin–lattice relaxation time T_1.

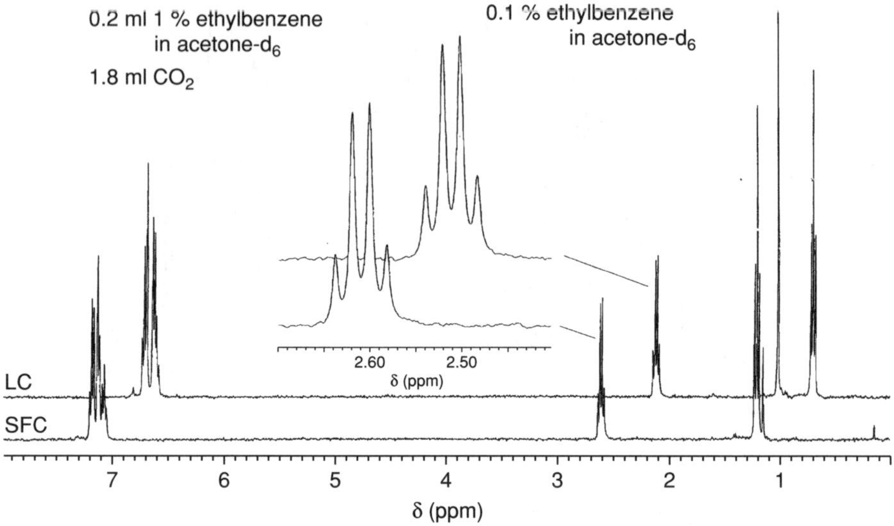

Figure 7.2.9 Comparison of NMR signal resolutions in the liquid and the supercritical state. The quartet of ethylbenzene shows no degradation in resolution in going from liquid to the supercritical state

7.2.6.4 SPIN–LATTICE RELAXATION TIMES

By using the inversion recovery method with the standard pulse sequence (180°–τ–90°–T), it is possible to determine the spin–lattice relaxation times in the supercritical state. The ^{1}H-T_1 values of benzyl-*n*-butylphthalate have been determined in the supercritical state at 250 bar and 321 K (Figure 7.2.10). Values of the order of 20 s result for the aromatic protons, whereas the methyl group exhibits a value of 8 s, which is two to three times higher than the corresponding values in the liquid state at the same temperature. This phenomenon can be explained by the decreased viscosity in the supercritical fluid when compared to the liquid state.

7.2.6.5 SOLVATION

The solvation power of a solvent, as a function of the temperature and pressure, changes significantly in the gaseous, as well as in the sub- and supercritical states. It is well known that the chemical shift depends very much on these solvating states, and therefore the strength of hydrogen bonding [25] and solute-solvent interactions [26,27] can be determined from NMR spectra, although this is also why severe chemical shift changes can be expected during a pressure or temperature gradient in SFC–NMR experiments.

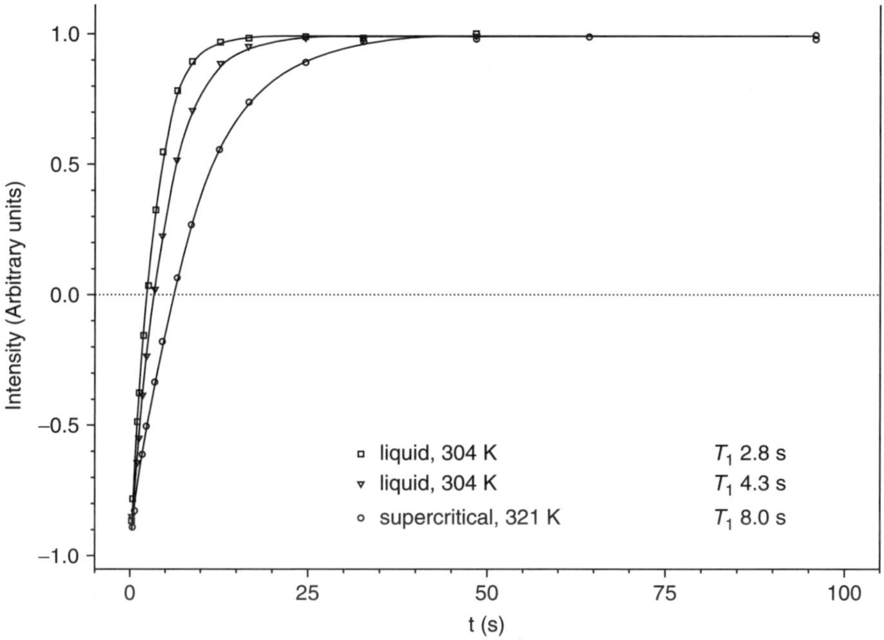

Figure 7.2.10 Values of the spin–lattice relaxation times, T_1, for the methyl protons of benzyl-*n*-butylphthalate in the liquid state at two different temperatures, and in the supercritical state

Separations in the supercritical state are very often performed with a pressure gradient, and very rarely with temperature gradients. Figure 7.2.11 shows two examples of the chemical shift changes which occur on applying pressure gradients in supercritical fluid NMR spectroscopy. Figure 7.2.11(a) shows the contour plot of the proton chemical shift of chloroform in supercritical CO_2 as a function of the applied pressure. Within a time increment of 12 s, a pressure gradient, starting at 90 bar and rising to 244 bar, at a temperature of 325 K, was applied. It is evident that, due to the increasing density of CO_2 with increasing pressure the ^{1}H NMR signals shift to higher field. Even worse is the behaviour of polar groups such as the OH group in methanol depicted in Figure 7.2.11(b). The signal of the OH group crosses the low-field-shifting methyl group several times, due to the phenomenon of cluster formation. This behaviour necessitates that only modest pressure changes can be made during acquisition of ^{1}H NMR spectra in supercritical fluids, or otherwise a severe signal line broadening would result. Taking into account a pressure gradient from 115 bar to 180 bar within 20 min (actually used during a separation), up to 16 transients can be co-added, thus introducing negligible line broadening. The drift of the signal, as can be seen in the contour plot, is, to a first approximation, proportional to the calculated density of the CO_2.

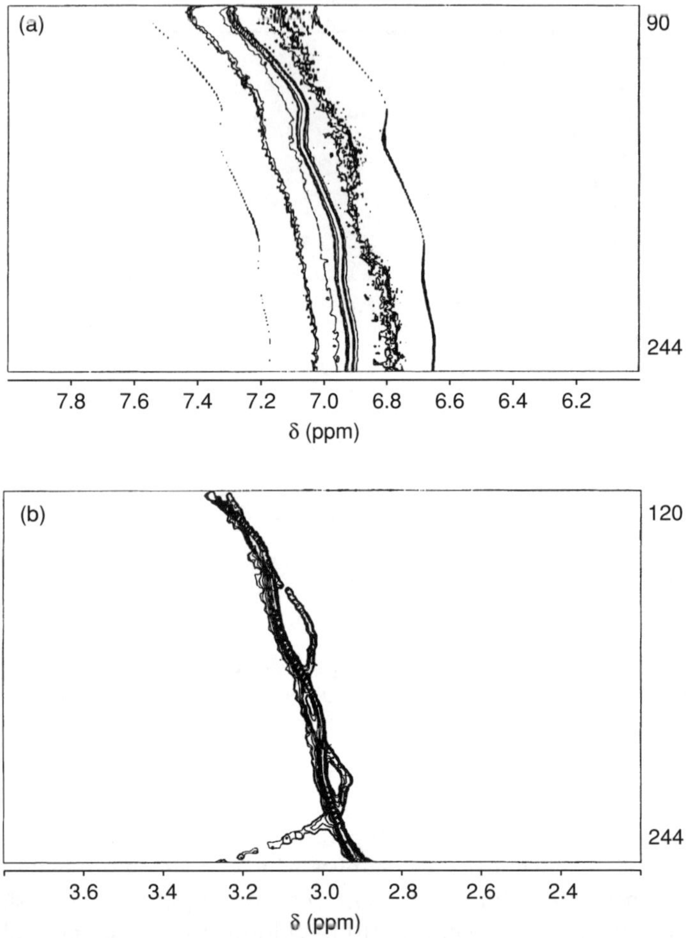

Figure 7.2.11 Contour plots of the signals of (a) chloroform, and (b) methanol in the supercritical state under a pressure gradient from 90 to 244 bar and 120 to 244 bar, respectively

7.2.6.6 REDUCING SPIN–LATTICE RELAXATION TIMES

The drawbacks already described must be taken into account in the acquisition process of SFC–NMR spectra. One is the increase in the spin–lattice relaxation times T_1 due to the reduced viscosity of the supercritical mobile phase. The other is the fact that with increasing pressure the susceptibility of the supercritical fluid changes, thus causing a high-field shift of the NMR signals. For the latter problem, up until now only the number of transients can be reduced to gain acceptable spectra, while for the problem of the long spin–lattice relaxation times T_1 this can be overcome in two ways. The first way, as carried out by Dorn, is to switch the premagnetization volume in front of the flow cell to improve the residence time in the magnetic field prior to detection, and there the Boltzmann distribution is established according to the normal T_1 values (see Figure 7.2.2 above). A second way is to influence the spin–lattice relaxation time by adding paramagnetic compounds to the solution – these materials can be free radicals, paramagnetic metal salts or paramagnetic chelate complexes. Then, the T_1 times are shortened by a factor of 10 up to 100, due to the paramagnetic relaxation effect. However, this affects not only the spin–lattice relaxation time, but also influences the spin–spin relaxation time T_2, which determines the line width of the NMR signals. Shortening the T_2 times leads to broad, and in the worst cases, to non-detectable NMR signals. This can only be negated by immobilizing the paramagnetic compounds on organic or inorganic matrices (e.g. silica) in front of the detection cell [22].

Figure 7.2.12 shows the principal set-up for this experiment. The column, containing the immobilized free radicals consists of an adapted PEEK tube, which fits into the flow probe below the detection cell. Figure 7.2.13 depicts a spectrum of d-*n*-butylphthalate recorded under the influence of a ('free-radical') column filled with 2,2,6,6-tetramethylpiperidin-1-oxyl (TEMPO), immobilized with an aminopropyl spacer on silica. The spectrum is taken from an on-line separation of a two-compound mixture. The line width is of the same order as

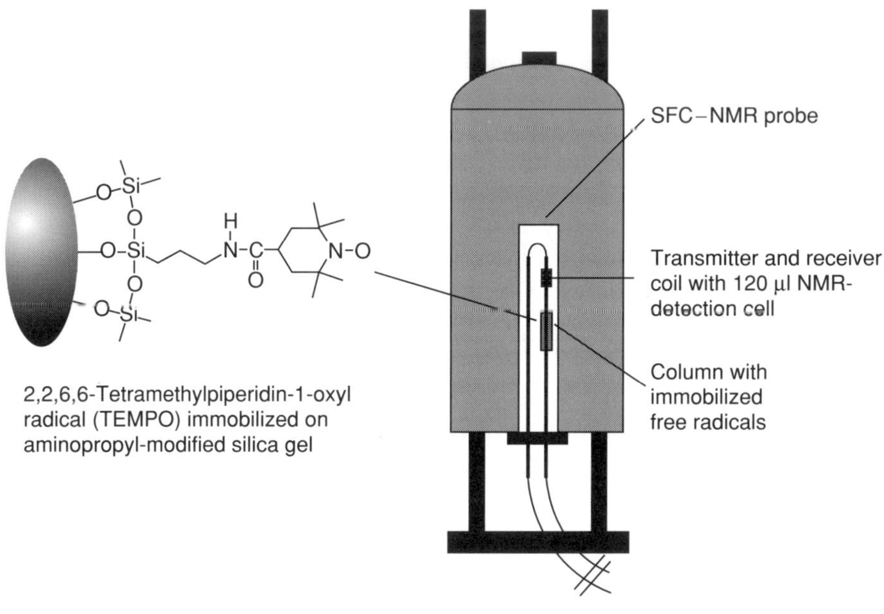

Figure 7.2.12 Immobilized free radicals placed in the high magnetic field for shortening the spin–lattice relaxation times T_1 for the fast magnetization of flowing analytes

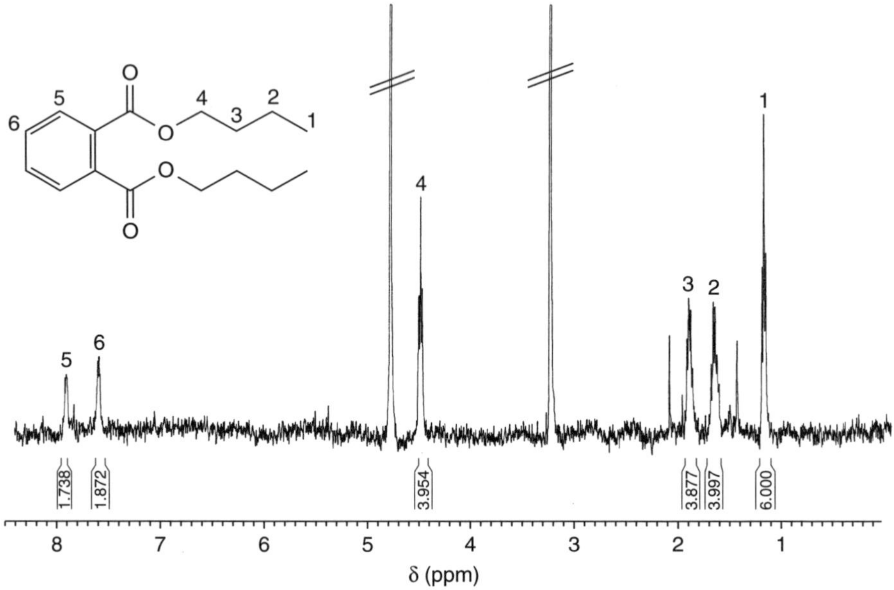

Figure 7.2.13 ^{1}H NMR spectrum of di-*n*-butylphthalate under the influence of immobilized free radicals

Table 7.2.1 Comparison of the integral values (ppm) of the aliphatic and aromatic protons in diallylphthalate and di-*n*-butylphthalate.

Diallylphthalate	Aliphatic protons	Aromatic protons
Theoretical value	10.0	4.0
TEMPO column	9.34	3.41
Dummy column	8.58	2.82
Di-*n*-butylphthalate	**Aliphatic protons**	**Aromatic protons**
Theoretical value	18.0	4.0
TEMPO column	17.83	3.61
Dummy column	17.45	2.82

that observed without the influence of the free-radical species, although the intensities, mainly of the aromatic protons, are much higher.

This is shown in Table 7.2.1, which compares the relative integrals of the compounds with and without the influence of the immobilized free radicals (IFRs). The ratio between the aliphatic and aromatic parts of the molecule corresponds much more with the theoretical values when the spectrum is recorded under the influence of the IFRs. One drawback here is the behaviour of the immobilized free radicals, which act like a separation phase, and so the selectivity, achieved by a conventional separation column, is destroyed by this 'phase'. Future developments will have to take in consideration the separation effect of these IFRs. With the development of capillary SFC–NMR coupling, it should be possible to prepare separation and IFR phases on the same matrix (mixed-mode phases) and place these in the magnetic field to achieve both separation and relaxation at the same time.

7.2.7 APPLICATIONS

7.2.7.1 FUEL DERIVATIVES

The first practical example of an on-line SFC–^{1}H NMR separation was recorded by Dorn and co-workers [16] (Figure 7.2.14). Since up to 90 % of a fuel is aliphatic, SFC–NMR on-line analysis is the matter of choice for separation and identification. Figure 7.2.14 shows a fuel mixture of isooctane, *n*-hexane, *n*-nonane, dodecane, and *n*-hexadecane, separated by SFC and detected by on-line ^{1}H NMR spectroscopy. The SFC separation was accomplished with a flow rate of 2.0 ml/min, a C18 250 × 4.6 mm column, operated at an isobaric pressure of 100 bar and a temperature of 323 K, using CO_2 with 1 % (w/w) CD_3CN as solvent. Each NMR spectrum consists of 20 co-added transients at an acquisition time of 1 s per transient. The total separation occurred within 5 min. The first eluting isooctane can be easily identified by the methylene-to-

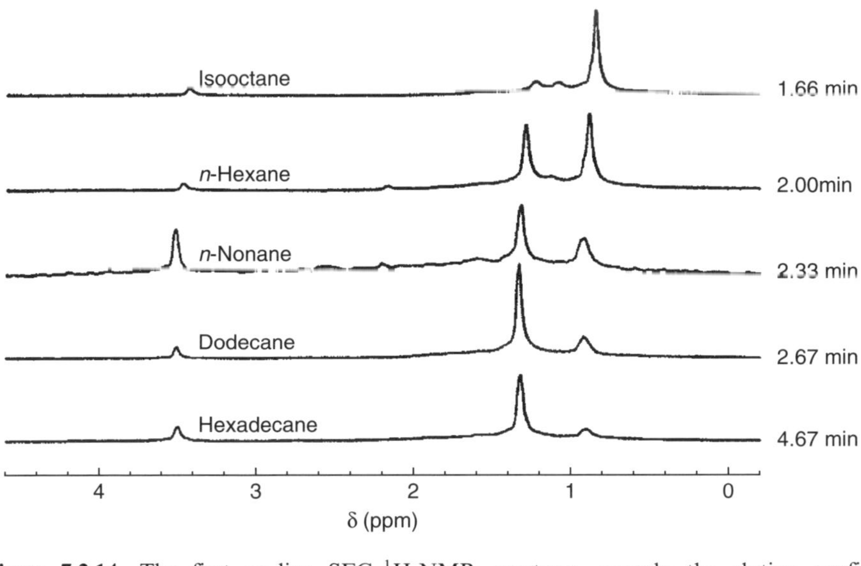

Figure 7.2.14 The first on-line SFC–^{1}H NMR spectrum, namely the elution profile (200 MHz) of a fuel mixture. Reprinted with permission from Allen, L. A., Glass, T. E. and Dorn, H. C., *Anal. Chem.*, **60**, 390–390 (1988). Copyright (1988) American Chemical Society

methyl signal ratio, which increases in the spectra of later-eluting compounds. Thus, longer elution times are consistent with an increasing chain length of the separated hydrocarbon.

7.2.7.2 ACRYLATES

In an on-line SFC–NMR separation employing a pressure gradient, up to 16 transients can be co-added without affecting the NMR resolution. In contrast to isobaric separations, the minimum number of transients is only determined by the concentration of the separated compound. Figure 7.2.15 shows the contour plot of the separation of four acrylates [17] in supercritical CO_2 with 1 % methanol-d_4 used as the polar modifier. The acrylates were separated by a 4.6 × 250 mm cyanopropyl column, at a flow rate of 1.0 ml/min, employing isobaric conditions (80 bar) for the first 7 min of the separation, a succeeding pressure ramp of 20 bar/min, and a final isobaric conditions at 100 bar. Throughout the separation, the temperature of the oven was kept constant at 333 K. Eight transients per row were co-added, thus resulting in a time resolution of 6.4 s. The residual ^{1}H NMR signals of methanol at 0.8 and 3.45 ppm show the prescribed high-field shift during the pressure ramp between retention times of 7–8 min.

At the left vertical axis of the contour plot, a one-dimensional chromatogram is reconstructed by co-adding all of the ^{1}H NMR signals that appear. This

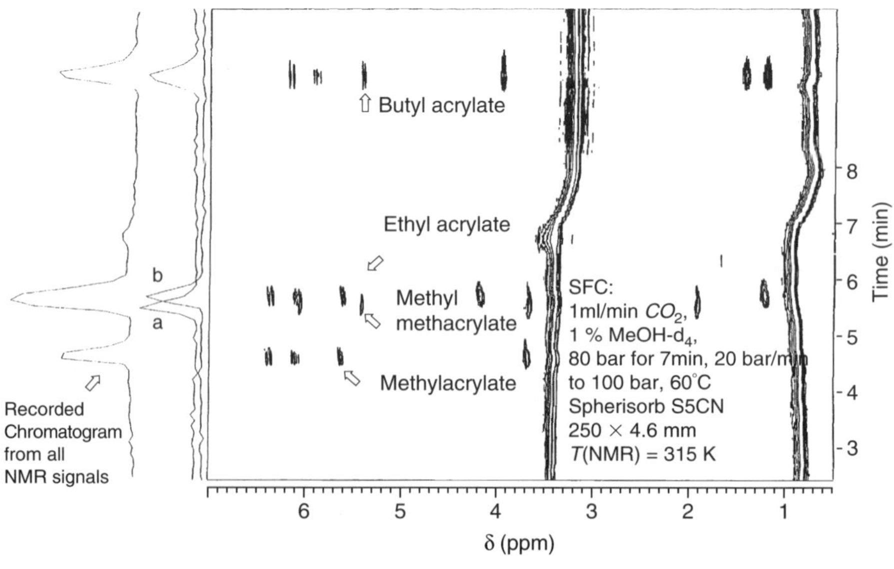

Figure 7.2.15 ^{1}H NMR chromatogram (contour plot, 400 MHz) of a SFC separation of acrylates in supercritical CO_2 with 1 % methanol-d_4 as the polar modifier

NMR mass chromatogram corresponds to a one-dimensional UV or refractive index chromatogram. The second peak consists of methyl methacrylate and ethyl acrylate, which is evident from the different chemical shifts provided from the on-line experiment. A clear differentiation is available from the chemical shift of the methyl group of both compounds. The chemical shift at 1.9 ppm indicates the methyl group at the double bond of methyl methacrylate, whereas the signal at 1.22 ppm is from the terminal methyl group of ethyl acrylate. Thus, the second dimension of the SFC–NMR run, provided by the ^{1}H chemical shifts, enables the separation of co-eluting compounds.

7.2.7.3 BIOMEDICAL COMPOUNDS

Biological and biomedical compounds are particularly temperature- and light-sensitive, and therefore hyphenated NMR spectroscopy provides a gentle way for analysis of such materials. In the spectrum of all-*trans* vitamin A acetate acquired during an HPLC–^{1}H NMR on-line run with *n*-heptane as the solvent, the aliphatic region between 0.6 and 1.5 ppm is covered by the remaining signals of *n*-heptane or distorted by the solvent suppression technique. Although the signals of the methyl groups at 1.7–1.8 ppm are observable, severe distortions from presaturation of the solvent signals lead to deviations in the intensity ratios. Therefore, the region that cannot be observed without distortions amounts to almost 2 ppm. In contrast, no such distortions occur in the spectra collected on-line during the SFC separation (see Figure 7.2.16). Once again, the

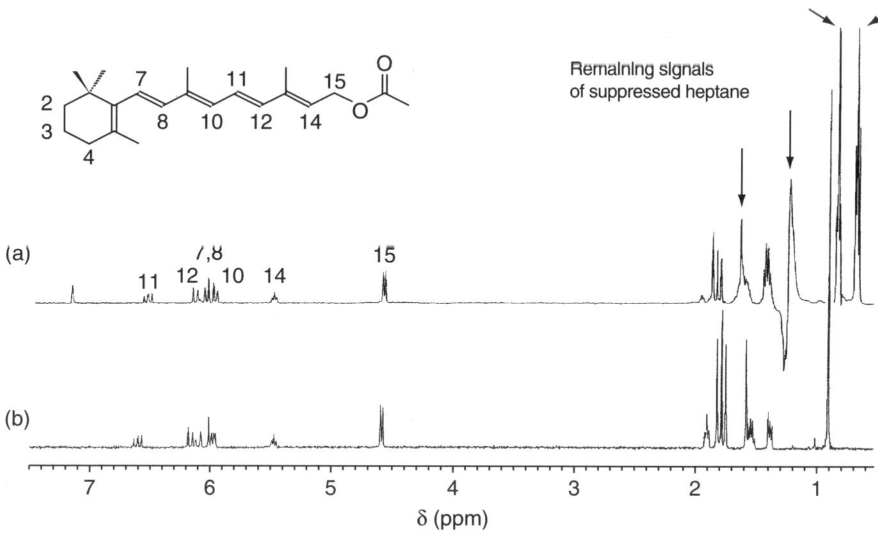

Figure 7.2.16 Comparison of continuous-flow 1H NMR spectra of all-*trans* vitamin A acetate, recorded (a) during an HPLC–NMR separation, and (b) during an SFC–NMR separation

strength of SFC NMR becomes clear. A line width of 1.7 Hz is observed, and no extensive baseline correction or further data processing is necessary.

7.2.8 SFE–NMR COUPLING

Extraction with supercritical CO_2 is a technical process of increasing importance. It provides a mild and rapid technique for the extraction of low- or medium-polarity substances. Supercritical CO_2 is used for supercritical fluid extraction (SFE) in important technical processes such as the decaffeination of coffee and the extraction of hops, as well as the extraction of naturally occurring compounds from biomaterials. As many applications are performed in the pharmaceutical, polymer, environmental and nutritional fields, direct on-line SFE–NMR would be an ideal tool to monitor the various extraction processes.

For the on-line SFE–NMR experiments, the set-up shown in Figure 7.2.17 can be used. A main pump serves an HP supercritical fluid chromatograph (G1205A), with analytical HPLC columns being used as extraction cells. The continuous-flow NMR cell is connected between the column outlet and the back-pressure regulator.

7.2.8.1 NATURAL COMPOUNDS

In Figure 7.2.18, the contour plot of an extraction of 820 mg of roasted coffee is shown [20]. The contour plot is similar to that given above Figure 7.2.11 where the horizontal axis represents the proton chemical shift and the vertical axis represents the pressure. A linear pressure gradient from 85 to 220 bar over 55 min, and a CO_2 flow of 1 ml/min at a temperature of 317 K was applied. One can clearly see the caffeine, which is extracted at a pressure of 105 bar, as

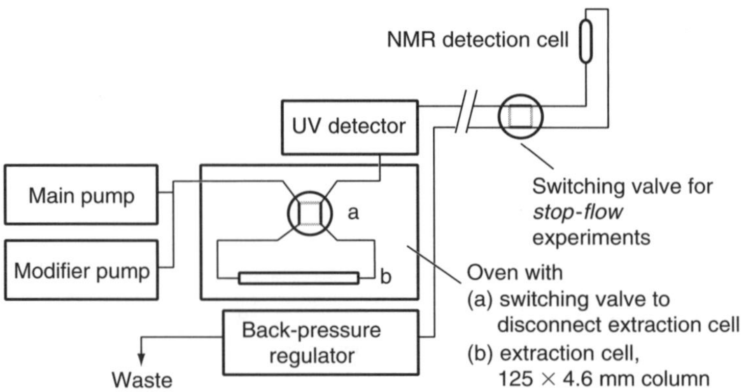

Figure 7.2.17 Schematic of the experimental set-up used for SFE–NMR coupling

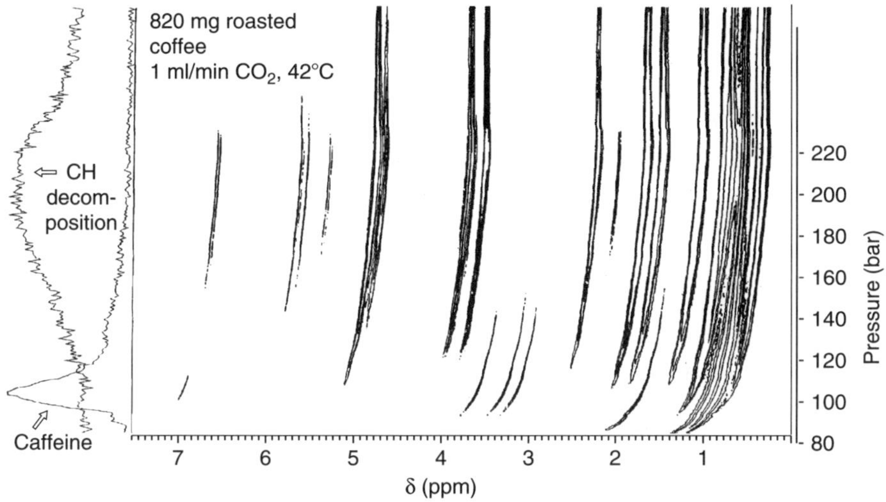

Figure 7.2.18 Extraction profile (contour plot) of roasted coffee using a linear pressure gradient from 85 to 220 bar

indicated in the contour plot. What also is conspicuous is the up-field shift of the proton signals caused by the increasing density. The projections shown on the vertical axis of the contour plot are generated by summing the intensities of the NMR signals of the indicated compound, and therefore represents an extraction profile. The NMR spectrum corresponding to a pressure of 105 bar is shown in Figure 7.2.19, and allows the identification of the caffeine.

At higher pressures, a complex mixture of triglycerides and other ingredients in the coffee, which are not separated by their solubility, are extracted over a very wide range of pressures. The spectrum corresponding to the maximum of the extraction profile at 200 bar is shown in Figure 7.2.20. The enlarged section in this figure shows the proton pattern corresponding to the outlined glycerol sub-unit. Examination of the contour plot of the extraction process shows that the sample composition at the extraction profile between 120 and 220 bar is nearly the same, with a maximum being reached at ca. 200 bar. This extraction profile can be explained by the presence of different, although chemically similar substances which show no substantial difference in the NMR spectrum. In addition, the extraction of the same compound from different sites in the matrix may lead to a broader extraction profile.

7.2.8.2 PLASTICIZER PLASTIFIER FROM PVC

SFE can be used to extract compounds from various polymers, such as poly (vinyl chloride) (PVC). As an example, 644 mg PVC pieces are packed into an

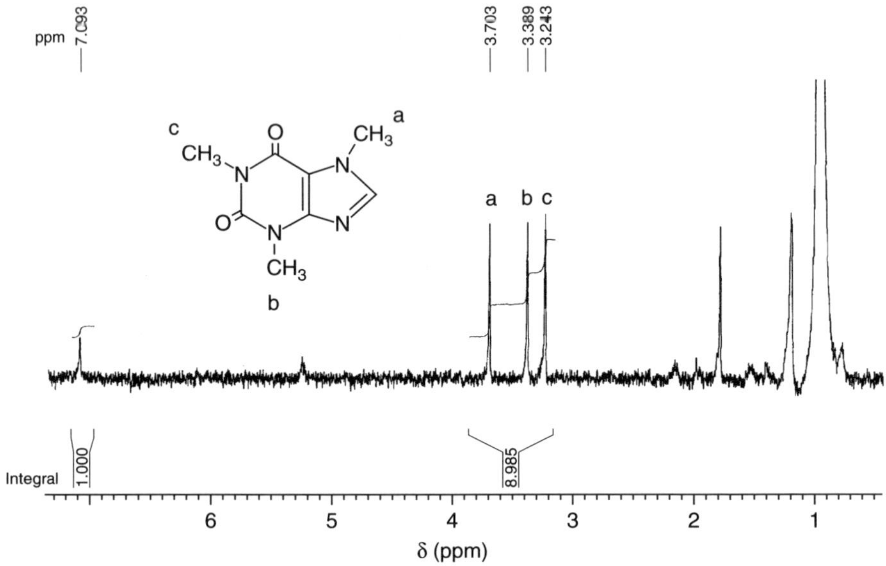

Figure 7.2.19 Continuous-flow ^{1}H NMR spectrum (400 MHz) of caffeine extracted from coffee with supercritical CO_2

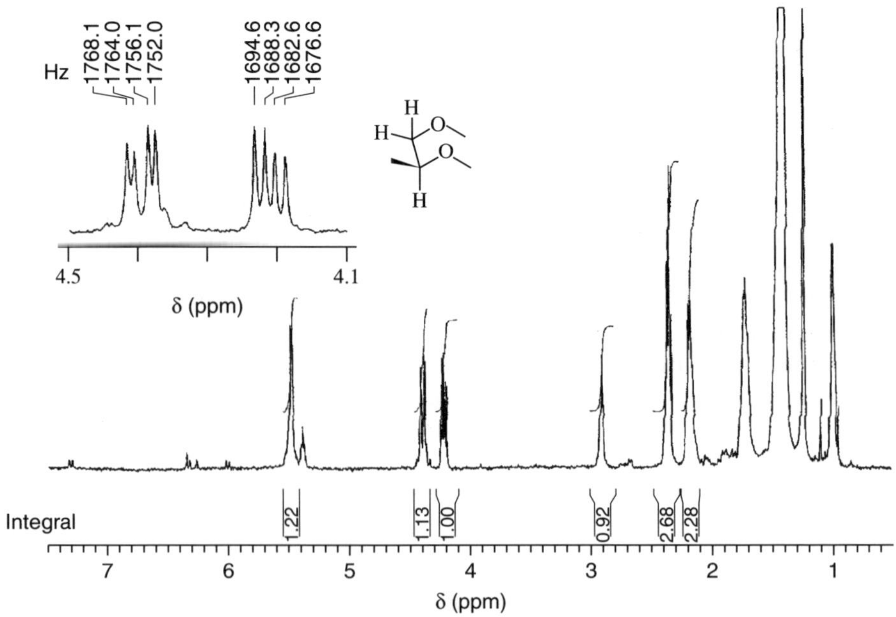

Figure 7.2.20 Continuous-flow ^{1}H NMR spectrum (400 MHz) of the triglycerides separated during the extraction process of coffee with supercritical CO_2

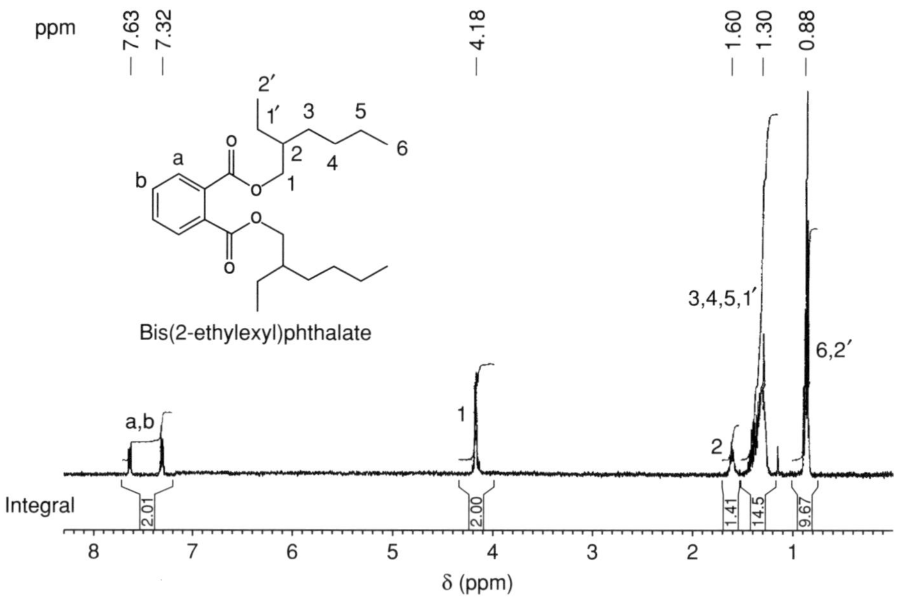

Figure 7.2.21 Stop-flow 1H NMR spectrum (400 MHz) of an extracted phthalate from PVC using supercritical CO_2

HPLC column which is closed with conventional frits on both sides. The extraction cell is kept at a temperature of 315 K, with a CO_2 flow rate of 1.0 ml/min, while for the extraction a linear pressure gradient from 79 to 317 bar over 50 min is employed. The weight reduction after the extraction is 119 mg, corresponding to 18.5 %. At the final extraction density of 0.9 g/ml (315 bar), the valve at the SFC probe is closed in order to switch to stop-flow conditions. The proton spectrum shown in Figure 7.2.21 is acquired under static conditions. From this, the structure of the extracted phthalate could be assigned as bis(2-ethylhexyl) phthalate.

7.2.9 CONCLUSIONS

The direct coupling of supercritical fluid chromatography and proton nuclear magnetic resonance spectroscopy is a very interesting tool for special cases of structure elucidation. The advantage of NMR detection in the supercritical state is the absence of ^{1}H NMR signals of solvents and impurities, while the drawbacks of this hyphenated technique are the pressure-dependence of the NMR signals and the increased spin–lattice relaxation times. Despite these problems, the spectral quality obtained together with the capability to acquire two-dimensional NMR spectra, may lead to SFC–NMR coupling becoming an established hyphenated technique. Furthermore, the enormous application power of supercritical fluid extraction with ^{1}H NMR detection will lead to new structural elucidation pathways in the search for new, natural occurring compounds which can be used in medical therapy.

REFERENCES

1. Brame Jr, E. G., *Anal. Chem.*, 1965, **37**, 1183.
2. Tsuda, T., Ojika, Y., Izuda, M., Fujishima, I. and Ishii, D., *J. Chromatogr.*, 1972, **69**, 194.
3. Buddrus, J. and Herzog, H., *Org. Magn. Reson.*, 1981, **15**, 211.
4. Herzog, H. and Buddrus, *J., Chromatographia*, 1984, **18**, 31.
5. Berger, T. A., *Packed Column SFC*, The Royal Society of Chemistry, Cambridge, UK, 1995.
6. Chester, T. L., Pinkston, J. D. and Raynie, D. E., *Anal. Chem.*, 1998, **70**, 301 R.
7. Anton, K. and Berger C. (Eds), *Supercritical Fluid Chromatography with Packed Columns: Techniques and Applications*, Marcel Dekker, New York, 1998.
8. Pal, R., Juhas, M. and Strumpf, A., *J. Chromatogr., A*, 1998, **819**, 249.
9. Berger, T. A., Wilson, W. H. and Deye, J. D., *J. Chromatogr. Sci.*, 1994, **32**, 179.
10. Strode, J. T. B., Taylor, L. T. and van Beek, T. A., *J. Chromatogr., A*, 1996, **783**, 115.
11. Williams, K. L. and Sander, L. C., *J. Chromatogr., A*, 1997, **785**, 149.
12. Bargman-Leyder, N., Thiébaut, D., Vergne, F., Tambuté, A. and Caude, M., *Chromatographia*, 1994, **39**, 673.
13. Benedek, G. B. and Purcell, E. M., *J. Chem. Phys.*, 1954, **22**, 2003.
14. Jonas, J., *Science*, 1982, **216**, 1179.
15. Jonas, J., Ballard, L. and Nash, D., *Biophys. J.*, 1998, **75**, 445.
16. Allen, L. A., Glass, T. E. and Dorn, H. C., *Anal. Chem.*, 1988, **60**, 390.
17. Albert, K., Braumann, U., Tseng, L.-H., Nicholson, G., Bayer, E., Spraul, M., Hofmann, M., Dowle, C. and Chippendale, M., *Anal. Chem.*, 1994, **66**, 3042.
18. Albert, K., Braumann, U., Steck, R., Spraul, M. and Ecker, R., *Fresenius' J. Anal. Chem.*, 1995, **352**, 521.
19. Braumann, U., Händel, H., Strohschein, S., Spraul, M., Krack, G., Ecker, R. and Albert K., *J. Chromatogr., A*, 1997, **761**, 336.
20. Braumann, U., Händel, H., Albert, K., Ecker, R. and Spraul, M., *Anal. Chem.*, 1995, **67**, 930.
21. Albert, K., *J. Chromatogr., A*, 1997, **785**, 65.
22. Fischer, H., Tseng, L.-H., Raitza, M. and Albert, K., *Magn. Reson. Chem.*, 2000, **38**, 336.
23. Yonker, C. R., Zemanian, T. S., Wallen, S. L., Lienhan, J. C. and Franz, J. A., *J. Magn. Res., A*, 1995, **113**, 102.
24. Bai, S., Taylor, C. M., Mayne, C. L., Pugmire, R. J. and Grant, D. M., *Rev. Sci. Instrum.*, 1996, **67**, 259.
25. Bai, S. and Yonker, C. R., *J. Phys Chem., A*, 1998, **102**, 8641.
26. Kanakubo, M., Aizawa, T., Kawakami, T., Sato, O., Ikushima, Y., Hatakeda, K. and Saito, N., *J. Phys. Chem., B*, 2000, **104**, 2749.
27. Bai, S., Taylor, C. M., Liu, F., Mayne, C. L., Pugmire, R. J. and Grant, D. M., *J. Phys. Chem., B*, 1997, **101**, 2923.

7.3 Nanoliter NMR

7.3.1 NMR Spectroscopy in Microdomains

MICHAEL E. LACEY†, ANDREW G. WEBB‡ and JONATHAN V. SWEEDLER†

University of Illinois at Urbana-Champaign, IL, USA
[†School of Chemical Sciences; ‡Department of Electrical and Computer Engineering]

7.3.1.1 EVOLUTION OF SMALL COILS FOR MAGNETIC RESONANCE

Since investigators continuously strive to extract increasing amounts of data from shrinking sample sizes, much commercial and academic research on NMR hardware focuses on optimizing RF probe sensitivity to allow the analysis of trace materials. Significant advances have been achieved through several distinct modifications of instrumentation. This present chapter provides a review of the characterization of nanoliter volumes via static NMR spectroscopy and emphasizes some of the more relevant developments in probe technology that have enabled such measurements.

To our knowledge, Odelblad provided the first detailed report of NMR spectroscopy using small, specialized RF coils for mass-limited analysis. In 1966, he described continuous-wave techniques to study the physical chemistry of mucus secreted from human cervical cells [1]. In this investigation, a set of solenoidal microcoils with 200–1000 μm diameters (and a fixed length of just more than 1 mm) were used. In a high-strength (3.7 T) permanent magnet, he recorded the proton chemical shifts and spin–lattice (T_1) relaxation times of cervical secretions with these probes. Shoolery followed in 1979 with the first high-resolution NMR spectra from a reduced-diameter RF coil [2]. With coils that were closely fitted to the analyte container, a decrease in the diameter of the sample tube from 10 mm (1.5 ml volume) to 1.7 mm (15 μl) shortened the NMR experiment time by a factor of 40 for a given signal-to-noise ratio (S/N) from a fixed sample mass of 5 mg of cholesterol. In this study, direct-observe ^{13}C spectra were also obtained from as little as 1 mg of sample in 16 h. For ^{1}H spectra, 1 μg of cortisone acetate was detected in a comparable experiment time. In still another early application of solenoids for NMR spectroscopy, spectra were collected in less than 3 min

On-Line LC–NMR and Related Techniques
Edited by K. Albert.

from the small superfused hind limb muscles of mice with 2-mm diameter coils [3]. These studies, as well as various others [4], laid the foundation for current high-resolution NMR spectroscopic capabilities in the nanoliter regime.

7.3.1.2 STRATEGIES TO INCREASE SIGNAL-TO-NOISE

The need for a higher mass sensitivity is obvious when volumes are reduced from 200 μl to 20 nl, as 10 000-fold fewer spins are present for a given concentration. How can this performance challenge be met?

7.3.1.2.1 SAMPLE TUBES AND PLUGS

Without manipulating the RF coil, one can improve the S/N of an NMR experiment via smaller sample tubes (which may even feature glass that is matched to the magnetic susceptibility of the solvent) or tube inserts to decrease the volume required for an experiment. For example, by reducing the total volume (V_{tot}) by a factor of three and maintaining the same total number of spins, the concentration can be correspondingly increased threefold [5]. For lossy samples, the decrease in V_{tot} can reduce the noise contribution from the sample, thereby increasing the S/N of the spectrum. Alternatively, the sample can be loaded into a spherical microcell that is positioned entirely within the active region of the RF coil. The spherical geometry minimizes the magnetic-susceptibility-induced line broadening that occurs since the sample/air and glass/air interfaces are located in or near the observe volume (V_{obs}). While this strategy can be implemented easily, the lower volume limit for commercial products is currently of the order of 20 μl [6]. In addition, gains in S/N are achieved only if the linewidth (LW) of the smaller sample is not degraded significantly.

Another approach to small-volume NMR spectroscopy relies upon plastic plugs (matched to the susceptibility of the solvent) within a tube initially made for larger volumes. The plug design restricts the sample to the V_{obs}, again optimizing the sample observation efficiency [6]. For a 3-mm o.d. tube, for example, plugs allow high-resolution spectra to be obtained from a minimum sample volume of $\sim$ 60 μl. No such susceptibility-matched inserts are commercially available for nanoliter volumes, although liquid perfluorocarbon susceptibility-matched plugs have recently been demonstrated for solenoidal microcoils with observe volumes less than 1 μl [7,8]. While these approaches improve sample handling, the overall probe performance can be further enhanced by making changes to the RF coil.

7.3.1.2.2 MINIATURIZATION OF SADDLE-TYPE COILS

In general, 'saddle'-shaped geometries (see diagram in Figure 7.3.1.1(a)) dominate the RF coils of standard high-resolution NMR probes. Consideration of the spatial configuration of magnetic vector potentials indicates that the sensitivity (as assessed by the RF field, B_1, per unit current, i) of this coil is given by the following [9]:

$$\frac{B_1}{i} = \frac{n\mu\sqrt{3}}{\pi}\left[\frac{2dh}{(d^2+h^2)^{\frac{3}{2}}} + \frac{2h}{d\sqrt{d^2+h^2}}\right] \tag{7.3.1.1}$$

where n is the number of turns, μ is the permeability of free space, d is the coil diameter, and h is its length. Clearly, reduction of the diameter and length of the coil increases the sensitivity. In addition to the theoretical relationship presented in the above equation, Hoult and Richards also established the principle of reciprocity which states that the duration of a 90° pulse provides a direct measure of sensitivity for a single RF coil system [9]. That is, shorter 90° pulse widths for a given transmitter power correspond to more sensitive RF coils. Virtually all commercial NMR probe manufacturers currently offer probes that utilize the increased sensitivity of small saddle coils. To our knowledge, the first major development in this area involved Varian's use of 1.7-mm tubes for both ^{1}H- and ^{13}C-NMR of mass-limited analytes [2,10–14]. The introduction of inverse detection microprobes designed for 3-mm diameter tubes with total sample volumes of $\sim$ 140 μl and observe volumes of $\sim$ 60 μl expanded the capabilities of small-volume NMR spectroscopy [15–19]. Inverse detection microprobes with 1.7-mm diameter tubes further decreased the requisite sample volumes and enabled rapid data acquisition on reduced sample amounts [20–22].

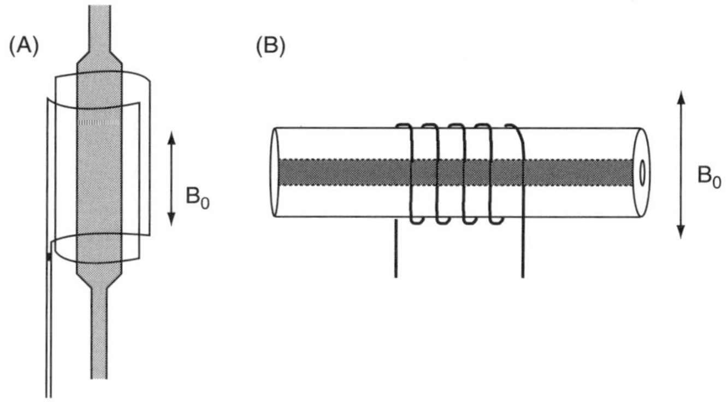

Figure 7.3.1.1 Schematics of two different RF coil geometries showing the coil and the sample tube position relative to the B_0 field: (a) saddle-type; (b) solenoidal

7.3.1.2.3 SOLENOIDAL COIL GEOMETRIES

For solenoidal RF coils (see Figure 7.3.1.1(b)), the theoretical relationship between sensitivity and coil dimensions is given by the following:

$$\frac{B_1}{i} = \frac{\mu n}{d\sqrt{1 + \left(\frac{h}{d}\right)^2}} \tag{7.3.1.2}$$

This equation indicates that the coil sensitivity varies inversely with the diameter of the coil (for a fixed length-to-diameter ratio). For diameters below $\sim$3 mm, the AC resistance of the coil itself acts as the major noise source, even for lossy biological samples. The resistance depends on both the winding geometry (including wire diameter, number of turns, and turn spacing) and the resistivity of the conductor.

7.3.1.2.4 MICROCOIL PROBES

In the size regime larger than $\sim$ 100 μm in diameter, the S/N per unit sample volume of solenoids can be expressed as follows:

$$S/N_{puv} = \frac{\omega_0^2\left[n/d\sqrt{1 + (h/d)^2}\right]}{\sqrt{n^2 d\omega_0^{1/2}/h}} \propto \frac{\omega_0^{7/4}}{d} \tag{7.3.1.3}$$

where ω_0 is the nuclear precession frequency [23]. As the mass sensitivity (S_m) depends on the S/N_{puv}, the above expression demonstrates that S_m improves for a microcoil with a fixed length-to-diameter ratio as the coil diameter decreases (i.e. $S_m \propto 1/d$). As thermal noise in the coil windings rather than sample noise is the primary factor in microscopic studies, minimizing the coil resistance also improves the performance. For coil diameters under 100 μm, the sensitivity dependence on size is reduced to a square-root relationship with the diameter [23]. Figure 7.3.1.2 illustrates the agreement between the actual and predicted improvement in S/N per unit volume with decreasing coil diameter [23]. Complementary to these results, Minard and Wind have recently published modeling studies of the spatial uniformity of the NMR response within solenoidal transceivers [24,25]. This work provides practical guidelines for maximizing the NMR signal-to-noise ratios for microcoils in terms of coil length, diameter, number of turns, operating frequency, and wire size.

To date, most NMR microcoils have been wound directly on to a capillary tube which functions as both sample container and coil form [26,27]. According to electromagnetic field theory, a sample enclosed by a perfectly uniform and infinitely long hollow cylinder (e.g. a fused silica capillary) experiences a uniform static magnetic field [28]. In reality, the susceptibility variation of the

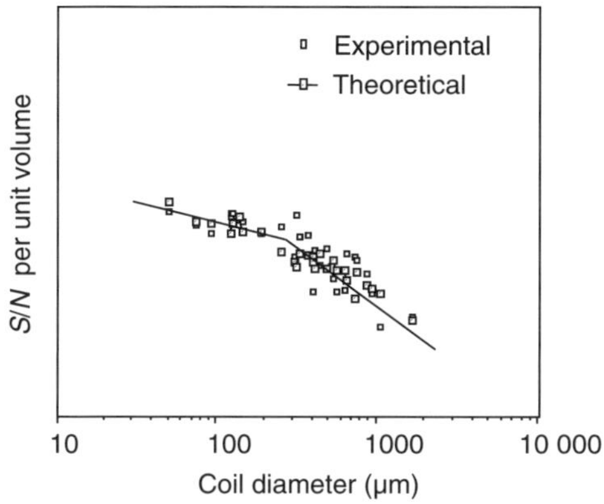

Figure 7.3.1.2 S/N per unit sample volume as a function of coil diameter for solenoidal coils. The $1/d$ and $(1/d)^{1/2}$ asymptotes do not account for the resistance of the lead and thus are not followed exactly by either theoretical or experimental data points. Adapted by permission of Academic Press, from Peck, T. L., Magin, R. L. and Lauterbur, P. C., *J. Magn. Reson., B.*, 1995, **108**, 114

materials near the sample (largely copper wire and its coating, adhesive and air) can locally distort the static magnetic field in the sample region. Furthermore, as the sample more closely approaches the coil windings, capillary coating and surrounding environment, magnetic field inhomogeneities become more pronounced. The resultant line broadening degrades the spectral resolution. Consequently, substantial effort has been devoted to studying the effects of coil proximity to the sample, especially considering the 'Lilliputian' dimensions of microcoil NMR when compared to traditional arrangements.

While conventional microliter-scale probes use fabrication materials with near-zero magnetic susceptibility or sample plugs matched to the susceptibility of the solvent, one distinctive approach undertaken for microcoils surrounds the coil and capillary region with a fluid of magnetic susceptibility that approximates copper, the coil material. As shown in Figure 7.3.1.3, this has proved very effective in decreasing the line width, improving the line shape, and increasing S/N. This inexpensive and easily implemented strategy avoids the difficulties posed by the fabrication of zero-magnetic-susceptibility materials with small dimensions and even smaller tolerances.

In order to carry out complete structural elucidation of unknown compounds (especially for complex molecules), the RF probe should enable a variety of heteronuclear NMR techniques to be performed. In particular, inverse detection ^{1}H–^{15}N and ^{1}H–^{13}C experiments such as heteronuclear multiple quantum coherence (HMQC) [29,30] and heteronuclear single quantum coherence (HSQC) [31] find almost ubiquitous application in myriad research environments. Although the microliter-scale probes described above feature both heteronuclear and homonuclear capabilities, no commerical product is

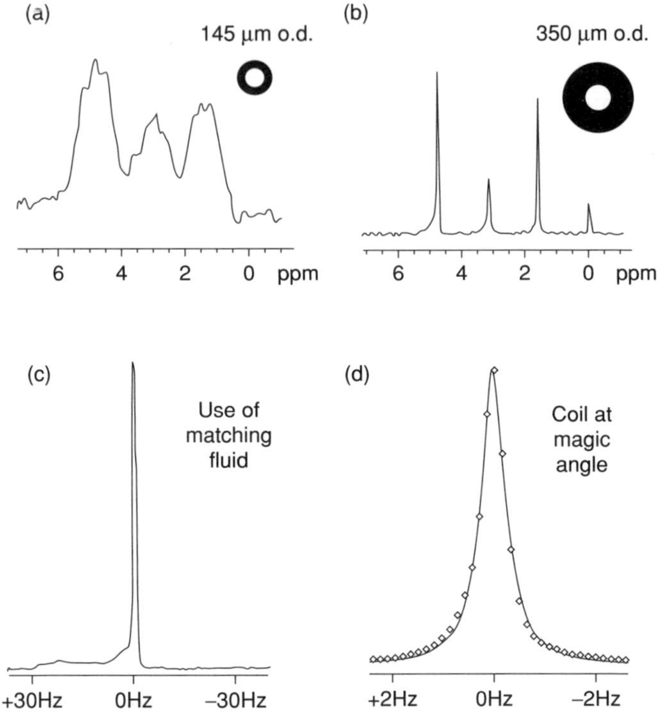

Figure 7.3.1.3 Progress in the improvement of spectral linewidth and line shape for NMR microcoils. (a) Without magnetic susceptibility matching fluid, the linewidths for arginine are 200 Hz for a thin-walled (75-μm i.d./145-μm o.d.) capillary. (b) For a thicker-walled (75-μm i.d./350-μm o.d.) capillary, without matching fluid, the linewidth is reduced to $\sim$ 11 Hz but the line shape is irregular [43]. (c) The effect of improved microcoil fabrication and the addition of magnetic susceptibility matching fluid surrounding the microcoil for a spectrum of 10 % H_2O/90 % D_2O. The line width decreases to less than 1 Hz, S/N significantly improves but the line shape has a baseline defect [4]. The effect of a microcoil configured at the magic angle inside the magnet. The linewidth is 0.6 Hz for 10 % H_2O/90 % D_2O and the data (diamonds) show no significant deviations in peak shape compared to the Lorentzian fit (line). All spectra are obtained for ^1H at 300 MHz

available for the nanoliter-volume regime. As a result, some relevant design parameters for multinuclear NMR microcoil probes are presented here.

These types of experiments call for efficient doubly tuned coils, ideally with a separate deuterium lock channel. For more complex molecules, such as proteins, considerably more intricate NMR pulse sequences, such as (HNCO) [32,33], require the probe to operate at three or four distinct frequencies. High efficiency is demanded from the proton observe channel. Ideally, the additional circuitry allowing multiple tuning should not interfere with the proton efficiency when compared to a singly tuned proton coil. In practice, some reduction is tolerated. The two most important design criteria for such inverse detection probes are (1) minimization of losses of the multiple-tuned circuitry such that all channels are maximally efficient, and (2) optimal isolation among the individual channels. High efficiency results in less power being deposited during decoupling, maximized decoupling bandwidth, and shorter pulse widths. Designs which provide geometric and/or electrical orthogonality improve isolation. Inverse detection experiments involve proton detection while simultaneously employing the heteronuclear channel for high-powered decoupling. Cross-talk between the channels causes severe S/N degradation. The second criterion poses a particular challenge for microcoils since the capacitors and inductors that comprise the multiple frequency impedance matching circuitries are typically located in much closer proximity than for conventional-scale NMR probes. Prior to ascertaining their spectroscopic performance, these probes are generally characterized electrically. Tests include the measurement of the S_{12} scattering parameter, defined as the reverse insertion voltage gain when source and load impedances are 50 Ω. That is, for 1 volt applied to the inputs of the proton channel, S_{12} is a measure of the electrical potential detected at the output of the carbon channel if both channels have been impedance matched to 50 Ω. For perfectly isolated channels, the scattering parameter will be 0. In practice, S_{12} values commonly vary from 0.1 % (−30 dB) to 0.01 % (−40 dB).

7.3.1.2.5 MICROFABRICATED RF COILS

Because of their high degree of control over small-scale geometries, microfabrication methods can prove useful in the effort to take advantage of the increased sensitivity of reduced-diameter RF coils. In this vein, Figure 7.3.1.4(a) illustrates a solenoidal microcoil product of microcontact printing [34]. While this work produced high-resolution ^{1}H-NMR spectra of ethylbenzene and acetone, the sensitivity of the printed microcoil was inferior to others made manually from copper wire. The decreased sensitivity was attributed to a higher coil resistance which resulted from a lower cross-sectional area and the higher resistivity of the electrodeposited copper wire (compared to annealed copper). While microcontact printing offers the capability to make solenoidal coils, alternative geometries are better suited to conventional lithographic techniques. Planar coils with inner diameters less than 50 μm have been fabricated for aluminum and gold deposited on glass substrates. Additionally, superconducting planar coils have been reported with dimensions of ∼ 100 μm for low-field continuous wave ^{3}He-NMR.[35] However, the potential for coupling dedicated preamplifiers and microcoils on a single platform makes gallium arsenide (GaAs) a more attractive substrate than glass [36–39]. As shown in Figure 7.3.1.4(b), planar receiver coils with 3.5 turns, a trace width of 10 μm, a trace separation of 10 μm, an i.d. of 60 μm, and an o.d. of 200 μm have been fabricated. A combination of lift-off and electroplating technologies

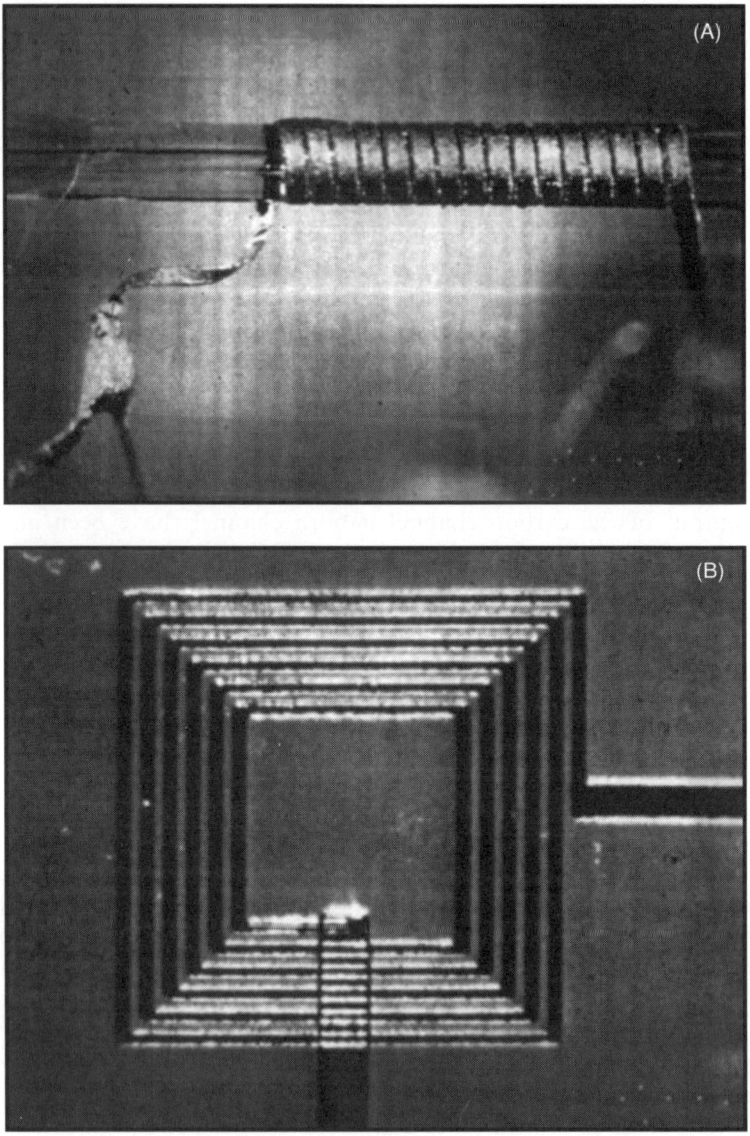

Figure 7.3.1.4 Microfabricated RF coils: (a) photograph of a 325 μm-diameter solenoidal microcoil fabricated by microcontact printing. Adapted by permission of the American Institute of Physics, from Rogers, J. A., Jackman, R. J., Whitesides, G. M., Olson, D. L. and Sweedler, J. V., *Appl. Phys. Lett.*, 1997, **70**, 2464. (b) Scanning electron micrograph of a planar microcoil with an o.d. of 200 μm and an i.d. of 60 μm

was used to transfer the pattern from a mask written by traditional electron-beam lithography to a GaAs substrate. A 1.5-cm RF transmitter coil, constructed from a single turn of 18-gauge copper wire, was mounted orthogonally to the microfabricated receiver coil on the circuit board.

With planar coils of n concentric turns and constant interturn spacing and trace width, the S/N per unit volume is given by the following:

$$S/N_{\text{puv}} = \frac{\sum_n \frac{\mu_0}{D[1+(2z/D)^{3/2}]}}{\sqrt{\sum_n \frac{4\rho D}{d_c^2}}} \tag{7.3.1.4}$$

where z refers to the on-axis distance from the coil, ρ corresponds to the resistivity, D is the o.d. of the spiral, and d_c is the diameter of the conductor [26]. Equation (7.3.1.4) demonstrates a weaker dependence of S/N on coil diameter for these planar coils when compared with solenoids. While fewer turns reduce space and fabrication effort, a higher number of turns produces an increased inductance and a lower capacitance necessary for impedance matching. Currently, the use of microfabrication techniques is in an embryonic stage. Although NMR spectra with linewidths under 2 Hz have been achieved [26,34] and microfabricated fluidic systems have been coupled to chip-based NMR detectors [40], magnetic-susceptibility-induced line broadening remains a major challenge. Additionally, planar microcoils must overcome the difficulties of extremely low signal strengths from sub-nanoliter observe volumes. Nevertheless, these techniques should prove instrumental as NMR spectroscopy examines continually smaller dimensions.

7.3.1.2.6 EVALUATIONS OF OVERALL PROBE PERFORMANCE

From a practical perspective, direct comparisons of probe sensitivity guide researchers in determining which probe to use for a particular application. Since a valid evaluation requires careful attention to sample preparation and NMR processing conditions for a particular analyte, a comprehensive survey of available commercial probes under uniform conditions is not available from the literature. Nevertheless, the overall effects of RF coil design on mass and concentration sensitivity can be garnered from a relatively few number of experiments. In a previous report [41], four different NMR probes (one Nano•NMR probe, one Nalorac SMIDG probe, and two microcoils of significantly different dimensions) were compared with respect to mass- and concentration-sensitivity from an equivalent sample mass and identical NMR experimental conditions. The results verified that the concentration-sensitivity improved with larger sample volumes and that the smaller diameter RF coils yielded the highest mass-sensitivity. This head-to-head comparison clearly showed that design parameters could be manipulated with relative ease to suit particular experimental requirements. In another recent publication, a direct comparison was made between 1.7-mm submicro and 3-mm micro gradient NMR probes from Nalorac for the acquisition of ^1H–^{13}C and ^1H–^{15}N heteronuclear shift correlation data on pharmaceutical standards [42].

In order to understand the implications of mass and concentration limitations on experiments, one can ponder a potent natural anti-cancer agent, of which only 10 pmol of material has been isolated from a marine sponge. If the sample is dissolved in a V_{obs} of 5 nl, the sample concentration is 2 mM. In contrast, dissolution of the material in 230 μl (V_{obs} for a typical 5-mm spinning tube) yields a concentration of only 44 nM. Although each NMR probe contains the same mass-limited amount of sample in their respective V_{obs} regions, the microcoil offers a significantly higher S/N for a given acquisition time (t_{acq}) [4]. Most mass-limited conditions require the probe with the smallest coil because of its enhanced mass sensitivity. Alternatively, consider NMR binding studies of a toxic compound to protein receptors that typically require low concentrations (μM) to determine the point at which strong interaction ceases. In this situation, one has to sacrifice S_m to gain overall S/N by using a larger-volume sample. A situation which is not mass-limited, or where the sample already exists in a relatively large volume and cannot be concentrated, benefits by using the probe with the highest concentration sensitivity, which usually also has the biggest V_{obs}. Clearly, no single probe fulfills all needs, and one must understand the subtleties of individual investigations. Considerations of probe sensitivity are weighed against the amount of sample available for analysis and the acceptable concentration range for a given experiment. Although the use of small sample volumes can be limited by such factors as sample loading and solubility, the microcoil approach definitely offers impressive performance gains [26,27,41]. Since microcoil probes have demonstrated superior mass sensitivity, their advantages in mass-limited applications are evident.

7.3.1.3 STATIC NMR SPECTROSCOPY WITH NANOLITER VOLUMES

Although the survey of microliter-volume NMR spectroscopy presented above demonstrates that commercial probe manufacturers have managed to deliver a wide range of capabilities at this scale, experiments on nanoliter volumes present additional challenges. While initial reports [43–45] illustrated the potential of sub-millimeter coils for nanoliter-volume NMR spectroscopy, the relatively poor linewidths (7–11 Hz) were too large to resolve most scalar couplings. In the first example of high-resolution ^1H-NMR of nanoliter-volume samples, a 130-fold improvement in mass sensitivity (relative to a conventional probe) was demonstrated for a microcoil with a 5-nl observe volume [4]. As a representative example of the performance of such a probe, Figure 7.3.1.5 shows a spectrum acquired from 3.3 nmol of α-bag cell peptide from the sea slug *Aplysia californica*; the limit of detection (LOD) is 124 pmol for an NMR experiment

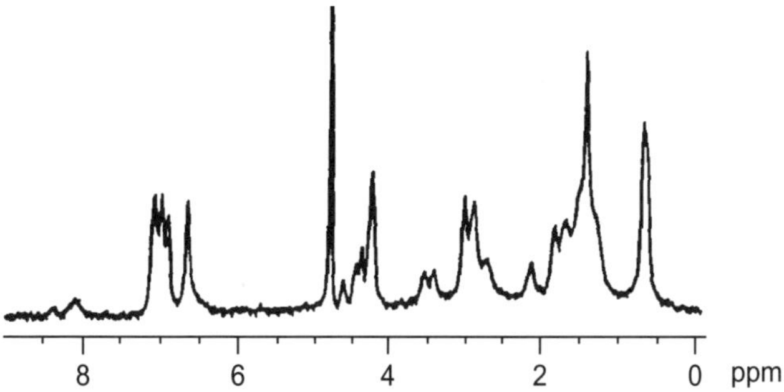

Figure 7.3.1.5 300 MHz ^{1}H spectrum of *Aplysia californica* α-bag cell peptide [1–7]. The 5-nl detection cell contains 3.0 μg (3.3 nmol) of peptide. The LOD is 112 ng (120 pmol) for an acquisition time of 11.8 min. Adapted with permission from Olson, D. L., Peck, T. L., Webb, A. G., Magin, R. L. and Sweedler, J. V., *Science*, 1995, **270**, 1967. Copyright (1995) American Association for the Advancement of Science

time of 11.8 min [4]. Since this first demonstration of high-resolution, nanoliter-volume NMR spectroscopy, efforts to understand and optimize probe design parameters have been underway for some time. More specifically, one report evaluated two microcoil probes that require a total sample volume of only 200 nl [46]. The performance of these probes was characterized with two model compounds and compared with a Nano•NMR probe. In addition, a strategy to minimize NMR background interference through the use of a spatially selective pulse sequence, combined with spectral subtraction, was presented. Another investigation [47] showed that, while larger filling factors generally resulted in better concentration sensitivity, this trend was difficult to achieve with the smallest-diameter microcoils because of spectral degradation due to the proximity of the coil to the sample. While most of these studies were performed at 300 MHz or below, further refinements of such microcoil probes have included experiments at higher fields and the incorporation of a lock channel [8,48].

Because of its high gyromagnetic ratio and natural abundance, the majority of microcoil work has focused on ^{1}H-NMR. However, nanoliter-volume probes have been constructed for both direct and inverse detection of ^{13}C [49,50]. Figure 7.3.1.6 depicts a representative circuit diagram of an inverse detection probe used for either ^{1}H–{^{15}N} or ^{1}H–{^{13}C} experiments. The inner solenoidal coil is tuned to protons, and the outer surface coil to the heteronucleus [49]. The surface coil consists of a single circular loop, 7 mm in diameter, and is mounted 1 mm from the microcoil edge. A chip capacitor is placed across the coil leads as close to the loop as possible. Although surface coils have high sensitivity and are simple to construct, they are traditionally not used in high-resolution NMR due to relatively poor B_1 homogeneity. However, with the reduced observe dimensions of microcoils, surface coils prove highly efficient both for pulsing and decoupling the observe volumes of microcoil probes.

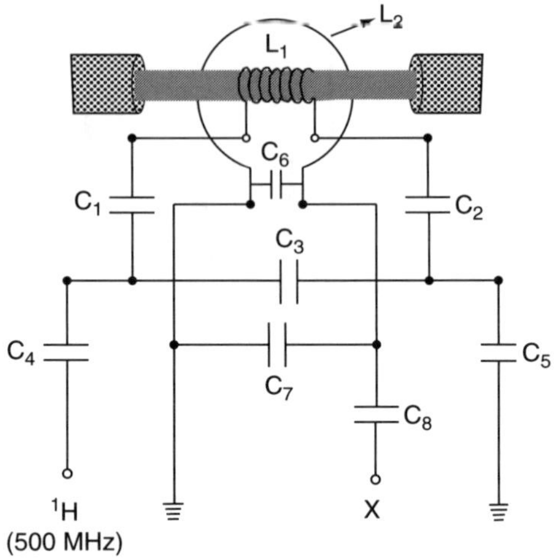

Figure 7.3.1.6 Representative circuit diagram for inverse detection probes showing the requisite RF coils and capacitor layout

Furthermore, the B_1 homogeneity is sufficiently high to produce a 180° pulse which inverts the magnetization cleanly during pulse-width-calibration procedures. Like most other microcoils made from copper wire, the surface coil and solenoidal microcoil are housed in a plastic container which also holds the magnetic susceptibility matching fluid. All circuit elements, including the coils, are positioned on a double-sided printed circuit board, with the impedance-matching circuitry for protons and heteronuclear frequencies on opposite sides. Because the S/N of inverse detection experiments is lower than that of homonuclear proton experiments, the typical sample volumes used are larger than those described earlier. With optimized coil volumes ranging from 700 nl to 1.2 µl, and concentrations of 30–100 mM, high-quality HMQC and HSQC spectra have been obtained in a few hours [50]. To demonstrate the performance of a typical inverse detection microcoil probe, a two-dimensional HMQC spectrum of chloroquine diphosphate in D_2O is presented in Figure 7.3.1.7. The unique inverse detection microcoil probes described in this section makes acquisition of two-dimensional ^1H–^{13}C correlation spectra possible for sample masses as low as tens of micrograms and data acquisition times of only a few hours. This represents an order of magnitude improvement compared to the commercial microliter-scale NMR probes described earlier.

In order to maximize sample handling efficiency, microcoil experiments can be performed on samples of limited volume with a setup similar to that shown in Figure 7.3.1.8. The liquid susceptibility-matching approach [7] described earlier allows observe factors of up to 70 % without broadening the linewidth. One recent example uses perfluorinated liquid plugs to restrict the total sample

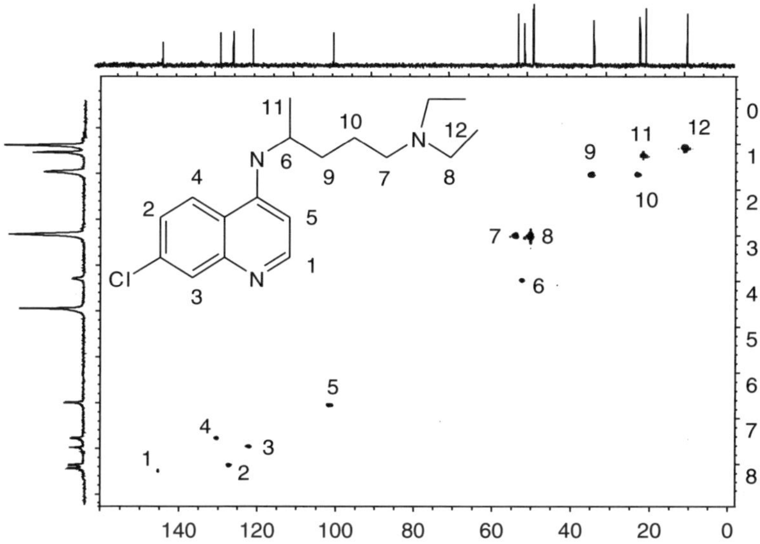

Figure 7.3.1.7 A ^{13}C-decoupled HMQC spectrum of 54 mM chloroquine diphosphate in D_2O acquired with an inverse detection microcoil probe. The 740-nl V_{obs} contained 40 nmol (13 μg) of chloroquine. The data, 32 transients per slice, 1024 × 128 (×2, hypercomplex) points, were acquired in 3.6 h. The data were zero-filled to 256 points in the ^{13}C dimension. A 40° shifted sinebell function was applied, followed by Gaussian multiplication prior to Fourier transformation

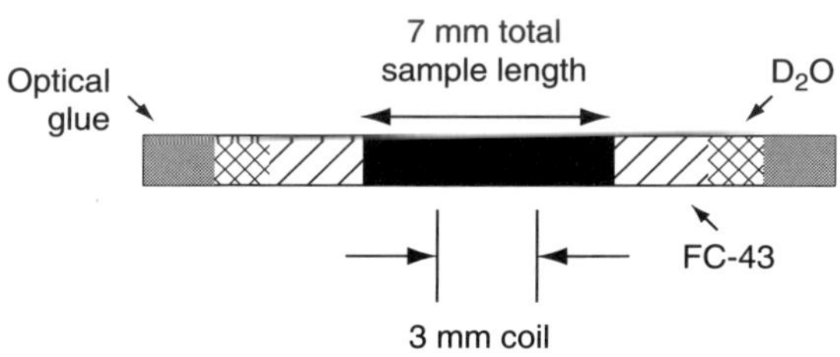

Figure 7.3.1.8 Instrumental configuration for mass-limited characterization by microcoil NMR. The analyte is bracketed on both sides by plugs of an immiscible perfluorinated organic liquid and sealed with an optically curable glue to prevent dilution during the data-collection period

volume of the product cleaved from individual solid-phase resin beads [8]. To emphasize the non-destructive nature of the technique, the product was collected and examined by mass spectrometry after microcoil NMR analysis. ^{1}H-NMR spectra were acquired at 600 MHz in 1 h of experiment time from approximately 540 pmol of total product. Through its enhanced sensitivity, this type of microcoil probe provides a means to ease the bottleneck associated with the need for high-throughout structural characterization of combinatorial products. Of course, the combination of multiple microcoils in a single probehead further increases sample throughput, and is the subject of another chapter.

7.3.1.4 COUPLING OF NMR SPECTROSCOPY AND MICROSEPARATIONS

In addition to relatively low sensitivity, NMR spectroscopy also typically experiences difficulties with the structural elucidation of mixtures. Due to the intrinsically high information content of the spectra, overlapping resonance signals are frequently challenging to interpret. Although augmented magnetic field strengths and multidimensional experimental methods increase data dispersion and aid in deconvolution, complex matrices can be more effectively treated in many situations by the combination of separations and NMR analysis. Moreover, the hyphenation of techniques allows researchers to address more sophisticated issues. As microcoils are generally wrapped directly around fused silica capillaries, the union of microseparations and nanoliter-volume NMR detection is a natural enhancement of probe development. During the past two decades, hyphenated analytical techniques have become rather commonplace. Multimode separation and detection strategies are requisite in such diverse areas as characterization of pharmaceutical candidates, understanding chemical communication within living organisms, and geochemical analysis of planetary samples. In comparison with conventional NMR probes, the enhanced mass sensitivity of reduced-diameter RF transceivers enables the coupling of one of the most structurally rich detection schemes to highly efficient capillary separations. The experimental versatility of NMR spectroscopy as a detector for separations has been applied successfully to solve a wide variety of problems. The chapters which follow in this book cover in detail the combination of NMR and microseparations.

7.3.1.5 CONCLUSIONS

As techniques for chemical analysis are used in continually smaller domains, experimental challenges for inherently insensitive methods such as NMR spectroscopy become increasingly severe. Among the various schemes to boost the intrinsic sensitivity of an NMR experiment, the development of small-volume RF probes has experienced a renaissance during the past decade. Commercial NMR probes now allow analyses of nanomole quantities in microliter volumes from natural product extracts and combinatorial chemical syntheses. Figure 7.3.1.9 illustrates the range of volumes that can be examined by NMR probes and accessories such as microsample tubes and inserts. With recently reported advances in sample preconcentration for microcoil NMR analysis [51], dilute microliter-volume samples can now be concentrated into nanoliter-volume

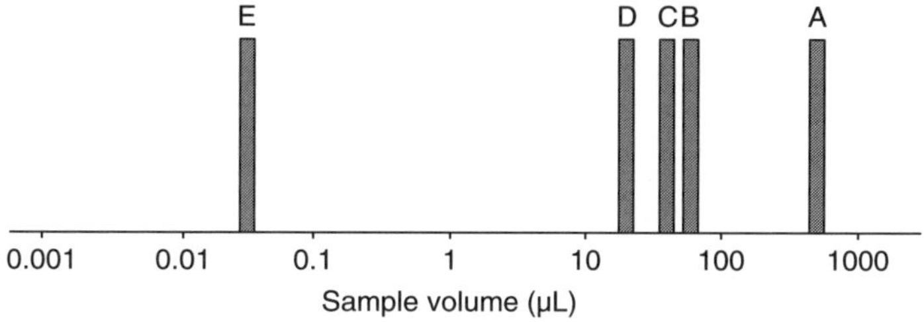

Figure 7.3.1.9 Relative sample volumes required for an assortment of NMR probes: (a) conventional 5-mm tubes; (b) 3-mm tubes with microcells; (c) Varian Nano•NMR; (d) 1.7-mm tubes with minimal solvent; (e) microcoil with a $V_{obs} \approx 30\,nl$

bands. Since such an approach aids sample handling for the smallest-volume (and correspondingly highest-mass-sensitivity) probes, the range of utility of microcoils is greatly expanded. Although the development of nanoliter-volume probes is still in the research phase, the enhancements in sensitivity already achieved are exciting. Through the fabrication of nanoliter-volume NMR probes, mass-limited analytes are becoming viable samples for NMR analysis.

REFERENCES

1. Odelblad, E., *Nordisk Forening for Obsterik och Gynekologi*, Karolinska Institute, Stockholm, 1966.
2. Shoolery, J. N., *Topics Carbon-13 NMR Spectrosc.*, 1979, **3**, 28.
3. Wiseman, R. W., Moerland, T. S. and Kushmerick, M. J., *NMR Biomed.*, 1993, **6**, 153.
4. Olson, D. L., Peck, T. L., Webb, A. G., Magin, R. L. and Sweedler, J. V., *Science*, 1995, **270**, 1967.
5. Reynolds, W. F., Yu, M. and Enriquez, R. G., *Magn. Reson. Chem.*, 1997, **35**, 614.
6. Wilmad Glass Product Catalogue, http://www.wilmad.com (accessed April 2001).
7. Behnia, B. and Webb, A. G., *Anal. Chem.*, 1998, **70**, 5326.
8. Lacey, M. E., Sweedler, J. V., Larive, C. K., Pipe, A. J. and Farrant, R. D., *J. Magn. Reson.*, 2001, **53**, 215.
9. Hoult, D. I. and Richards, R. E., *J. Magn. Reson.*, 1976, **24**, 71.
10. Shoolery, J. N., *Varian Instrum. Appl.*, 1976, **10**, 18.
11. Shoolery, J. N., Wehrli, F. and Wirthlin, T., *Chem.-Tech. (Heidelberg)*, 1977, **6**, 55.
12. Shoolery, J. N. and Majors, R. E., *Am. Lab.*, 1977, **9**, 51.
13. Chatterjee, A., Mukhopadhyay, S. and Shoolery, J. N., *Ind. J. Chem.*, *B*, 1978, **16**, 67.
14. Shoolery, J. N. and Southwick, E. W., *J. Agric. Food Chem.*, 1979, **27**, 1400.
15. Crouch, R. C. and Martin, G. E., *Magn. Reson. Chem.*, 1992, **30**, 566.
16. Shockcor, J. P., Wurm, R. M., Silver, I. S., Crouch, R. C. and Martin, G. E., *Tetrahedron Lett.*, 1994, **35**, 4919.
17. Crouch, R. C. and Martin, G. E., *J. Nat. Prod.*, 1992, **55**, 1343.
18. Crouch, R. C., Martin, G. E., Musser, S. M., Grenade, H. R. and Dickey, R. W., *Tetrahedron Lett.*, 1995, **36**, 6827.

19. Sharaf, M. H. M., Schiff, Jr, P. L., Tackie, A. N., Phoebe, Jr, C. H., Howard, L., Meyers, C., Hadden, C. E., Wrenn, S. K., Davis, A. O., Andrews, C. W., Minick, D., Johnson, R. L., Shockcor, J. P., Crouch, R. C. and Martin, G. E., *Magn. Reson. Chem.*, 1995, **33**, 767.
20. Martin, G. E., Crouch, R. C. and Zens, A. P., *Magn. Reson. Chem.*, 1998, **36**, 551.
21. Martin, G. E., Guido, J. E., Robins, R. H., Sharaf, M. H. M., Schiff, P. L. and Tackie, A. N., *J. Nat. Prod.*, 1998, **61**, 555.
22. Hadden, C. E. and Martin, G. E., *J. Nat. Prod.*, 1998, **61**, 969.
23. Peck, T. L., Magin, R. L. and Lauterbur, P. C., *J. Magn. Reson., B*, 1995, **108**, 114.
24. Minard, K. R. and Wind, R. A., *Concepts Magn. Reson.*, 2001, **13**, 128.
25. Minard, K. R. and Wind, R. A., *Concepts Magn. Reson.*, 2001, **13**, 190.
26. Webb, A. G., *Prog. NMR Spectrosc.*, 1997, **31**, 1.
27. Olson, D. L., Lacey, M. E. and Sweedler, J. V., *Anal. Chem.*, 1998, **70**, 257A.
28. Rao, N. N., *Electromagnetics*, Prentice-Hall, New York, 1992.
29. Mueller, L., *J. Am. Chem. Soc.*, 1979, **101**, 4481.
30. Bax, A., Griffey, R. H. and Hawkins, B. L., *J. Magn. Reson.*, 1983, **55**, 301.
31. Bodenhausen, G. and Ruben, D. J., *Chem. Phys. Lett.*, 1980, **6**, 185.
32. Kay, L. E., Ikura, M., Tschudin, R. and Bax, A., *J. Magn. Reson.*, 1990, **89**, 496.
33. Grzesiek, S. and Bax, A., *J. Magn. Reson.*, 1992, **96**, 432.
34. Rogers, J. A., Jackman, R. J., Whitesides, G. M., Olson, D. L. and Sweedler, J. V., *Appl. Phys. Lett.*, 1997, **70**, 2464.
35. Koivuniemi, J., Kiviranta, M., Seppa, H. and Krusius, M., *J. Low Temp. Phys.*, 1998, **110**, 255.
36. Peck, T. L., Magin, R. L., Kruse, J. and Feng, M., *IEEE Trans. Biomed. Eng.*, 1994, **41**, 706.
37. Stocker, J. E., Peck, T. L., Webb, A. G., Feng, M. and Magin, R. L., *IEEE Trans. Biomed. Eng.*, 1997, **44**, 1122.
38. Dechow, J., Forchel, A., Lanz, T. and Haase, A., *Microelectron. Eng.*, 2000, **53**, 517.
39. Dechow, J., Forchel, A. W. B., Lanz, T. and Haase, A., *Proc. SPIE-Int. Soc. Opt. Eng.*, 1999, **3857**, 98.
40. Trumbull, J. D., Glasgow, I. K., Beebe, D. J. and Magin, R. L., *IEEE Trans. Biomed. Eng.*, 2000, **47**, 3.
41. Lacey, M. E., Subramanian, R., Olson, D. L., Webb, A. G. and Sweedler, J. V., *Chem. Rev.*, 1999, **10**, 3133.
42. Martin, G. E. and Hadden, C. E. *Magn. Reson. Chem.*, 1999, **37**, 721.
43. Wu, N., Peck, T. L., Webb, A. G., Magin, R. L. and Sweedler, J. V., *J. Am. Chem. Soc.*, 1994, **116**, 7929.
44. Wu, N., Peck, T. L., Webb, A. G., Magin, R. L. and Sweedler, J. V., *Anal. Chem.*, 1994, **66**, 3849.
45. Wu, N., Webb, A., Peck, T. L. and Sweedler, J. V., *Anal. Chem.*, 1995, **67**, 3101.
46. Olson, D. L., Lacey, M. E. and Sweedler, J. V., *Anal. Chem.*, 1998, **70**, 645.
47. Webb, A. G. and Grant, S. C., *J. Magn. Reson., B*, 1996, **113**, 83.
48. Subramanian, R., Lam, M. M. and Webb, A. G., *J. Magn. Reson.*, 1998, **133**, 227.
49. Subramanian, R. and Webb, A. G., *Anal. Chem.*, 1998, **70**, 2454.
50. Subramanian, R., Sweedler, J. V. and Webb, A. G., *J. Am. Chem. Soc.*, 1999, **121**, 2333.
51. Kautz, R. A., Lacey, M. E., Wolters, A. M., Foret, F., Webb, A. G., Karger, B. L. and Sweedler, J. V., J. *Am. Chem. Soc.*, 2001, **123**, 3159.

7.3.2 Capillary Separation Techniques

ALEXANDRE BEZERRA SCHEFER and KLAUS ALBERT
Institut für Organische Chemie, Universität Tübingen, Tübingen, Germany

It is evident from the results outlined in the preceding chapters that the direct on-line coupling of HPLC and related techniques together with NMR spectroscopy yields a wealth of unequivocal structural information. The main pitfalls of the current NMR-hyphenated techniques are residual solvent resonances together with the occurrence of signals of solvent impurities. The spectroscopic information in the solvent and impurity signal range is lost, thus rendering structural elucidation as a very difficult task. Therefore, the use of deuterated solvents, as in routine NMR spectroscopy, would be favourable. Low solvent consumption of deuterated solvents can be achieved when capillary separation techniques are performed. They would also allow the investigation of limited amounts of sample.

These two reasons are the main driving forces for the current on-going attempts for the development of a feasible arrangement for the on-line coupling of capillary separation techniques together with NMR spectroscopy. Whereas a solenoidal microcoil configuration is used by the group of Sweedler and Webb [1–4] (see Chapters 7.3.1 and 8.2), a double-saddle Helmholtz coil is employed by the Tübingen research team [5–16]. Here, the capillary is positioned in the z-direction parallel to the magnetic field direction of the cryomagnet. Thus, no NMR signal disturbances due to additional magnetic field effects occur in electrodriven separations.

Moreover, an easy exchange of the detection capillary is possible. This feature is of enormous importance, because in the current status of prototype development cracking of the capillaries is a routine occurrence.

7.3.2.1 CAPILLARY HPLC–NMR COUPLING

Figure 7.3.2.1 shows the current experimental arrangement employed in Tübingen for capillary separations. A capillary HPLC pump is placed at a distance of about 2.5 m from the 14 T (proton resonance frequency, 600 MHz) cryomagnet,

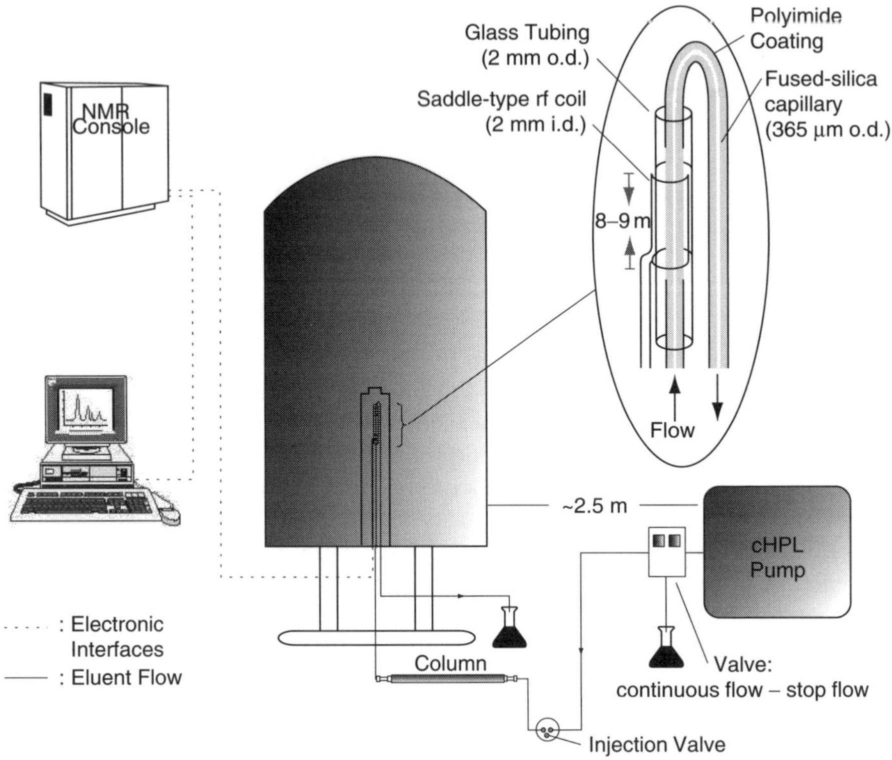

Figure 7.3.2.1 Experimental set-up for capillary HPLC-NMR coupling

whereas the separation capillary and the injection valve is located below the cryomagnet. Capillary separation is performed at a flow rate of 1.5 μl/min on a C_{18} packed fused silica capillary (length 250 mm, i. d. 250 μm, o. d. 350 μm). NMR detection is performed with a microprobe equipped with a Helmholtz microcoil with an internal diameter of 2 mm. Within the detection coil of the microprobe (overall length 9 mm), the polyimide coating of the capillary (i. d. 180 μm, o. d. 365 μm) is removed, so resulting in a detection volume of 750 nl. The HPLC system is composed of a micro-pump, a Valco Two-Position-Actuator Control Module and a Valco injection device with a 5 μl loop. The connections are made by means of fused silica capillaries: 50 μm i. d./365 μm o. d. from the HPLC micro-pump up to the NMR detection cell, and 100 μm i. d./365 μm o. d. from the detection cell up to the waste.

Figure 7.3.2.2 shows the current sensitivity obtained with a capillary (c)HPLC–NMR probe with a 2.0 mm Helmholtz microcoil and a detection volume of 750 nl (detection capillary, i. d. 180 μm). The signal of the anomeric protons of 0.2 M sucrose in D_2O yields a signal-to-noise ratio (S/N) of 50:1.

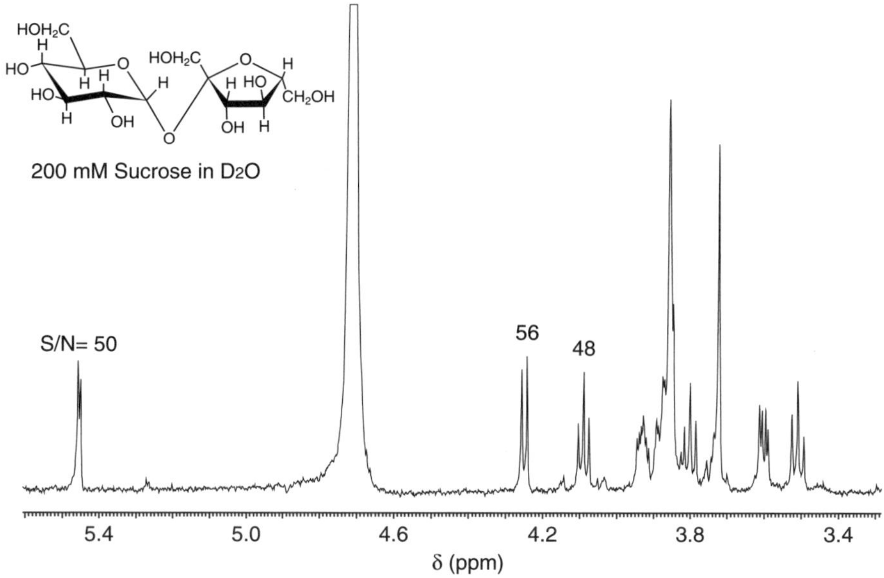

Figure 7.3.2.2 Current sensitivity test of the cHPLC–NMR probe (sample, 200 mM sucrose in D_2O; ^1H resonance frequency, 600 MHz; detection volume, 750 nl)

7.3.2.2 cHPLC–NMR SEPARATION OF A REAL-LIFE SAMPLE OF NATURAL COMPOUNDS

As a typical example, the cHPLC–NMR separation of an extract of black beetles is shown in Figure 7.3.2.3. The methanolic extract was prepared from the vessels of one hundred black beetles (*Palembus ozularis*) by Marit Wahrendorf at the Institute of Pharmaceutical Biology (Prof. Dr. Michael Wink) in Heidelberg, Germany [16]. These insects are found in South America and their vessels contain anti-inflammatory compounds which can act against asthma. As the extract was received with a non-deuterated solvent, before the injection into the chromatographic system, the solvent was evaporated and then methanol-d_4 was used to dissolve the dry extract.

The separation was carried out at ambient temperature with a solution of D_2O/acetonitrile-d_3 (60:40, v/v) as eluent. A flow rate of 1.5 μl/min was applied and the injection was performed by filling up the 5 μl injection loop with the black beetles extract and switching on the valve for 90 s under a back-pressure of about 478 psi.

For the continuous-flow measurements, the pseudo-2D spectrum was recorded with a spectral width of 9616 Hz and 64 transients with 8K complex data points, thus resulting in an acquisition time of 0.42 s/transient along the 128 t_1 increments. A relaxation delay of 1.2 s was used and the time resolution

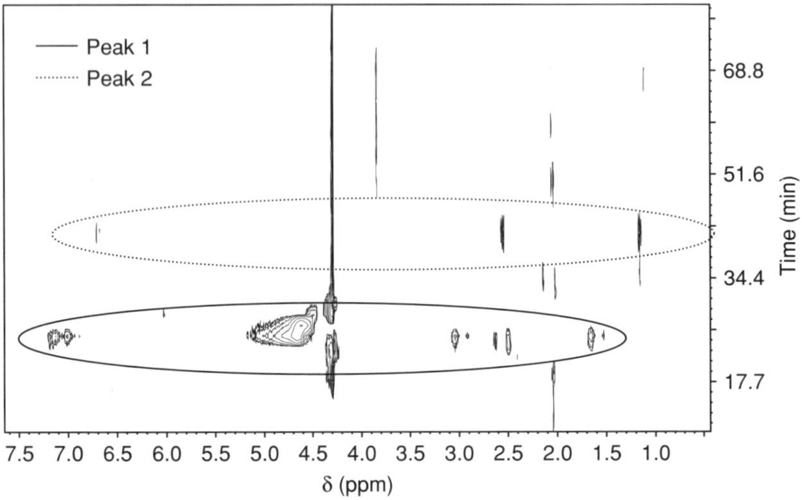

Figure 7.3.2.3 Continuous-flow cHPLC–NMR chromatogram (600 MHz) obtained for the separation of an extract of black beetles

along the F1 dimension was 87 s/increment. The pulse angle was set to 90° and the spectrum was referenced to 4.9 ppm by using the HDO residual signal. The acquired data were treated as a 2D-NMR matrix (t_1, retention time) and processed with XWINNMR software. Prior to the Fourier transformation, the data was apodized with a shifted squared sinebell window function (shift 2.1) in the F_2 dimension only.

Stop-flow ^{1}H NMR and ^{1}H–^{1}H TOCSY spectra were recorded over the peaks at 25 min (peak 1) and 41 min (peak 2). The maximum of each chromatographic peak was determined by monitoring the run through a continuous-flow experiment. The peak at 25 min was recorded without suppression of solvent peaks since the separation was carried out in deuterated solvents, and the acquirement was performed with the spectrometer gain set to its maximum value. Thus, while intense peaks were observed from residual protonated water and acetonitrile, no spectral distortion or clipping resulted from these resonances. However, the second stop-flow spectrum on the chromatographic peak at 36 min was recorded with suppression of the HDO peak. Although not really necessary, the irradiation of such a peak was performed to show the possibility of suppressing a weak signal without any effects on the chemical shift range of interest. For the first peak, a total of 258 transients was acquired over a spectral width of 7143 Hz and 32K complex data points, which gave rise to an acquisition time of 2.9 s (Figure 7.3.2.4). For the second peak, a total of 512 transients was acquired over a spectral width of 48 087 Hz and 32K complex data points, hence resulting in an acquisition time of 3.4 s. Both of the 1D spectra were processed by applying an exponential window function prior to the Fourier transformation with 1.0 Hz line-broadening.

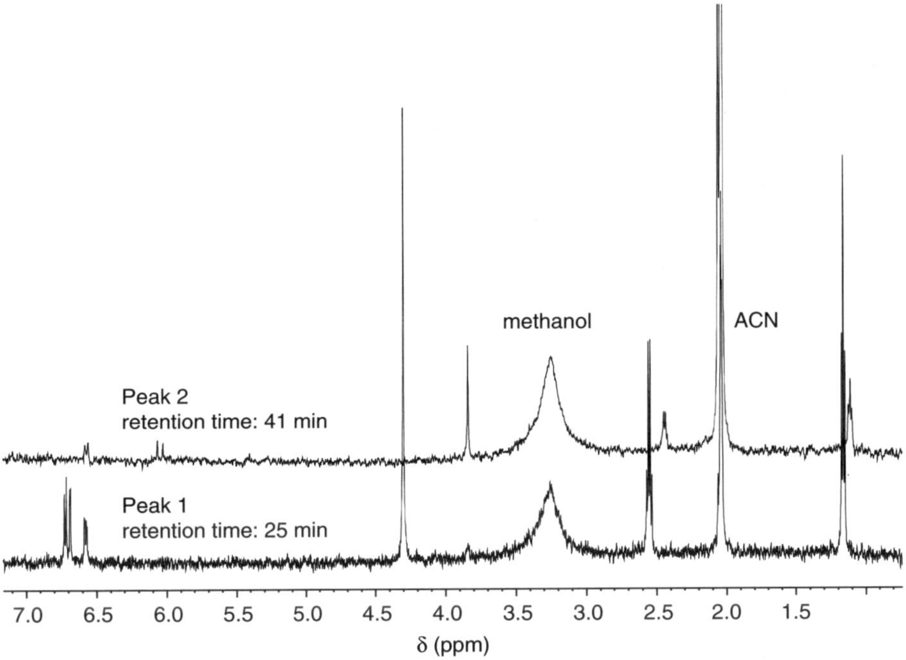

Figure 7.3.2.4 Stop-flow ^{1}H NMR spectra (600 MHz) of peaks 1 and 2 from the black beetles cHPLC–NMR separation

Suppression of the HDO signal was applied during the 2D-TOCSY experiments for both compounds since the 1D-spectrum of the second peak did not show distortions due to the irradiation. For peak 1, a total of 223t_1 increments with 16 transients and 1K complex data points was acquired in the simultaneous mode with a spectral width of 7143 Hz in both dimensions. With an acquisition time of 71.7 ms and an applied mixing time of the MLEV spin lock of 139.5 μs, the total acquisition time amounts to 3 h 47 min. The data were apodized with a shifted squared sinebell window function in both dimensions and zero-filled in the F_1 dimension to 512 data points. For peak 2, a total of 512t_1 increments with 64 transients and 1K complex data points were acquired in the simultaneous mode with a spectral width in both dimensions of 4808 Hz. With an acquisition time of 106.5 ms and an applied mixing time of the MLEV spin lock of 139.5 μs, the total acquisition time amounts to 15 h 29 min. The data were apodized with a shifted squared sinebell window function in both dimensions and zero-filled in the F_1 dimension to 1K data points. Before every stop-flow spectrum, optimization of shimming on the D_2O lock signal, together with tuning of the transmitter channel, was performed.

The continuous-flow measurements showed the presence of two major compounds. They were afterwards assigned by stop-flow measurements. Peak 1 eluted at ~25 min, and peak 2 at ~41 min (see Figure 7.3.2.3). Comparison between the ^{1}H-NMR spectra recorded during the stop-flow experiments showed that the two compounds are very similar (Figure 7.3.2.5). Peak 1 could be assigned to 2-ethylbenzo-1,4-quinone and peak 2 to 2-ethyl-3-methoxy-benzo-1,4-quinone.

This example clearly shows that with further improvement in sensitivity, on-line cHPLC–NMR, in conjunction with cHPLC–MS, is likely to be used as the major tool for the successful treatment of problems encountered with genomics, proteomics and drug metabolomics.

7.3.2.3 ELECTRODRIVEN SEPARATIONS COUPLED TO NMR

Electrodriven separations, such as capillary electrophoresis (CE) and capillary electrochromatography (CEC), are based on the different electrophoretic mobilities in an electric field of the molecules to be separated. They provide a higher separation efficiency then conventional HPLC since the electrophoretic flow (EOF) has a plug-flow profile. Whereas the mobile phase in CE is driven only by the electro-osmotic flow, it is generated in CEC by a combination of EOF and pressure. CEC has a high sample capacity which favours its hyphenation with NMR.

A similar system to the arrangement outlined in Figure 7.3.2.1 can be successfully employed for the hyphenation of electrodriven separations with NMR. Here, the cHPLC pump is substituted by a power supply for voltage generation. Special attention has to be given to the grinding of the (capillary) connections between the separation and the detection capillaries. A disadvantage may be the long length of capillaries between the power supply and the injection device. Here, in the future, shielded cryomagnets will likely lead to a considerable improvement, because then the supply of the electrodriven separations can be positioned at a distance close to the cryomagnet. Nevertheless, even with this outline prototype design several applications have been reported for the analysis of natural compounds, drugs and drug metabolites [10–15]. One example shows the separation of paracetamol metabolites from a urine extract performed with CE–NMR and pressurized CEC–NMR coupling. The shorter retention times, demonstrated in the on-flow contour plot (Figure 7.3.2.6), together with the enhanced sensitivity (Figure 7.3.2.7) are clearly visible in the CEC separation. With further improvements in coil design and sensitivity levels, the hyphenation of both CE and CEC with NMR has the potential to solve more real-world problems.

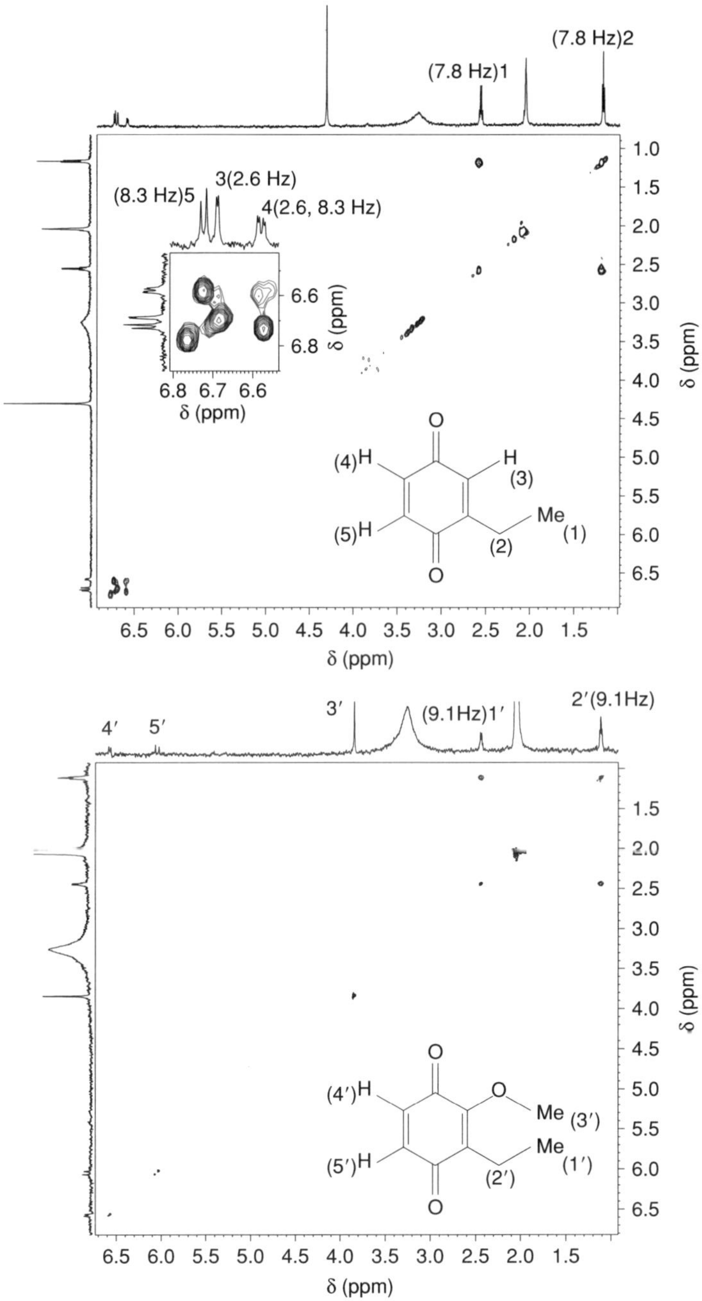

Figure 7.3.2.5 Stop-flow ^{1}H–^{1}H TOCSY spectra of (a) peak 1, and (b) peak 2 from the black beetles cHPLC–NMR separation

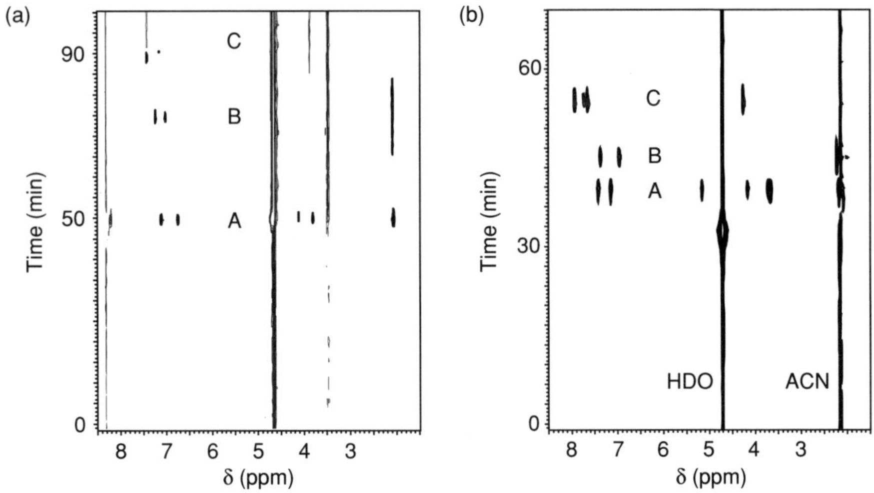

Figure 7.3.2.6 On-flow contour plots (600 MHz) of the (a) CE–NMR, and (b) pressurized CEC–NMR separations of the paracetamol metabolites obtained from a human urine extract: A, paracetamol glucoronide; B, paracetamol sulfate; C, hippuric acid

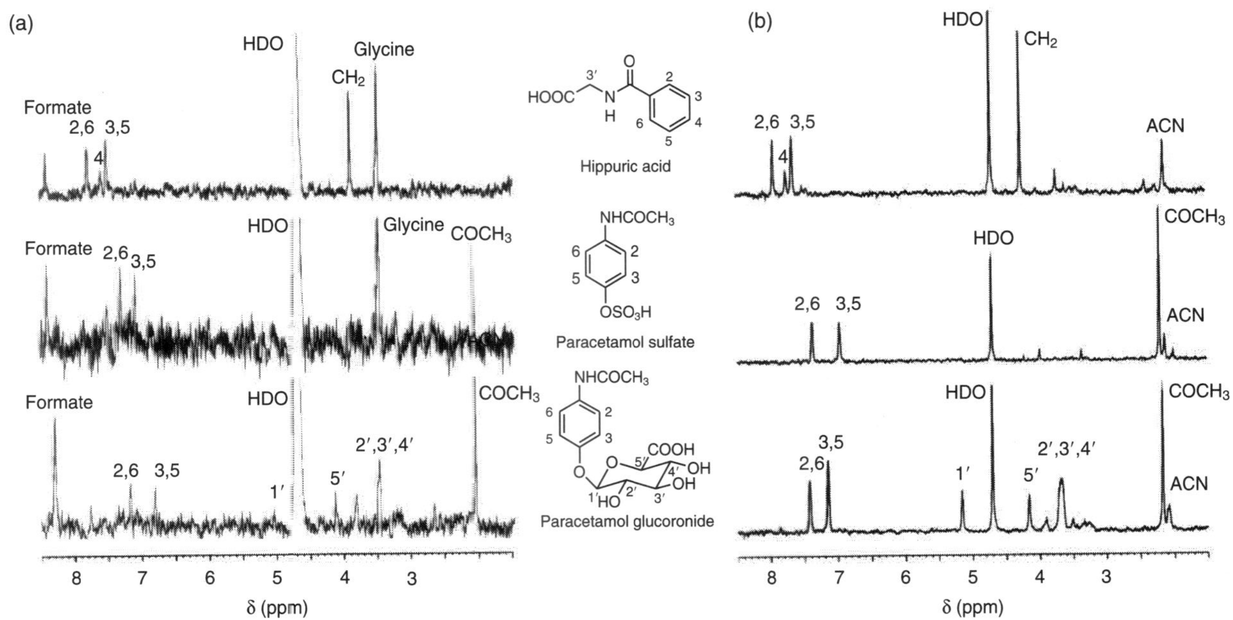

Figure 7.3.2.7 Single rows (64 accumulated transients) extracted from the continuous-flow ^{1}H-NMR chromatograms shown in Figure 7.3.2.6 of the (a) CE–NMR, and (b) pressurized CEC–NMR separations of the paracetamol metabolites obtained from a human urine extract

REFERENCES

1. Wu, N., Peck, T. L., Webb, A. G., Magin, R. L. and Sweedler, J. V., *J. Am. Chem. Soc.*, 1994, **116**, 7929.
2. Wu, N., Peck, T. L., Webb, A. G., Magin, R. L. and Sweedler, J. V., *Anal. Chem.*, 1994, **22**, 3849.
3. Olson, L., Peck, T. K., Webb, A. G., Magin, R. L. and Sweedler, J. V., *Science*, 1995, **270**, 1967.
4. Webb, A. G., *Prog. Nucl. Magn. Reson. Spectrosc.*, 1997, **31**, 1.
5. Albert, K., *Angew. Chem. Int. Ed. Engl.*, 1995, **34**, 641.
6. Behnke, B., Schlotterbeck, G., Tallarek, U., Strohschein, S., Tseng, L.-H., Keller, T., Albert, K. and Bayer, E., *Anal. Chem.*, 1996, **68**, 1110.
7. Albert, K., Schlotterbeck, G., Tseng, L.-H. and Braumann, U., *J. Chromatogr., A*, 1996, **750**, 303.
8. Schlotterbeck, G., Tseng, L.-H., Händel, H., Braumann, U. and Albert, K., *Anal. Chem.*, 1997, **69**, 1421.
9. Albert, K., Braumann, U., Tseng, L.-H. and Schlotterbeck, G., *Biomed. Chromatogr.*, 1998, **12**, 158.
10. Pusecker, K., Schewitz, J., Gförer, P., Tseng, L.-H., Albert, K. and Bayer, E., *Anal. Chem.*, 1998, **70**, 3280.
11. Pusecker, K., Schewitz, J., Gfrörer, P., Tseng, L.-H., Albert, K., Bayer, E., Wilson, I. D., Bailey, N. J., Scarfe, G. B., Nicholson, J. K. and Lindon, J. C., *Anal. Comm.*, 1998, **35**, 213.
12. Schewitz, J., Gförer, P., Pusecker, K., Tseng, L.-H. Albert, K., Bayer, E., Wilson, I. D., Bailey, N. J., Scarfe, G. B., Nicholson, J. K. and Lindon, J. C., *Analyst*, 1998, **123**, 2835.
13. Gfrörer, P., Schewitz, J., Pusecker, K., Tseng, L.-H., Albert, K. and Bayer, E., *Electrophoresis*, 1999, **20**, 3.
14. Schewitz, J., Pusecker, K., Gförer, P., Götz, U., Tseng, L.-H., Albert, K. and Bayer, E., *Chromatographia*, 1999, **50**, 333.
15. Gförer, P., Tseng, L.-H., Rapp, E., Albert, K. and Bayer, E., *Anal. Chem.*, 2001, **73**, 3234.
16. Schefer, A. B., Rapp, E., Wahrendorf, M. S., Wink, M., Ferreira, A. G., Bayer, E. and Albert, K., unpublished results.

8 Future Developments – Introduction

Developments in LC–NMR over the past twenty years have transformed the technique from its academic inception to integration as an important component of chemical and biochemical analysis in almost every pharmaceutical company. To a large extent, this highly successful development has come about because of the great increases in sensitivity of NMR spectroscopy, and the carefully designed coupling of separation science to NMR detection. Standard 4.6 mm separation columns coupled to 5 mm NMR coils are used at frequencies up to 800 MHz, and have found extensive applications in the assay of biofluids. Very high separation efficiency and NMR mass sensitivity can be achieved by using capillary separations coupled to sub-millimeter NMR detectors. It seems likely that future developments in the field of LC–NMR will come about as a result of continued increases in sensitivity and throughput.

Two initial approaches towards this goal are outlined in this present section. Presently, the vast majority of experiments are performed by using proton detection, either direct or indirect. Much valuable information could be gained by direct heteronuclear detection, and in particular, ^{13}C NMR. Sensitivity factors currently preclude the direct detection of ^{13}C resonances, although the polarization transfer techniques outlined in the first part of this section offer potential sensitivity gains of orders of magnitude. A second development concentrates on increasing the throughput of the NMR experiment, an ever increasingly important consideration in the pharmaceutical industry. This approach incorporates multiple, small, highly sensitive NMR detectors in a single NMR probe head, and is described in the second part of this section.

On-Line LC–NMR and Related Techniques
Edited by K. Albert.

8.1 HPLC–^{13}C NMR

KLAUS ALBERT
Institut für Organische Chemie, Universität Tübingen, Tübingen, Germany

Structure elucidation of unknown organic compounds is usually performed by the combined use of ^{1}H and ^{13}C NMR spectroscopy. The latter spectroscopy provides direct information about the carbon skeleton of the investigated molecule, thus revealing valuable structural features such as carbonyl and carboxyl moieties, which cannot be deduced by ^{1}H NMR spectroscopy.

The big disadvantage of ^{13}C NMR spectroscopy is its low sensitivity. Due to the natural abundance of 1.1% of the ^{13}C isotope and due to long spin–lattice relaxation times (T_1) of the order of seconds, the acquisition of a routine ^{13}C NMR spectrum of a 0.1 M solution of an organic compound takes at least one minute.

In on-line HPLC–NMR coupling, the commonly recorded nuclei are ^{1}H and ^{19}F, because their natural abundances are 99.9 and 100%, respectively. Thus, a direct monitoring of chromatographic separations is possible, as outlined earlier in Chapter 1. Indirect access to the information content of ^{13}C NMR spectra is obtained in the stop-flow mode, where 'inverse' detected ^{1}H, ^{13}C correlation spectra can be recorded. The acquisition of these type of 2D-spectra relies on the fact that a direct proton carbon connectivity via scalar coupling is present. Quartenary carbons without any directly attached protons are not detected. Thus, it is of major interest to record ^{13}C NMR spectra which reveal all possible information within a coupled LC experiment.

Continuous-flow ^{13}C NMR spectra can be recorded with a flow-probe design similar to that outlined above in Figure 1.8 in Chapter 1 [1,2]. A ^{13}C RF coil, with a length of 15 mm, is directly attached to the glass tube, so resulting in a detection volume of about 190 μl. An outer RF coil, mounted co-axially to the first, is tuned to ^{1}H for ^{1}H decoupling and to ^{2}H for field/frequency stabilization (lock) (Figure 8.1.1). In principal, continuous-flow ^{13}C NMR spectra of neat liquid samples or highly concentrated solutions of organic molecules can be obtained with the same quality as static ^{13}C NMR spectra. As outlined in Chapter 1, the flow-inherent decrease in the apparent spin–lattice relaxation time, $T_{1\text{flow}}$, leads to an increase in intensity versus acquisition time (compare Figure 1.2 and 1.4 in Chapter 1).

On-Line LC–NMR and Related Techniques
Edited by K. Albert.

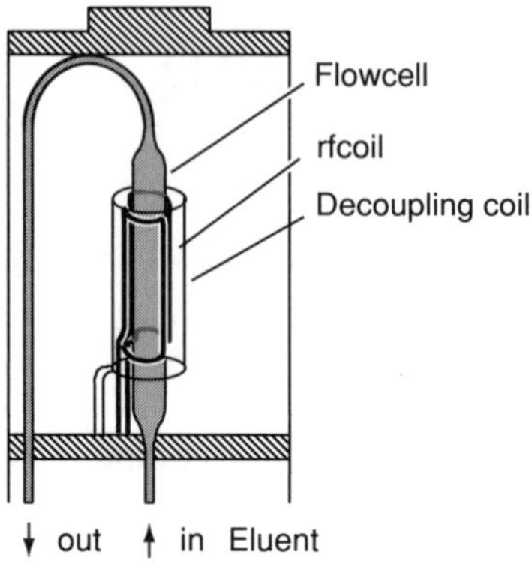

Figure 8.1.1 Schematic of a continuous-flow ^{13}C NMR probe with a detection volume of 190 μl

Figure 8.1.2 shows the ^{13}C NMR spectra of 0.1 M cholesterol acetate recorded under static-flow and different continuous-flow conditions. All NMR spectra have been recorded under 'inverse-gated decoupling conditions', so eliminating the effect of different carbon signal intensities due to $^{1}H^{13}C$ dipolar relaxation effects (nuclear Overhauser effect (NOE)). The residence time τ within the detector volume varies with flow rate (see Table 1.1 in Chapter 1). The maximum signal intensity can be obtained by adjusting the relaxation delay to the difference between residence time and acquisition time, according to the following equation:

$$\text{residence time } \tau = \text{relaxation delay} + \text{acquisition time} \tag{8.1.1}$$

Thus, at a given acquisition time of 0.344 s, a flow rate of 1 ml/min and a detection volume of 190 μl, the residence time is 11.3 s and the relaxation delay 10.966 s. At a flow rate of 10 ml/min, the residence time is 1.13 s while the relaxation delay is 0.787 s. Thus, higher flow rates result in shorter total acquisition times (number of transients multiplied by the sum of relaxation delay and acquisition times) when recording a ^{13}C NMR spectrum. Due to the different spin–lattice relaxation times of the carbon nuclei, they exhibit a different flow-rate-dependent behaviour. The signal of C-9 of cholesterol acetate with a T_1 of the order of 0.5 s shows an increase in signal intensity of 67 % with a flow rate increase from 1 ml to 20 ml/min, whereas signals for carbon atoms with longer relaxation times of the order of 10 s (e.g. C-5) are decreased at higher flow rates (see Figure 8.1.2).

This example demonstrates that high-quality ^{13}C NMR spectra can be obtained in the continuous-flow mode. Thus, it is possible to monitor electrochemical reactions of high-concentration compounds (0.1 M solutions) [3], or even to use diluted samples with ^{13}C-labelled positions [4–6].

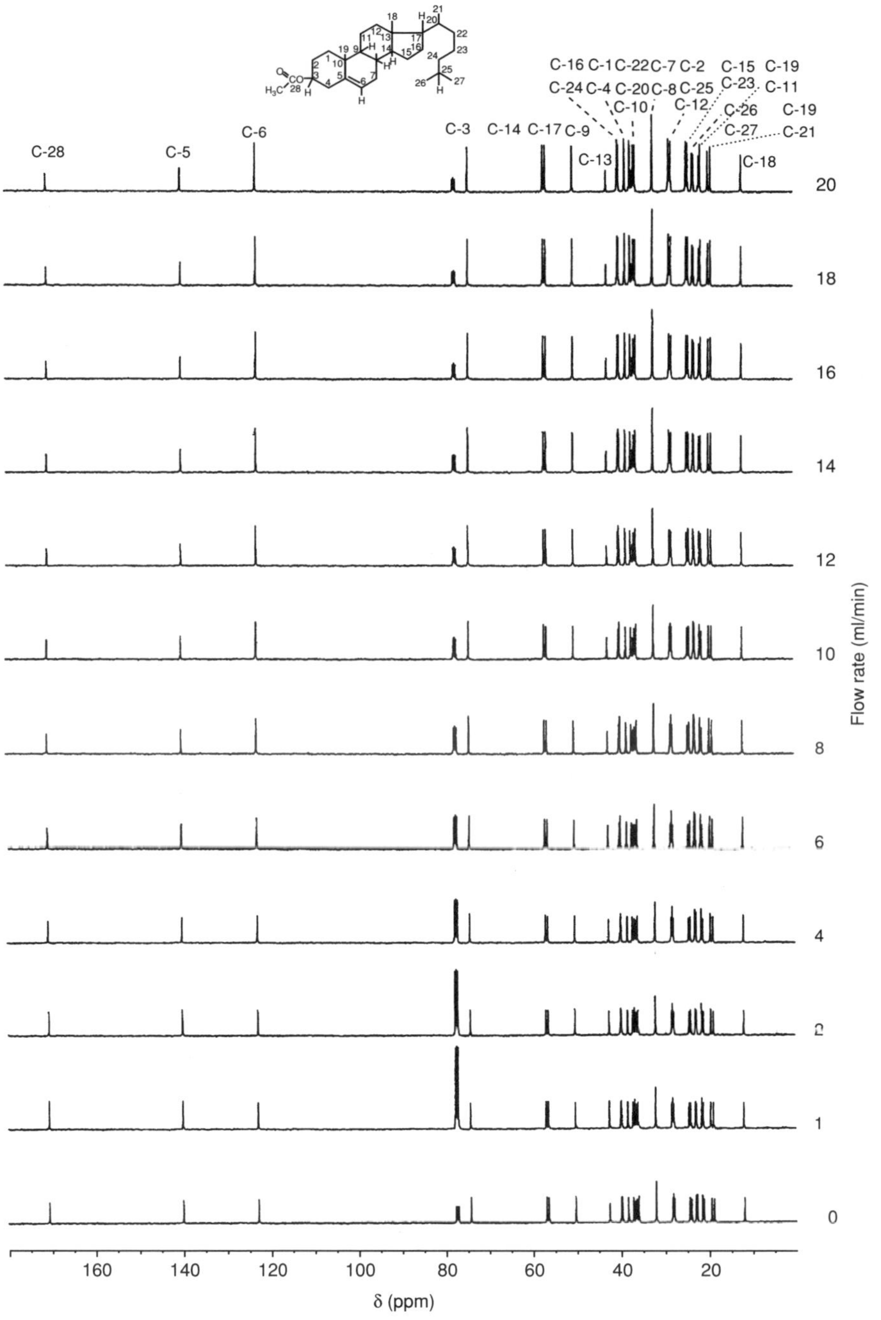

Figure 8.1.2 ^{13}C NMR (100 MHz) spectra of cholesterol acetate under static- and continuous-flow conditions

However, this technique is not feasible for recording continuous-flow ^{13}C NMR spectra of chromatographic peaks. For this purpose, a major increase in sensitivity has to be achieved. Two approaches are currently under development to achieve this aim. One is the use of immobilized free radicals to reduce the apparent T_1 relaxation times, while the other is to use the benefit of electron ^{13}C magnetization transfer by employing dynamic nuclear polarization. It is well-known in NMR spectroscopy that the addition of paramagnetic relaxation agents, such as chromium tris(acetylacetonate) $Cr(acac)_3$ leads to a significant reduction of T_1, but also causes major line broadening due to the reduction of T_2. The benefitial effect of the reduction of T_1 can now be separated from the T_2 drawback by employment of the continuous-flow procedure. Here, a pre-column, with immobilized paramagnetic relaxation agents covalently attached to silica, is placed before the continuous-flow detector cell [2]. Flowing nuclei entering the flow cell take the 'memory' of the paramagnetic-reduced T_1 relaxation times with them, whereas T_2 is not effected due to the absence of a direct contact between the paramagnetic species and the investigated molecule (Figure 8.1.3). Different paramagnetic species can be covalently attached to silica and show different relaxation activities [2]. An increase in sensitivity of the order of 70:1 can be obtained by the application of a gadolinium society (Gd-1, 4,7,

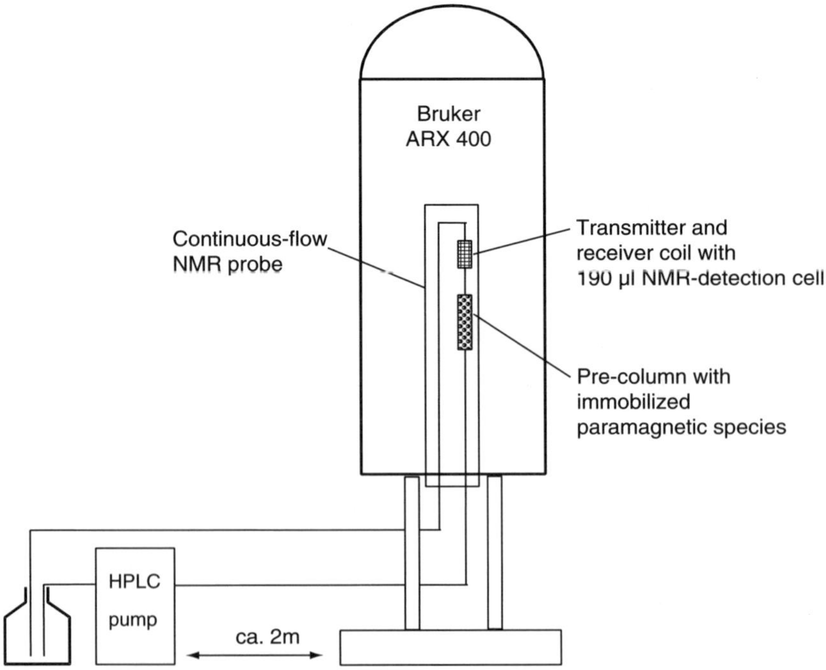

Figure 8.1.3 Schematic of the experimental arrangement used for continuous-flow ^{13}C NMR experiments employing a pre-column with immobilized paramagnetic species

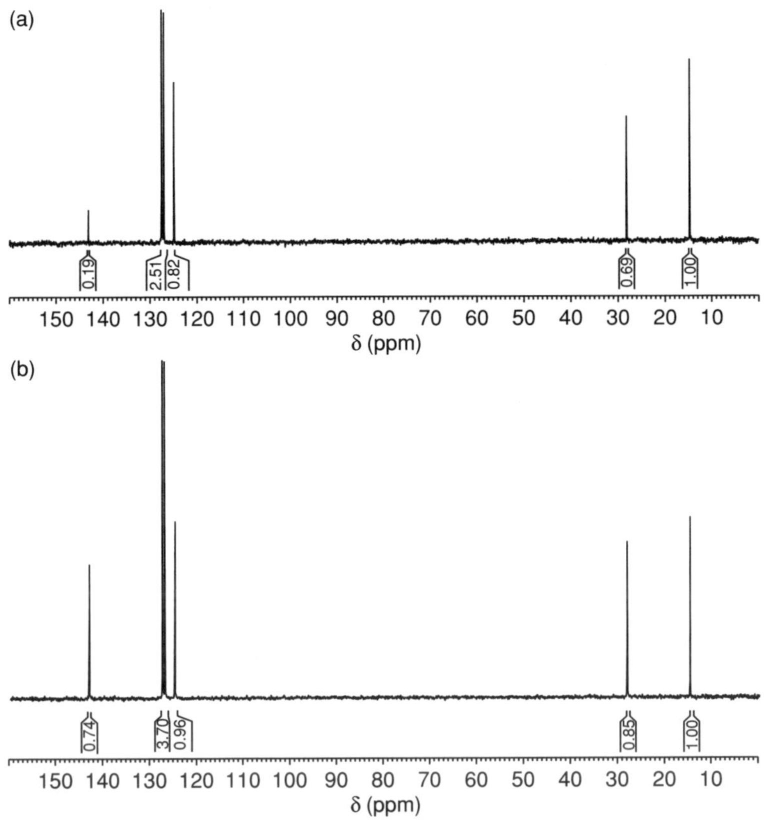

Figure 8.1.4 Continuous-flow ^{13}C NMR spectra of ethylbenzene (at a flow rate of 10 ml/min): (a) without immobilized paramagnetic species; (b) with immobilized paramagnetic species (Gd-DOTA) in a pre-column

10-tetraazacyclododecane N, N′, N″, N‴-tetraacetic acid (Gd-DOTA) as the paramagnetic relaxation agent. Figure 8.1.4 illustrates the gain in sensitivity by comparing two continuous-flow ^{13}C NMR spectra of ethylbenzene recorded with and without immobilized paramagnetic species. Ongoing research in our laboratory deals with the development of mixed-mode stationary phases by combining paramagnetic species together with *n*-alkyl moieties for reversed-phase separations.

A modification of the design of the flow cell can also lead to a further increase in sensitivity [7,8]. The addition of a proton RF coil for selective excitation of the protons before entering the ^{13}C flow cell enables the full build-up of the nuclear Overhauser effect in a continuous-flow system (Figure 8.1.5). The combination of this NMR flow cell design, together with the use of immobilized paramagnetic species, could lead to the registration of continuous-flow ^{13}C NMR spectra of separated species.

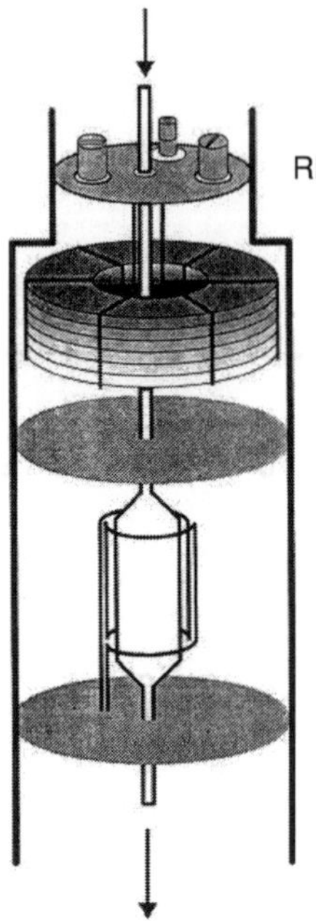

Figure 8.1.5 Schematic of a continuous-flow ^{13}C probe employing an additional proton toroidal coil

The second approach is the use of the dynamic nuclear polarization (DNP) detection principle. Dorn and co-workers have pioneered the application of this technique [9,10]. Whereas the NOE enhancement of ^{13}C nuclei in the conventional $^{13}C\{^1H\}$ recording is dependent upon the γ_H/γ_C ratio (NOE $= \gamma_H/2\gamma_C$ $= 2{:}\,1$), the DNP enhancement relates to the γ_e/γ_{13C} ratio (2640:1). In an electron–nucleus spinsystem, the electron–electron transitions are saturated by microwave irradiation and magnetization transfer from electron to nucleus (Overhauser effect) occurs via a scalar and/or dipolar mechanism. The DNP enhancement, A, is described by the following equation:

$$A = \rho f s(\gamma_e/\gamma_C) \qquad (8.1.2)$$

where ρ is the so-called coupling factor, showing its highest positive value of $+\,0.5$ for dipolar coupled electron–nuclear interactions and, on the other hand, its highest negative number of -1 for scalar coupled interactions; f is the leakage factor, a measure for the nucleus relaxation due to nuclear–electron interactions, s represents the saturation factor, a measure of the extent of saturation of the electronic transitions at the Larmor frequency ω_e, and γ_e and γ_C are the gyromagnetic ratios of the electron and the ^{13}C nucleus,

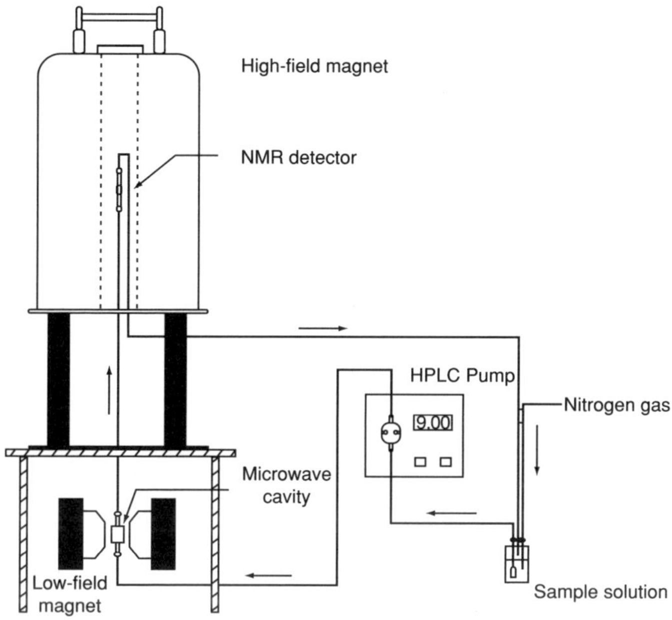

Figure 8.1.6 Schematic of the experimental arrangement used for performing DNP experiments

respectively. According to the above equation, the theoretical DNP enhancement factors are $A_{\text{dipolar}} = -1320$ and $A_{\text{scalar}} = +2640$, respectively.

Thus, it is desirable to establish an experimental arrangement where electron transmission can be performed in the presence of flowing liquids. This arrangement is outlined in Figure 8.1.6, which shows the set-up which Dorn and co-workers used to perform DNP experiments. In this way, polarized ^{13}C NMR signals, with either positive or negative phases, can be obtained. Figure 8.1.7 shows the ^{13}C NMR chromatogram of a model mixture of benzene and several halogenated compounds.

In this figure, the different signal intensities of chloroform, tetrachloroethylene, hexafluorobenzene, 1,1,1-trichlorotrifluoroethane and benzene can be clearly seen. The drawback of this technique is that not all organic molecules undergo DNP enhancement, although the principle works extremely well with halogenated compounds.

Overall, both types of experiment clearly show that the application of polarization transfer experiments has great potential for recording continuous-flow ^{13}C NMR spectra. The further refining and fine-tuning of these types of experiments could lead to practical related applications of on-line HPLC–^{13}C NMR experiments.

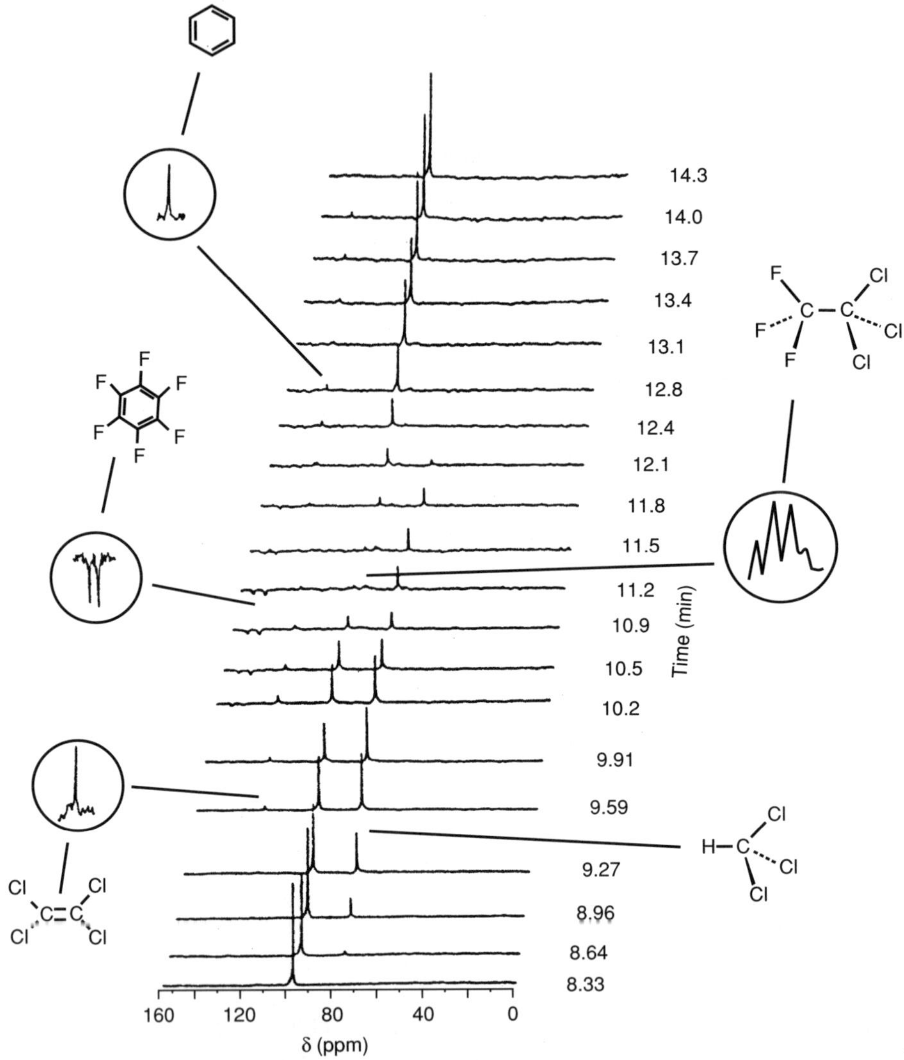

Figure 8.1.7 LC–^{13}C DNP profile of a mixture of halogenated compounds (chloroform, tetrachloroethylene, hexafluorobenzene and 1,1,1-trichlorotrifluoroethane), together with benzene, in CCl_4 recorded at a flow rate of 2.5 ml/min

REFERENCES

1. Albert, K., Nieder, M., Bayer, E. and Spraul, M., *J. Chromatogr.*, 1985, **346**, 17.
2. Fischer, H., Bertagnolli, H., Seiler, M., Eberhardinger, U., Schmitt-Willich, H. and Albert, K., *J. Phys. Chem.*, submitted for publication.

3. Albert, K., Dreher, E., Straub, H. and Rieker, A., *Magn. Reson. Chem.*, 1987, **25**, 919.
4. Albert, K., Kruppa, G., Zeller, K.-P., Bayer, E. and Hartmann, F., *Z. Naturforsch., C*, 1984, **89**, 859.
5. Hütter, P., Albert, K., Bayer, E., Zeller, K.-P. and Hartmann, F., *Biochem. Pharmacol.*, 1987, **36**, 2729.
6. Albert, K., Sudmeier, J. L., Anwer, M. and Bachovchin, W. W., *Magn. Reson. Med.*, 1989, **11**, 309.
7. Günther, U., Sudmeier, J., Albert, K. and Bachovchin, W., *J. Magn. Reson., A*, 1995, **117**, 73.
8. Sudmeier, J., Günther, U., Albert, K. and Bachovchin, W., *J. Magn. Reson., A*, 1996, **118**, 145.
9. Stevenson, S. and Dorn, H. C., *Anal. Chem.*, 1994, **66**, 2993.
10. Stevenson, S., Glass, T. and Dorn, H. C., *Anal. Chem.*, 1998, **70**, 2623.

8.2 Parallel NMR Detection

ANDREW G. WEBB†, JONATHAN V. SWEEDLER‡ and DANIEL RAFTERY¶

[†Department of Electrical and Computer Engineering, University of Illinois at Urbana-Champaign, IL, USA; ‡School of Chemical Sciences, University of Urbana-Champaign, IL, USA; ¶Department of Chemistry, Purdue University, IN, USA]

8.2.1 INTRODUCTION

The vast majority of nuclear magnetic resonance (NMR) experiments are performed by using a probehead in which a single sample is contained inside a single radiofrequency (RF) coil. Data acquisition can be described as a single-input, single-output system. For example, in order to obtain structural information on a complex sample, a series of homo- and heteronuclear multi-dimensional experiments are run in a purely serial fashion. The total data acquisition time for a protein sample may be several weeks. In studies of more simple molecules, a single one- or two-dimensional experiment may be all that is required, in which case a number of samples can be run rapidly one after another, but again in serial mode. An important example of this latter mode of operation is the analysis of combinatorial chemistry products where large numbers of compounds are produced in parallel syntheses [1]. Existing approaches to high-throughput NMR analysis use either robotic sample changers or combine an autosampler, which injects samples directly into the probe, with a flow-through probe design. Other analytical chemistry techniques, such as mass spectrometry, have introduced a considerable degree of parallelism into data acquisition in order to increase the throughput [2]. Very often, separations are coupled with these detection methods, such as in liquid chromatography/mass spectrometry (LC–MS) or liquid chromatography/tandem mass spectrometry (LC–MS–MS). These configurations have increasingly been integrated with samples preparation in standard 96-well titre plates. Massive parallelism has also been introduced into microseparation techniques such as capillary electrophoresis (CE) [3].

Introducing parallelism into NMR data acquisition requires simultaneous acquisition of signals from more than one sample. The simplest way to achieve this is to use multiple samples within a single radiofrequency (RF) coil, and to design either pulse sequences or post-processing routines to separate the signals

On-Line LC–NMR and Related Techniques
Edited by K. Albert.

from each sample. However, this approach suffers from a less than optimal sensitivity due to the low effective filling factor for each small sample with respect to the large coil. A more sensitive approach uses one RF coil for each sample. The fundamental problem with NMR is that each coil must be placed in the homogeneous region of the magnet in order to record a high-resolution spectrum. In addition, inhomogeneities in the magnetic field due to the magnetic susceptibility effects from additional coils must be compensated. With conventionally sized coils it is not possible to place more than one in a highly homogeneous region, and so substantial parallelism requires the use of RF 'microcoils' [4]. The detection efficiency can then be increased further, as described elsewhere in this book, by hyphenating microcoil NMR detection with microscale separation techniques such as CE and capillary electrophoretic chromatography (CEC) [5] or capillary high performance liquid chromatography (CHPLC) [6].

8.2.2 MULTIPLE SAMPLES WITHIN A SINGLE COIL

The simplest approach of using a single coil and multiple samples was first demonstrated by Banas in 1969 [7]. Although high-resolution spectra were obtained from seven samples simultaneously, there was no way of separating the individual spectra from the composite spectrum. An improvement upon this approach, in which spectra could be separated, was demonstrated recently [8]. Nineteen samples were placed inside a coil and the spectra separated by using a two-dimensional chemical-shift imaging (CSI) technique [9], as shown in Figure 8.2.1.

Two-dimensional Fourier transformation of the signal with respect to the x- and y-gradient values gives the spatial distribution of individual spectra. However, only a finite number of phase-encoding steps are acquired, and the point

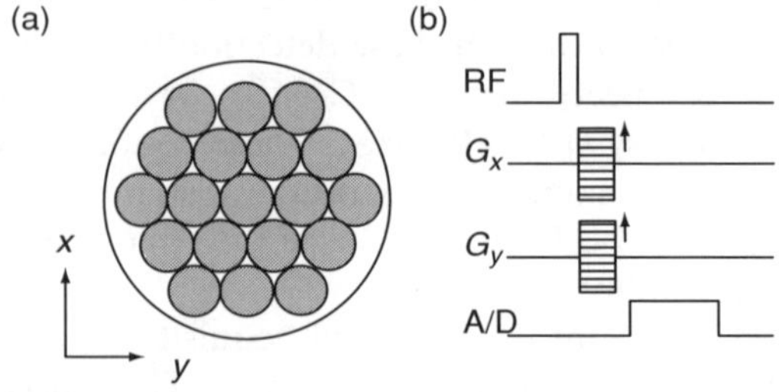

Figure 8.2.1 (a) Schematic of nineteen capillaries arranged within a single coil. The coil has a Helmholz geometry with the long axis oriented along the B_o/z-direction. (b) A CSI sequence was used to separate the spectra from the individual samples. After application of the RF pulse, magnetic field gradients are applied in the x- and y-directions, and the signal detected in the analog-to-digital (A/D) converter after the gradients are switched off. The x- and y-gradients are incremented sequentially

spread function (PSF) of the resulting spatial distribution of spectra is given by the Fourier transform of a truncated Fourier series. This is a sinc function, with the width of the main lobe, as measured by the distance between its two zero-crossings, inversely proportional to the number of gradient increments. The width of the PSF effectively determines the amount of spectral bleedthrough from sample to sample. The larger the number of gradient increments, then the narrower the PSF and the more accurate the spectral localization: however, clearly the data acquisition time is increased. Notice that the gradient increment Δk is effectively fixed, since the field-of-view (FOV) is given by $1/\Delta k$, and this parameter should be set to just encompass all of the samples. In order to obtain clean spectra from M samples, it is necessary to collect many more than M spectra, and thus CSI is a fundamentally slow technique of data acquisition. It should be noted, however, that the number of spectra can be reduced considerably by incorporating prior information about the spatial locations of the samples [10,11]. An additional problem is that the signal-to-noise ratio (S/N) for each sample is intrinsically low due to the small effective filling factor. For this example of nineteen samples arranged in a hexagonally close-packed geometry, the diameter of the coil is approximately five times larger than would be required if one coil were used for each sample. Since the sensitivity of the coil is inversely proportional to its diameter, this means that the S/N for each sample (assuming that the coil is the dominant noise source) is approximately one-fifth of the maximum value.

8.2.3 MULTIPLE COILS CONNECTED IN PARALLEL

The S/N of individual spectra is optimized when each sample is contained within a size-matched RF coil. As mentioned in the introduction to this chapter, this implies that the coils must be small in order for a number of them to fit inside the homogeneous region of the magnet. The first implementation using multiple coils was by MacNamara and co-workers [12], who connected four coils in parallel to make the circuit shown in Figure 8.2.2.

The NMR coils were solenoids constructed from four turns of polyurethane-coated 36-gauge copper wire wrapped around fused silica capillaries (20 mm long, 1.6 mm o.d., 0.8 mm i.d.) The coil length was 0.7 mm, thus giving an observe volume of 1.4 μl. The intercoil spacing was 3.2 mm, centre-to-centre. The entire coil array was surrounded by FC-43, an inert perfluorocarbon with a very similar volume magnetic susceptibility to that of copper. The circuit had a tuning range of ∼2 MHz and a quality (Q) value of 60. The assembly was mounted in a narrow-bore (39 mm diameter) probe body and Teflon tubes were attached to the capillaries to allow flow introduction of samples using a syringe. A single transmitter and receiver were used. Since the detected signal is a combination of all of the samples, some method must be devised to separate the individual signals. The first technique adopted involved applying a small

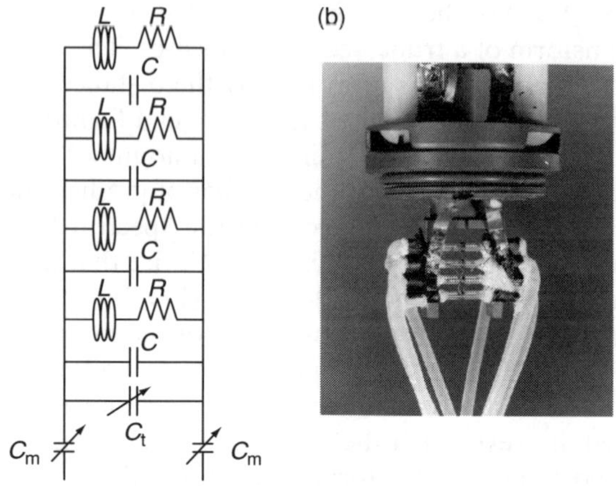

Figure 8.2.2 (a) Schematic of the four-coil probehead introduced in Reference [12]. The four individual solenoidal coils are represented by an inductance (L), series resistance (R), and inter-turn capacitance (C). The entire circuit was impedance-matched to 50 Ω at a frequency of 300 MHz by using the variable capacitors C_t and C_m. (b) Photograph of the four-coil assembly. Reprinted from MacNamara, E., Hou, T., Fisher, G., Williams, S. and Raftery, D., 'Multiplex Sample NMR: an approach to high-throughput NMR using a parallel coil probe', *Anal. Chem. Acta*, **397**, 9–16, copyright (1999), with permission of Elsevier Science

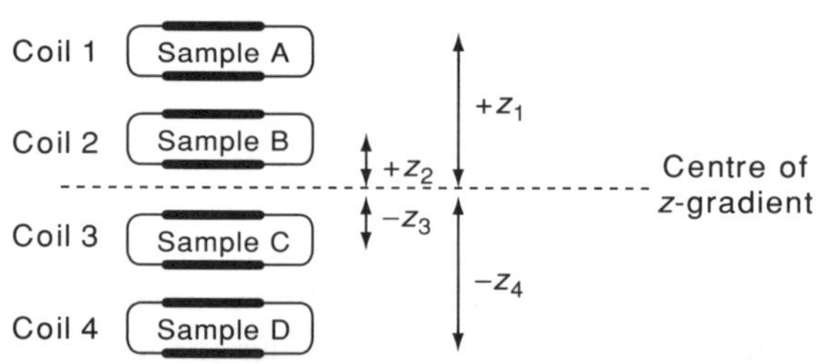

Figure 8.2.3 Schematic of the arrangement of the four RF coils within the z-gradient. The distances between the centres of the coils and the gradient are denoted by z_{1-4}

gradient in the z-direction during data acquisition, using the arrangement shown in Figure 8.2.3.

The resonance frequencies of the peaks in the four samples will shift by different amounts when the gradient is applied. Suppose that sample A is placed in coil 1, and that it has a single NMR resonance. The resonant frequency, ω_A, is then given by the following:

$$\omega_A = \gamma B_0(1 - \sigma_A) \tag{8.2.1}$$

where σ_A is the electronic shielding constant of compound A. If a gradient of strength G_1 is applied in the z-direction, and coil 1 is a distance of $+z_1$ cm from the centre of the gradient coil (see Figure 8.2.3), then the resonant frequency, ω_{A1}, is given by the following:

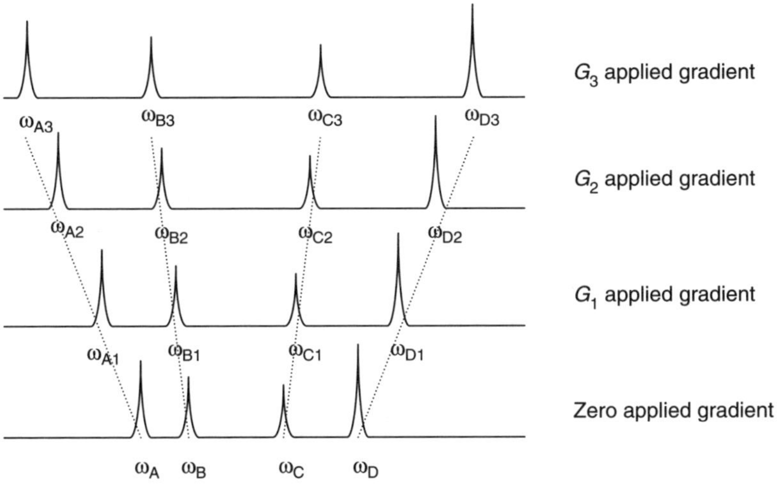

Figure 8.2.4 Effect on the resonant frequencies of samples A to D of applying a *z*-gradient during data acquisition. The gradient strengths are in the order $G_3 > G_2 > G_1 > 0$. The line-broadening effects on individual resonances are not shown for clarity

$$\omega_{A1} = \gamma B_0(1 - \sigma_A) + \gamma G_1 z_1 \tag{8.2.2}$$

The process is shown schematically in Figure 8.2.4 for four different values (including zero) of the applied gradient, and four compounds A–D contained in coils 1–4, respectively.

The spectra can be processed in different ways in order to produce signals from individual samples. In the first method, results from a 'peak-picking' routine in each acquired spectrum are correlated with a calculation of the shifts in resonant frequency expected for each sample from the values of the applied gradients and the positions of the coils with respect to the centre of the *z*-gradient. A second, more robust method involving spectral 'masking' and subtraction, is shown in Figure 8.2.5.

Figure 8.2.6 shows the results of this subtraction data processing method using four samples, i.e. H_2O, methanol, acetonitrile, and *t*-butanol, each in D_2O. The linewidths obtained after shimming were 3.1, 2.8, 3.6 and 3.4 Hz, respectively. Two values of the *z*-gradient were used, i.e. 0 and 48 mG/cm. The processed spectra showed very clean spectral subtraction, with signal 'bleedthrough' between coils below the noise level. As noted by the authors [12], one disadvantage of the subtraction technique is that it compromises quantitative information. The resonances are also broadened by the application of the *z*-gradient during data acquisition, meaning that the S/N is reduced.

One method of reducing the effects of the line-broadening due to the *z*-gradient present during data acquisition is to use 'reference deconvolution'

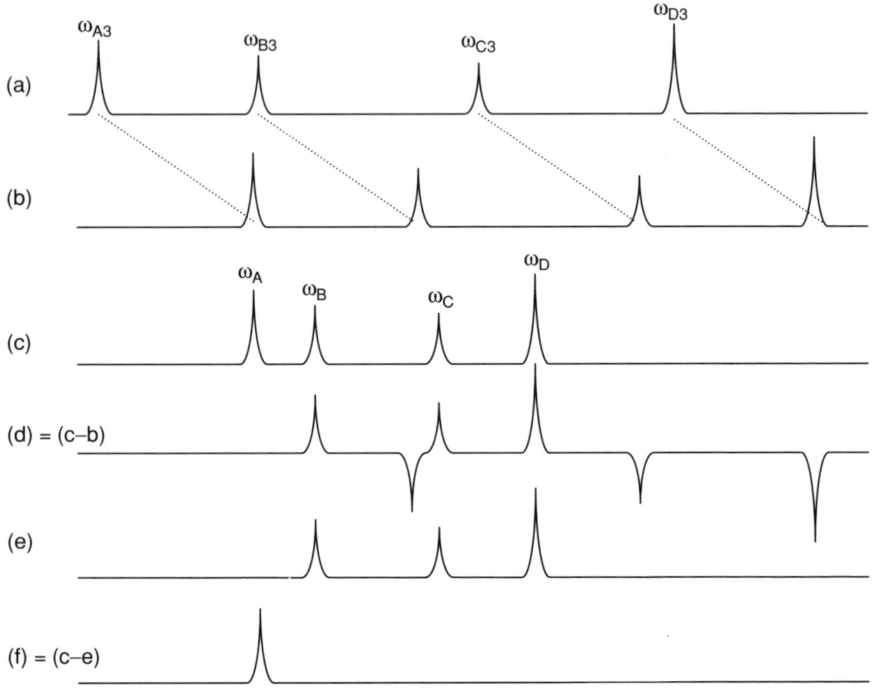

Figure 8.2.5 (a) The spectrum obtained from application of gradient G_3 (see Figure 8.2.4). (b) The spectrum from (a) is shifted in frequency by an amount equal to the calculated shift of sample A in coil 1 at position z_1 with respect to the centre of the gradient coil. (c) The spectrum from zero applied gradient. (d) Result of subtraction of the spectrum in (a) from that in (c). (e) Peaks from spectrum (d) are thresholded to remove all negative peaks. (f) Result of subtraction of the spectrum in (e) from that in (c). This corresponds to the spectrum from sample A in coil 1 only. The process can be repeated to obtain spectra from sample B in coil 2, etc. by varying the frequency shift in step (b)

[13]. The process of reference deconvolution [14] can be described mathematically as follows:

$$S_{\text{comp}}(f) = FT\left[\frac{S_{\text{exp}}(t)S_{\text{ideal}}(t)}{S_{\text{ref}}(t)}\right] \tag{8.2.3}$$

where $S_{\text{comp}}(f)$ is the spectrum after deconvolution, $s_{\text{exp}}(t)$ is the acquired time-domain signal, $s_{\text{ideal}}(\text{t})$ is the time-domain signal corresponding to the chosen, deconvolved Lorenzian linewidth, and $s_{\text{ref}}(t)$ is the time-domain experimental signal from only the particular resonance used as the reference for deconvolution. Hou *et al.* [13] showed that reference deconvolution could be incorporated easily into the subtraction procedure, by deconvolving both acquired spectra (with and without gradients) to the same linewidth.

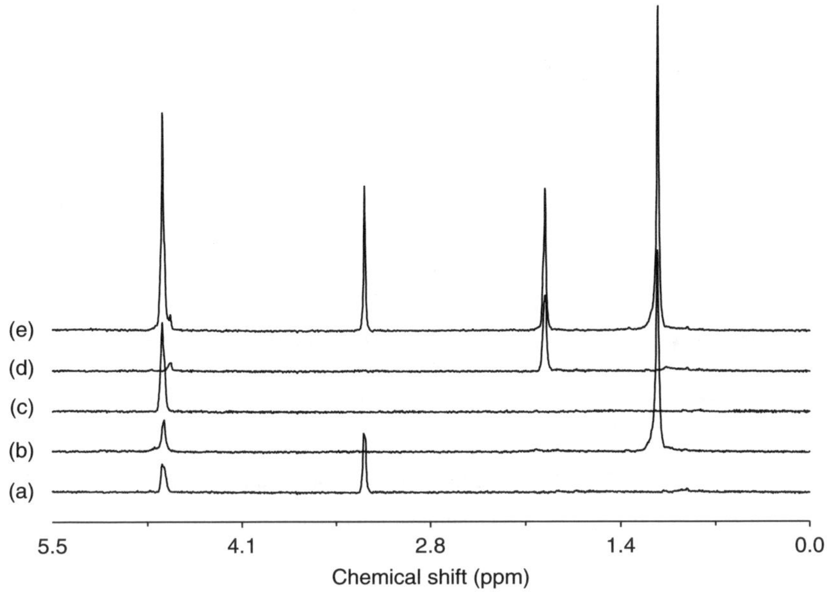

Figure 8.2.6 ^1H spectra of four different samples (500 mM in D_2O) obtained by using the subtraction method. These represent the post-processed spectra of (a) methanol (4.7 and 3.2 ppm), (b) *t*-butanol (4.7 and 1.1 ppm), (c) water (4.7 ppm) and (d) acetonitrile (1.9 ppm), and (e) the spectrum of the four different samples before data processing. Reprinted from MacNamara, E., Hou, T., Fisher, G., Williams, S. and Raftery, D., 'Multiplex Sample NMR: an approach to high-throughput NMR using a parallel coil probe', *Anal. Chim. Acta*, **397**, 9–16, copyright (1999), with permission of Elsevier Science

Figure 8.2.7 shows the results obtained for methanol using the same probe as described previously.

Two-dimensional spectroscopy can also be carried out by using this 'Multiplexed NMR' approach. For example, correlated spectroscopy (COSY) experiments have been performed with a *z*-gradient applied during the data acquisition time. Peak picking or subtraction methods can be used for post-processing the data. For the latter process, the authors suggested zeroing all the peaks within a certain bandwidth from the diagonal, with this bandwidth set to be less than the smallest *J*-coupling observed in the spectrum. Using four samples of 0.5 M ethanol, 1-propanol, dichloroacetic acid and acetaldehyde in D_2O, it was shown that the appropriate individual sub-spectra could be generated. The sub-spectra of 1-propanol and ethanol are shown in Figure 8.2.8.

These methods work well for spectra in which there is no significant spectral overlap between compounds, or for situations where the samples to be analysed are chemically quite similar. For example, parallel process monitoring could be accomplished by using such an approach. In cases where spectra contain many resonances that overlap, it is much more difficult to assign resonances based on frequency shifts. Two other approaches have been used in this case [15,16]. One

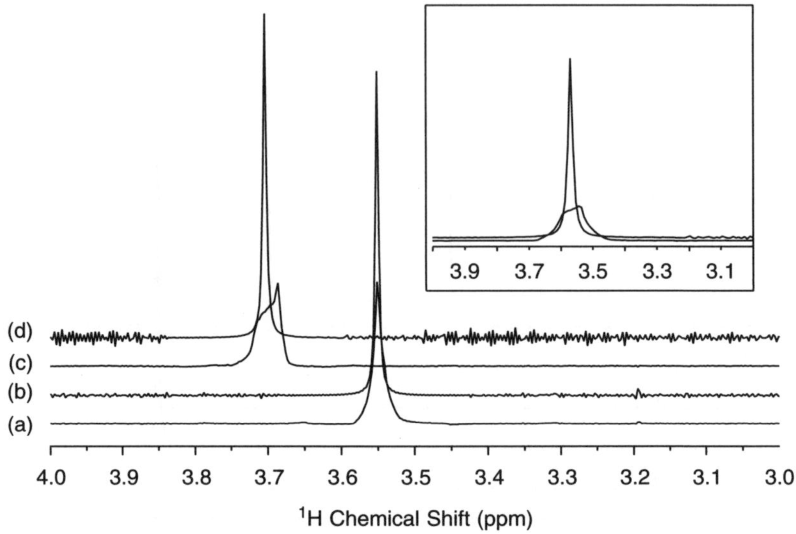

Figure 8.2.7 Reference deconvolution applied to the methanol peak: (a) original peak; (b) methanol peak deconvolved to a 2 Hz Lorenzian lineshape; (c) gradient-shifted methanol peak; (d) gradient-shifted methanol peak deconvolved to the same 2 Hz Lorenzian lineshape; (e) comparison of the sub-spectrum of the methanol sample using the subtraction algorithm with (sharp peak) and without (flattened peak) reference deconvolution. Reprinted from Hou, T., MacNamara, E. and Raftery, D., 'NMR analysis of multiple samples using parallel coils: improved performance using reference deconvolution and multi-dimensional methods', *Anal. Chem. Acta*, **400**, 297–305, copyright (1999), with permission of Elsevier Science

is to employ frequency-selective excitation, and the other is the CSI method already outlined above in Section 8.2.2. As shown in Figure 8.2.9, one-dimensional CSI acquisition of four samples results in a frequency versus spatial correlation which separates the frequency components based on their spatial position in the probe. Taking slices along the frequency dimension allows one to observe individual spectra from each sample coil. In this experiment, 12 gradient increments were required, along with 8 acquisitions per increment to cancel artifacts due to the poor RF homogeneity of the short solenoid coils used in the experiment. Such long acquisition times need to be reduced before high-throughput can be achieved when using CSI methods. Nevertheless, for samples at low concentration and with fast T_1 relaxation times, the CSI method is appropriate.

A second approach involving selective excitation, as used for slice selection in magnetic resonance imaging, is described by the pulse sequence shown in Figure 8.2.10. Notice that the gradient is switched off during data acquisition so that no line-broadening occurs. Application of the gradient causes the individual spectra from the samples in coils 1–4 to be centred at different frequencies, as described by Equation (8.2.2). The applied gradient should be large enough such that the bandwidth of each spectrum (typically 10 ppm for proton

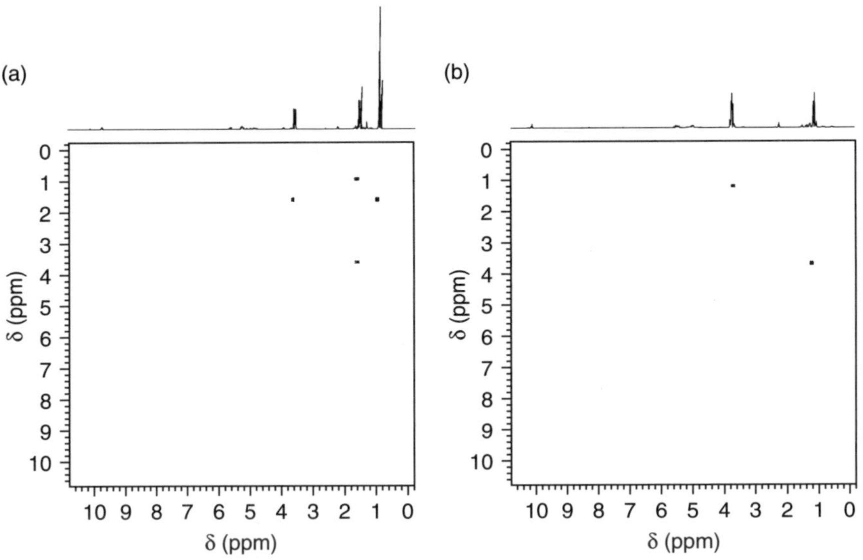

Figure 8.2.8 Two-dimensional sub-spectra generated from the original and gradient-shifted 2D-COSY spectra of four coils using the subtraction method and nulling of the diagonal peaks: (a) 1-propanol; (b) ethanol. Reprinted from Hou, T., MacNamara, E. and Raftery, D., 'NMR analysis of multiple samples using parallel coils: improved performance using reference deconvolution and multidimensional methods', *Anal. Chem. Acta*, **400**, 297–305, copyright (1999), with permission of Elsevier Science

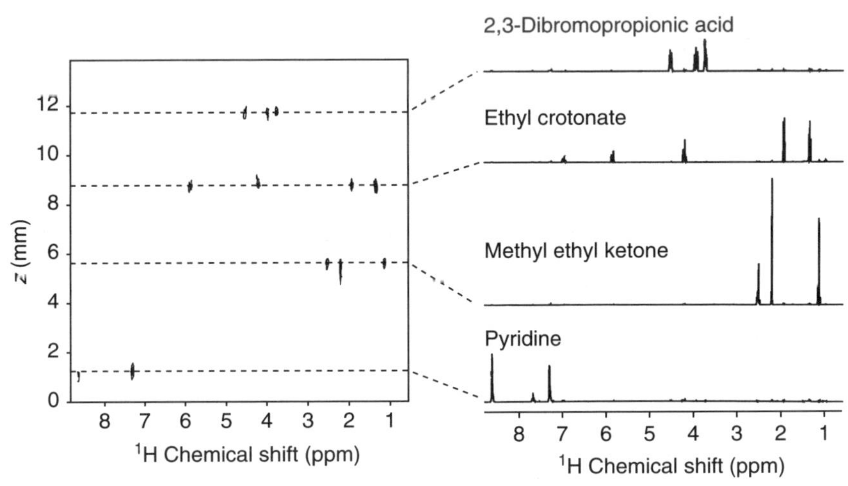

Figure 8.2.9 CSI experiment using four samples stacked vertically. The resulting contour plot is a spatial/chemical-shift map, which can be analysed by taking slices along the frequency axis to yield individual spectra for each sample position. Reprinted with permission from Hou, T., Smith, J., MacNamara, E., Macnaughtan, M. and Raftery, D., *Anal. Chem.*, **73**, 2541–2546 (2001). Copyright (2001) American Chemical Society

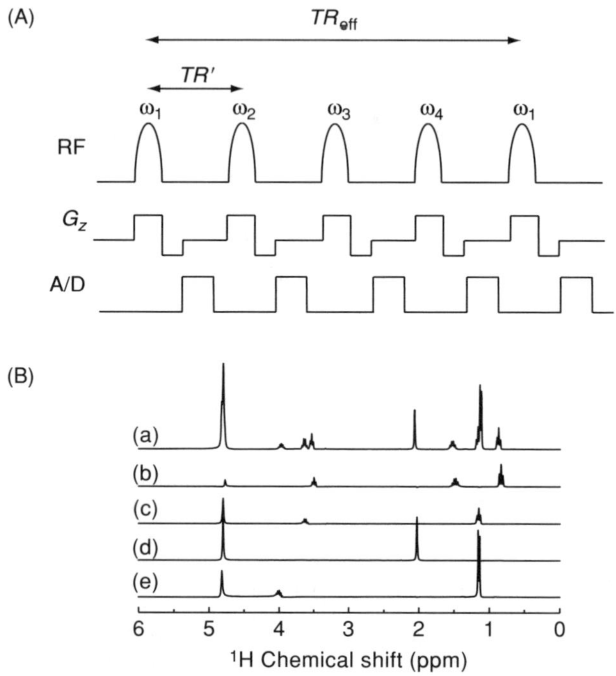

Figure 8.2.10 (A) Pulse sequence used for selective excitation of each of the four samples in turn. The RF pulses are frequency-selective and applied at different resonant offsets via phase modulation in the time domain. (B) Normal ^1H spectrum of (a) the four samples, and the resulting sub-spectra of 0.5 M (b) 1-propanol, (c) 2-propanol, (d) acetic acid and (e) ethanol in D_2O. Since the four spectra are acquired within the relaxation time TR_{eff}, which is typically set to be $3T_1$, there is an increase in efficiency by a factor of four. Reprinted with permission from Hou, T., Smith, J., MacNamara, E., Macnaughton, M. and Raftery, D., *Anal. Chem.*, **73**, 2541–2546 (2001). Copyright (2001) American Chemical Society

spectroscopy) is smaller than the value of the frequency separation between the samples. The RF pulse is frequency-selective, with a bandwidth greater or equal to 10 ppm, and centred at frequency ω_1 for the first pulse, ω_2 for the second, and so on.

Selective excitation is an efficient method of data collection for samples in which the value of the relaxation time (TR_{eff}) is less than three times the T_1 value of the sample. The smallest possible value of TR_{eff} is given by n times the data acquisition time for each sample, where n is the number of coils. In cases where very high spectral resolution is required, a large number of coils are used, and/or where the samples have short T_1 values, this method may be of limited use. The use of multiple receivers interfaced with a number of arrays, each containing a smaller number of coils, can potentially overcome this limitation. Alternatively, a one-dimensional CSI method can be used, with the proviso of losses in S/N or potentially longer acquisition times, as mentioned earlier.

8.2.4 MULTIPLE ELECTRICALLY DECOUPLED COILS

An alternative approach to parallel detection uses a separate RF coil and impedance-matching circuitry for each sample. The coil and the circuitry for each sample must be electrically decoupled from all of the other coils and circuits. The advantages of this approach include the potential for running different pulse sequences for each sample, the simple differentiation of the signals from individual samples which can be routed into separate receivers or time-multiplexed into a single receiver using an RF switch, and maintaining maximum NMR sensitivity. The major disadvantage is the increased complexity of the probe geometry and circuitry and the need for shielding to establish high electrical isolation between coils. These issues are exacerbated the more coils are incorporated.

The first NMR application was in solid-state spectroscopy, carried out by Oldfield [17]. For solids, the lineshapes are intrinsically large, and only ~ 1 ppm magnetic field homogeneity is needed. In Oldfield's design, two side-by-side 5 mm solenoid coils occupied the upper region of the magnet, while a third solenoid (which could be spinning) occupied the lower part of the magnet, with about 3 cm vertical separation between the upper and lower coils. For isolation, these latter two probes were separated by a Faraday shield. Operating at 11.7 T, the lower coil was tuned to ^{17}O at 67.77 MHz, and the top two to ^{2}H (76.74 MHz) and ^{27}Al (130.27 MHz), respectively (Figure 8.2.11(a)). Linewidths of ~1 ppm were obtained for all three coils simultaneously by using samples of $^{2}H_2O$, $H_2{}^{17}O$ and an aqueous solution of $^{27}Al(NO_3)_3$. The results obtained are shown in Figure 8.2.11(b).

The first liquid-state application, reported by Fisher *et al.* [18], used two coils, both tuned to the carbon frequency of 75.44 MHz at 7.4 T. The coils were

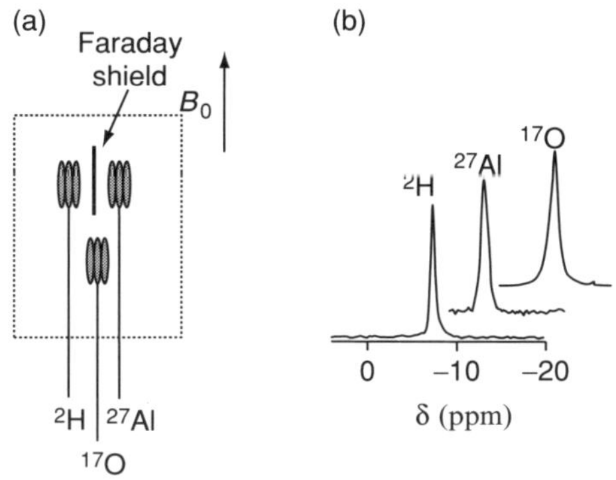

Figure 8.2.11 (a) Schematic of the three-coil solid-state probehead. (b) Spectra obtained from this probe: $^{2}H_2O$, $H_2{}^{17}O$ and $[^{27}Al(H_2O)_6]^{3+}$ resonances showing ~ 1 ppm linewidths. Data were obtained during the same time-period. Reproduced by permission of Academic Press from Oldfield, E., *J. Magn. Reson., A*, 1994, **107**, 255

four-turn solenoids, with diameter and length both of 4 mm, arranged vertically and separated by a horizontal piece of copper-plated circuit board grounded to the probe body. An aluminium plate separated the variable capacitors used for impedance-matching the two circuits. A commercial spectrometer was used as the transmitter, and the RF excitation pulse was channelled through a power splitter to separate duplexer circuits (consisting of crossed-diodes and quarter-wavelength cables). In addition to the receive channel of the commercial spectrometer, a second receive channel was constructed by the authors (Figure 8.2.12(a)). Samples of ^{13}C-labelled methanol and carbon tetrachloride were each placed in one of the coils. The NMR spectra, shown in Figure 8.2.12(b), showed no cross-talk between the two coils.

The concept of using decoupled coils has been developed further by Li *et al.* [19]. In this case, an RF switch was used for time-domain multiplexing of the signals into a single receiver channel. A four-coil system was constructed for operation at 6 T in a wide-bore (89 mm) magnet. Each coil was fabricated by using 17 turns of 50 μm diameter copper wire wrapped around a 180 μm i.d., 355 μm o.d., polyimide-coated fused silica capillary, giving an observe volume

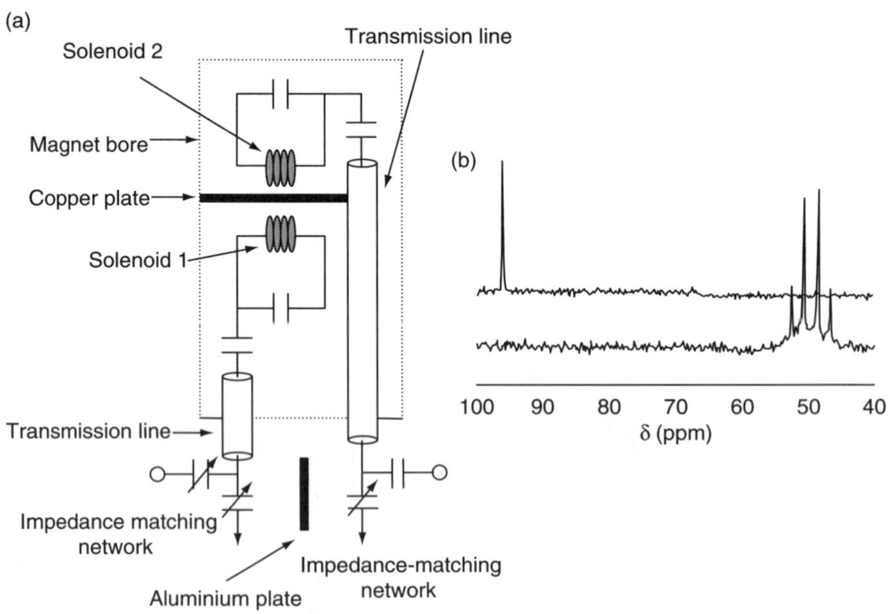

Figure 8.2.12 (a) RF circuitry and layout of a dual-channel ^{13}C probe. (b) Proton-coupled ^{13}C NMR spectra acquired simultaneously by using this probe. Both spectra were acquired by using single 90° pulses and 50 kHz bandwidths. The top trace is the spectrum obtained from 4 μl of ^{13}C-enriched methanol using a commercial receiver, while the bottom trace is obtained from 4 μl of ^{13}C-labelled carbon tetrachloride from a home-built receiver. Reproduced by permission of Academic Press from Oldfield, E., *J. Magn. Reson., A*, 1994, **107**, 255

(V_{obs}) of 28 nl. Teflon flow-tubes were attached to both ends of the capillary for sample loading. The coils were mounted on printed circuit boards and impedence-matching capacitors were added in a balanced configuration. The microcoils were mounted one above the other with a vertical spacing of 5 mm between adjacent coils. Alternate coils were rotated 90 degrees with respect to each other in order to reduce the coupling. The matching networks were also placed at 90 degrees to each other, again to reduce coupling. The whole system was surrounded by a container filled with FC-43. Figure 8.2.13 shows a schematic of the four-coil assembly.

Although the original implementation required four preamplifiers in the receiver chain, this was due to a relatively high-loss (2.3 dB) switch being used. In subsequent realizations, a much lower loss (0.1 dB) switch has been substituted, and this means that the receiver chain can be simplified considerably to give that shown in Figure 8.2.14. In experiments carried out using a Varian Inova console, five spare lines (SP1–SP5) can be set to high (+5 V) or low (0 V) from within the pulse program. The TTL signal is fed into an integrated circuit consisting of open-collector Darlington transistors. Five of these open-collectors are connected to the five inputs of the switch.

For experiments at 250 MHz, the four coils were initially filled with a 10 % H_2O/90 % D_2O mixture for shimming. The signal used for shimming consisted of a concatenation of the four individual free induction decays (FIDs). A simplex routine was used for optimization of the shim currents, with the time required for shimming being typically less than 30 min. The linewidths for the four coils were very similar, lying between 2 and 4 Hz. As shown in Figure 8.2.15, one-dimensional spectra were obtained from four solutions of 250 mM fructose, chloroquine, galactose and adenosine triphosphate, all in D_2O; 7 nmol of each sample was in the observe volume of the coil. The spectral width and transmitter frequency were optimized for each sample (using an Applescript

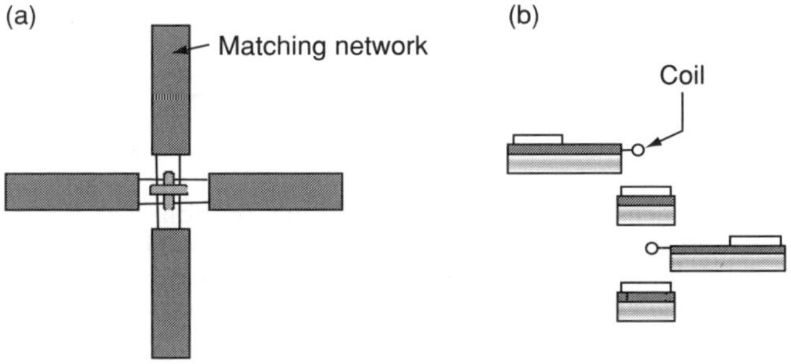

Figure 8.2.13 Schematic of the probehead using four decoupled coils, as reported in Reference [19]: (a) top–down view; (b) side view. Each coil and corresponding matching network is oriented at 90° with respect to its nearest neighbour in order to minimize coupling. Reprinted with permission From Li, Y., Wolters, A., Malaway, P., Sweedler, J. V. and Webb, A. G., *Anal. Chem.*, **71**, 4815–4820 (1999). Copyright (1999) American Chemical Society

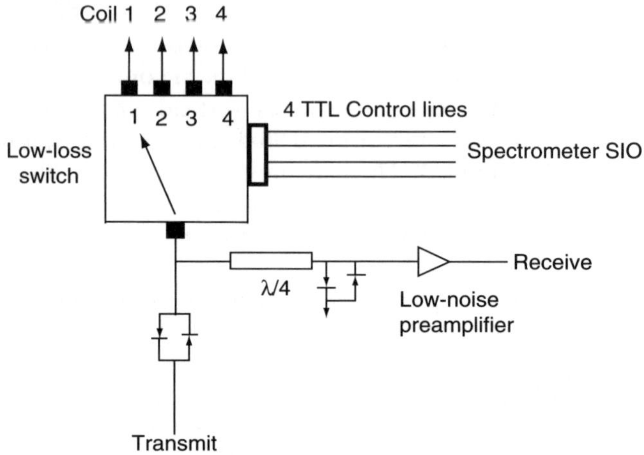

Figure 8.2.14 Schematic of the hardware set-up for acquisition of multiple signals with a single receiver channel. The TTL lines from the spectrometer are controlled from within the pulse programme and determine the position of the switch

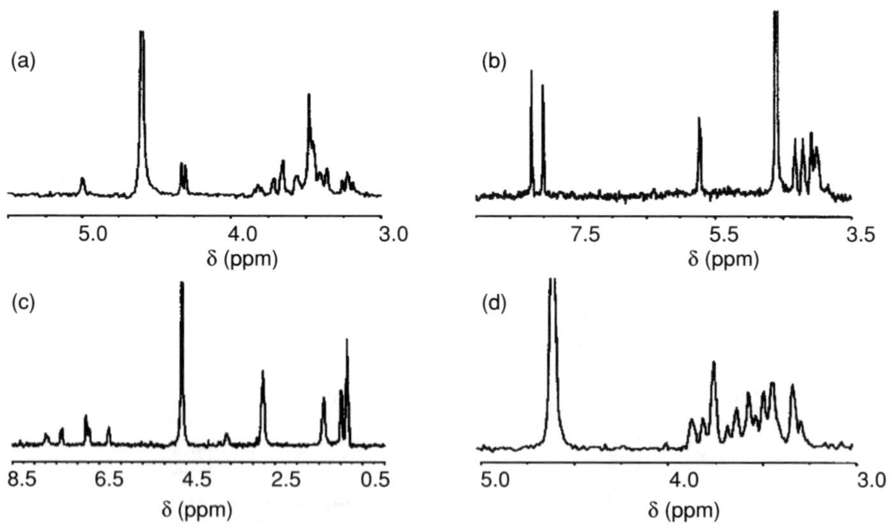

Figure 8.2.15 One-dimensional spectra acquired with the four-coil probe. Each sample (250 mM in D_2O) was loaded into the coil via the attached Teflon tubes; 32 scans were acquired for each spectrum, with no delay between excitations of successive coils. Concurrent with the switch position being incremented, the spectral width was optimized for each compound; 1 Hz line-broadening was applied before Fourier transformation and baseline correction. The spectral widths were: (a) 600 Hz (galactose); (b) 1400 Hz (adenosine triphosphate); (c) 2000 Hz (chloroquine); (d) 500 Hz (fructose). 2048 complex data points were acquired for each spectrum, giving data acquisition times of approximately 1.7, 0.7, 0.5 and 2.0 s, respectively. The delay between successive 90 degree excitations was 4.9 s for each sample. Reprinted with permission From Li, Y., Walters, A., Malaway, P., Sweedler, J. V. and Webb, A. G., *Anal. Chem.*, **71**, 4815–4820 (1999). Copyright (1999) American Chemical Society

macro for the Tecmag Libra system), with spectral widths of 600, 1400, 2000 and 500 Hz for galactose, adenosine triphosphate, chloroquine and fructose, respectively.

Standard correlation spectroscopy (COSY) experiments were run on the same four samples and the results are displayed in Figure 8.2.16. Since the acquisition time was approximately an order of magnitude less than the recycle delay, a full factor of four for improvement in throughput was achieved; in fact, the number of coils could be increased for yet further improvements in temporal efficiency. No signal bleedthrough was observed from one spectrum to another. Similar results were reported in this paper with a two-coil probe used at 500 MHz [19].

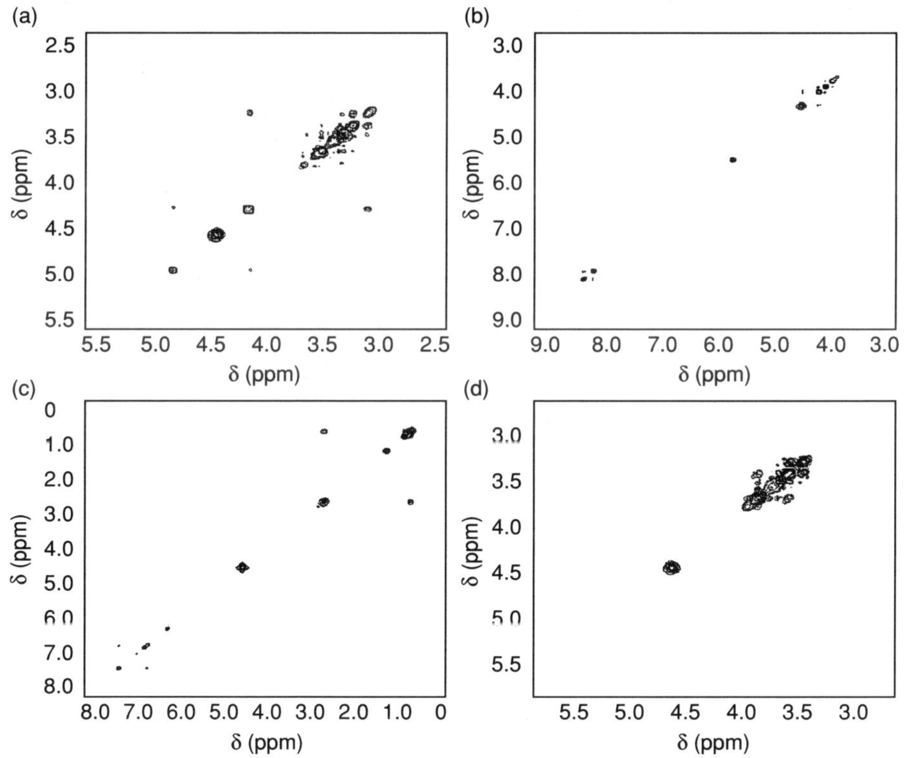

Figure 8.2.16 COSY spectra acquired with the four-coil probe, where the compounds and concentrations were the same as those of the one-dimensional spectra. Data acquisition parameters: spectral width, 2000 Hz; data matrix, 512 × 128 (complex); 16 signal averages; delay between successive coil excitations, 400 ms; effective recycle delay for each sample, ca. 1.7 s. Data were processed by using shifted sine-bell multiplication in both dimensions and displayed in magnitude mode. Reprinted with permission from Li, Y., Walters, A., Malaway, P., Sweedlar, J. V. and Webb, A. G., *Anal. Chem.*, **71**, 4815–4820 (1999). Copyright (1999) American Chemical Society

Recent developments have included the design of a two-coil probehead, with each coil having a much larger volume (15 μl), thus allowing much smaller concentrations of metabolites, i.e. less than 10 mM, to be studied [20]. These solenoids can either be separated vertically (Figure 8.2.17), or horizontally, and have electrical isolations greater than 30 dB. In addition, both of these coils have been double-tuned to proton and nitrogen frequencies, so allowing multi-nuclear multi-sample experiments to be carried out.

Figure 8.2.17 Photograph of a two-coil, multi-tuned probehead, with two solenoids arranged vertically. The impedance matching networks are shielded in copper boxes

Probeheads containing multiple coils can also be used to enhance NMR data acquisition in ways other than by simply running multiple samples. One such case is the hyphenation of CE separations with NMR data acquisition. The ideal NMR detector is a solenoid, due to its intrinsic higher sensitivity when compared to a Helmholz coil, and the smaller size scale at which these can be constructed, again leading to increased sensitivity. However, the horizontal geometry of the solenoid means that line-broadening of the resonances occurs as a result of gradients in the local magnetic field induced by current flow perpendicular to B_0 [21]. One way to avoid this is to carry out stop-flow CE–NMR, but this may compromise the separation efficiency, particularly if the data acquisition time is long. The scheme shown in Figure 8.2.18 uses multiple coils to overcome some of these problems [20]. Three coils are used, with the outlet capillary from each connected either to the negative potential of the voltage generator (allowing electro-osmotic flow) or to ground (stopping electro-osmotic flow), depending upon the position of the switch. The principle is that continuous CE separation can occur, but the capillary splitter can be used to divert a sample of interest into one of the NMR coils in which the capillary is grounded. Then, no line-broadening occurs, and extended NMR data acquisition times can be used to perform, for example, multidimensional experiments.

Some initial results obtained by using this probehead are shown in Figure 8.2.19, although no separations have yet been implemented. The coils are separated by approximately 1 cm. The results show that electro-osmotic flow is occuring in the sample connected to ground and the spectrum is correspondingly broadened, while a second peak, which has been parked in the second RF coil, undergoes no broadening and is isolated completely from electro-osmotic flow.

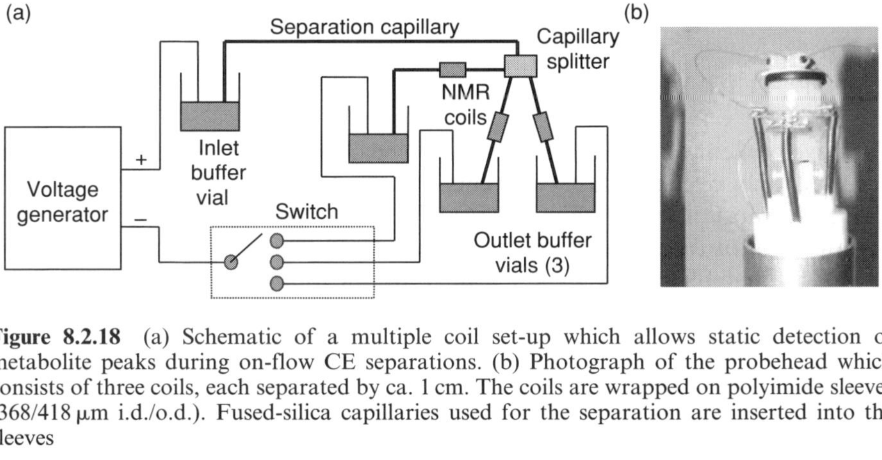

Figure 8.2.18 (a) Schematic of a multiple coil set-up which allows static detection of metabolite peaks during on-flow CE separations. (b) Photograph of the probehead which consists of three coils, each separated by ca. 1 cm. The coils are wrapped on polyimide sleeves (368/418 μm i.d./o.d.). Fused-silica capillaries used for the separation are inserted into the sleeves

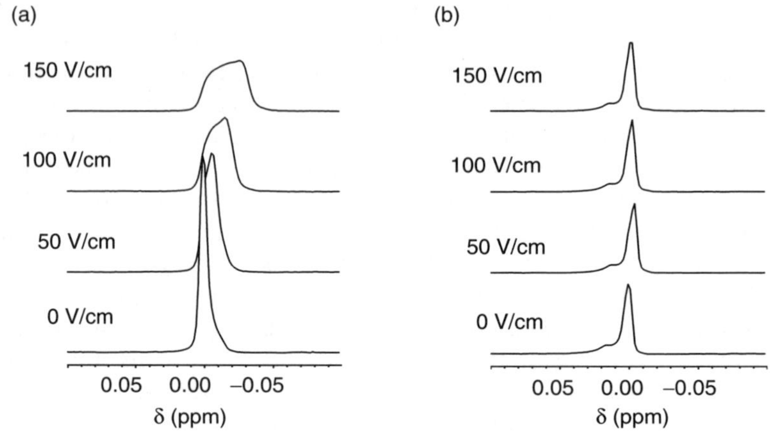

Figure 8.2.19 Spectra acquired simultaneously from two coils: (a) electro-osmotic flow causes increased broadening of the lineshapes with increasing applied voltage; (b) the lineshapes of the resonance in the second coil, where the sample is effectively electrically grounded and thus undergoes no electro-osmotic flow, is not affected by the value of the applied voltage.

Another CE separation method that has been adapted to on-line NMR detection for trace level separations is capillary isotachophoresis [22]. In this case, after the separation, the analyte bands are slowly moved through the capillary until they lie directly within the coil. Precise positioning of the analyte bands in the NMR detection coil can be difficult. A recent enhancement is the use of several NMR detection coils on a single separation capillary [23]. In this way, the first coil acts as a 'scout' coil and is optimized for sensitivity (not necessarily linewidth) to locate the analyte band as it moves through the coil. After an analyte band is detected, the flow is stopped after the appropriate time-interval so that the analyte bands are now located in the second coil, which is used to acquire high-resolution NMR spectra.

8.2.5 FUTURE DIRECTIONS

The development of different techniques for parallel NMR data acquisition has occurred only over the past two years, but is clearly a rapidly growing area. Currently, most of the development is in the area of probe hardware design, and optimization of different sequences and data processing techniques. Although one is ultimately limited by the extent of the homogeneous region of the magnet, one major thrust is to increase the number of coils. This may well involve the incorporation of microfabrication methods, as have already been prototyped for coil production [24]. The ultimate embodiment may well be to emulate other techniques, such as mass spectrometry, in interfacing to a 96-well plate with one coil per sample. This obviously also requires the development of

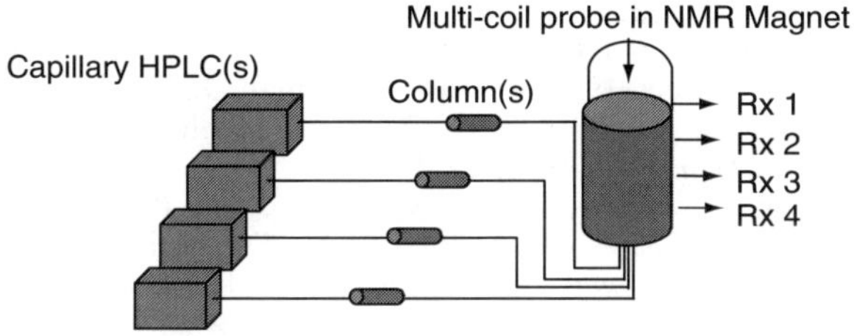

Figure 8.2.20 Schematic of the potential development of parallel CHPLC/NMR hyphenation systems. CHPLC is used for high separation efficiency, with a number of separations carried out simultaneously. Each separation is monitored either by using on-flow or stop-flow NMR detection using an RF microcoil, with each coil feeding data to a separate receiver

commercial spectrometers with an increased number of receiver channels, but manufacturers are already offering up to four receivers, with future development of many more. Hand-in-hand with an increase in the number of coils is the requirement for high spectral resolution, and this may involve the development of zero-susceptibility materials for very small coils. In the area of applications, clearly the tremendous advances in hyphenated microseparation/NMR detection can benefit from the incorporation of parallelism, and experiments such as those envisioned in Figure 8.2.20 are already in the trial stage.

The use of capillary separations, an NMR probe that contains multiple coils, and the associated capillary fluidics to deliver the samples to and from the coils is the next step in probe development. A future exciting development will be the interfacing of such 'intelligent' NMR probe and fluidic systems with other integrated detection modalities such as fluorescence, absorbance and mass spectrometry to provide an integrated system capable of delivering unprecedented structural information from complex samples.

ACKNOWLEDGEMENTS

Contributions from Andrew Wolters, Yu Li, Xiaozhong Zhang, Ting Hou, Jay Smith and Megan Macnaughtan are included in this article. This work was supported by grants from the National Institute of Health (PHS 2 R01 GM53030), the National Science Foundation (CHE CHE 95–31693), the Purdue TRASK fund, and the Sloan Foundation. This contribution was written during the sabbatical leaves of AGW at the Physikalisches Institut at the University of Würzburg, and DR at the Universidad Autonoma in Madrid, Spain.

REFERENCES

1. (a) Wilson, S. R. and Czarnik, A. W., *Combinatorial Chemistry: Synthesis and Application*, Wiley, New York, 1997. (b) Gorlach, E., Richmond, R. and Lewis, I.,

Anal. Chem., 1998, **70**, 3227. (c) Czarnik, A. W., *Anal. Chem.*, 1998, **70**, 378A. (d) Kyranos, J. N. and Hogan Jr, J., *A*, 1998, **70**, 389A. (e) Lam, K., Lebl, M. and Krchnak, V., *Chem. Rev.*, 1998, **97**, 411. (f) Ellman, J., Stoddard, B. and Wells, J., *Proc. Natl. Acad. Sci. USA*, 1997, **94**, 2779. (g) Hsieh-Wilson, L., Xiang, X. and Schultz, P., *Acc. Chem. Res.*, 1996, **29**, 164.

2. (a) Liu, H., Felten, C., Xue, Q., Zhang, B., Jedrzejewski, P., Kargar, B. L. and Foret, F., *Anal. Chem.*, 2000, **72**, 3303. (b) Bayliss, M. K., Little, D., Mallett, D. N. and Plumb, R. S., *Rapid Commun. Mass Spectrom.*, 2000, **14**, 2039. (c) Onorato, J. M., Henion, J. D., Lefebvre, P. M. and Kiplinger, J. P., *Anal. Chem.*, 2001, **73**, 119. (d) Zweigenbaum, J. and Henion, J., *Anal. Chem.*, 2000, **72**, 2446. (e) Jemal, M., *Biomed. Chromatogr.*, 2000, **14**, 422.
3. (a) Kang, S. H., Gong, X. and Yeung, E. S., *Anal. Chem.*, 2000, **72**, 3014. (b) Gaao, Q. and Yeung, E. S., *Anal. Chem.*, 2000, **72**, 2499. (c) Zhang, Y., Gong, X., Zhang, H., Larock, R. C. and Yeung, E. S., *J. Comb. Chem.*, 2000, **2**, 450. (d) Ma, L., Gong, X. and Yeung, E. S., *Anal. Chem.*, 2000, **72**, 3383.
4. (a) Odeblad, E., *Micro-NMR in High Permanent Magnetic Fields*, Nordisk Forening for Obsetrik och Gynekologi, Lund, Sweden, 1966. (b) Shoolery, J. N., *Topics Carbon-13 NMR Spectrosc.*, 1979, **3**, 28. (c) Peck, T. L., Magin, R. L. and Lauterbur, P. C., *J. Magn. Reson., B*, 1995, **108**, 114. (d) Olson, D. L., Peck, T. L., Webb, A. G., Magin, R. L. and Sweedler, J. V., *Science*, 1995, **270**, 1967. (e) Webb, A. G., *Prog. Nucl. Magn. Reson. Spectrosc.*, 1997, **31**, 1. (f) Webb, A. G. and Grant, S. C., *J. Magn. Reson., B*, 1996, **113**, 83. (g) Olson, D. L., Lacey, M. E. and Sweedler, J. V., *Anal. Chem.*, 1998, **70**, 645. (h) Lacey, M. E., Subramanian, R., Olson, D. L., Webb, A. G. and Sweedler, J. V., *Chem. Rev.*, 1999, **99**, 3133.
5. (a) Wu, N., Peck, T. L., Webb, A. G., Magin, R. L. and Sweedler, J. V., *J. Am. Chem. Soc.*, 1994, **116**, 7929. (b) Wu, N., Peck, T. L., Webb, A. G., Magin, R. L. and Sweedler, J. V., *Anal. Chem.*, 1994, **66**, 3849. (c) Olson, D. L., Lacey, M. E. and Sweedler, J. V., *Anal. Chem.*, 1998, **70**, 257A. (d) Pusecker, K., Schewitz, J., Gfrörer, P., Tseng, L.-H., Albert, K. and Bayer, E., *Anal. Chem.*, 1998, **70**, 3280. (e) Pusecker, K., Schewitz, J., Gfrörer, P., Tseng, L.-H., Albert, K., Bayer, E., Wilson, I. D., Bailey, N. J., Scarfe, G. B., Nicholson, J. K. and Lindon, J. C., *Anal. Commun.*, 1998, **35**, 213. (f) Schewitz, J., Gfrörer, P., Pusecker, K., Tseng, L.-H., Albert, K., Bayer, E., Wilson, I. D., Bailey, N. J., Scarfe, G. B., Nicholson, J. K. and Lindon, J. C., *Analyst*, 1998, **123**, 2835. (g) Gfrörer, P., Schewitz, J., Pusecker, K., Tseng, L.-H., Albert, K. and Bayer, E., *Electrophoresis*, 1999, **20**, 3.
6. (a) Wu, N., Webb, A. G., Peck, T. L., and Sweedler, J. V., *Anal. Chem.*, 1995, **67**, 3101. (b) Behnke, B., Schlotterbeck, G., Tallarek, U., Strohschein, S., Tseng, L.-H., Keller, T., Albert, K. and Bayer, E., *Anal. Chem.*, 1996, **68**, 1110. (c) Albert, K., Schlotterbeck, G., Tseng, L.-H. and Braumann, U., *J. Chromatogr., A*, 1996, **750**, 303. (d) Schlotterbeck, G., Tseng, L.-H., Handel, H., Braumann, U. and Albert, K., *Anal. Chem.*, 1997, **69**, 1421.
7. Banas, E. M., *Appl. Spectrosc.*, 1969, **23**, 281.
8. Ross, A., Schlotterbeck, G., Senn, H. and von Kienlin, M., in *Proceedings of the 42nd Experimental NMR Conference*, Orlando, FL, March 11–16, 2001, p. 102.
9. Brown, T. R., Kincaid, B. M. and Ugurbil, K., *Proc. Natl. Acad. Sci. USA*, 1982, **79**, 3523.
10. von Kienlin, M. and Mejia, R., *J. Magn. Reson.*, 1991, **94**, 268.
11. Hu, X., Levin, D. N., Lauterbur, P. C. and Spraggins, T., *Magn. Reson. Med.*, 1988, **8**, 314.
12. MacNamara, E., Hou, T., Fisher, G., Williams, S. and Raftery, D., *Anal. Chim. Acta*, 1999, **397**, 9.
13. Hou, T., MacNamara, E. and Raftery, D., *Anal. Chim. Acta*, 1999, **400**, 297.

14. (a) Wouters, J. M. and Petersson, G. A., *J. Magn. Reson.*, 1977, **28**, 81. (b) Wouters, J. M., Petersson, G. A., Agosta, W. C., Field, F. H., Gibbons, W. A., Wyssbrod, H. and Cowburn, D., *J. Magn. Reson.*, 1977, **28**, 93. (c) Taquin, J., *Rev. Phys. Appl.*, 1979, **14**, 669. (d) Morris, G. A., *J. Magn. Reson.*, 1988, **80**, 547. (e) Gibbs A. and Morris, G. A., *J. Magn. Reson.*, 1991, **91**, 77. (f) Morris, G. A., in *Signal Treatment and Signal Analysis in NMR*, D. N. Rutledge (Ed.), Elsevier, Amsterdam, The Netherlands, 1996, pp. 346–361. (g) Morris, G. A., Barjat, H. and Horne, T. J., *Progr. NMR Spectrosc.*, 1997, **31**, 197. (h) Metz, K. R., Lam, M. M. and Webb, A. G., *Concepts Magn. Reson.*, 2000, **12**, 21.
15. Raftery, D., Hou, T., Macnaughtan, M. and Smith, J., in *Proceedings of the 42nd Experimental NMR Conference*, Orlando, FL, March, 2001, 11–16, p. 66.
16. Hou, T., Smith, J., MacNamara, E., Macnaughtan M. and Raftery, D., *Anal. Chem.*, 2001, **73**, 2541–6.
17. Oldfield, E., *J. Magn. Reson., A*, 1994, **107**, 255.
18. Fisher, G., Petucci, C., MacNamara, E. and Raftery, D., *J. Magn. Reson.*, 1999, **138**, 160.
19. Li, Y., Wolters, A., Malaway, P., Sweedler, J. V. and Webb, A. G., *Anal. Chem.*, 1999, **71**, 4815.
20. Zhang, X., Wolters, A., Li, Y., Kelley, W., Sweedler, J. and Webb, A., in *Proceedings of the 41st Experimental NMR Conference*, Asilomar, CA, April 9–14, 2000, pp. 169–170.
21. Olson, D. L., Lacey, M. E., Webb, A. G. and Sweedler, J. V., *Anal. Chem.*, 1999, **71**, 3070.
22. Kautz, R. A., Lacey, M. E., Wolters, A. M., Foret, F., Webb, A. G., Karger, B. L. and Sweedler, J. V., *J. Am. Chem. Soc.*, 2001, **123**, 3159–3160.
23. Wolters, A. M., Lacey, M. E., Kautz, R. A., Foret, F., Karger, B. L., Webb, A. G. and Sweedler, J. V., in *Proceedings of the 42nd Experimental NMR Conference*, Orlando, FL, March 11–16, 2001, pp. 170–171.
24. (a) Peck, T. L., Magin, R. L., Kruse, J. and Feng, M., *IEEE Trans. Biomed. Eng.*, 1994, **41**, 760. (b) Stocker, J. E., Peck, T. L., Webb, A. G., Feng, M. and Magin, R. L., *IEEE Trans. Biomed. Eng*, 1997, **44**, 1122. (c) Koivuniemi, J., Kiviranta, M., Seppa, H. and Krusius, M., *J. Low. Temp. Phys.*, 1998, **110**, 255. (d) Rodgers, J. A., Jackman, R. J., Whitesides, G. M., Olson, D. L., Sweedler, J. V., *Appl. Phys. Lett.*, 1997, **70**, 2464. (e) Dechow, J., Forchel, A., and Lanz, T. and Haase, A., *Microelectron. Eng.*, 2000, **53**, 519.

Concluding Remarks

KLAUS ALBERT
Institut für Organische Chemie, Universität Tübingen, Tübingen, Germany

When Ernst Bayer asked me in 1976 to develop a NMR probe for the on-line NMR detection of chromatographic peaks we did not dare to apply for funding to the German Research Council. At this time, the rotation of the NMR tube was considered to be a prerequisite for high-resolution NMR and the NMR sensitivity required seemed to be beyond any valuable figure for on-line coupling experiments. Therefore, most of our colleagues in academia and industry were then laughing at this apparently foolish approach. However, the constant support and encouragement of Ernst Bayer finally let to the development of continuous-flow probes for cryomagnets which are now used in more than 200 laboratories throughout the world. Therefore, Ernst Bayer, who sadly died in January 2002, has to be considered as the father of modern LC–NMR hyphenation. His name will always be remembered.

Besides the development work camed out in Tübingen, Joachim Buddruss in Dortmund, Germany, as well as Harry Dorn in Blacksburg, USA and Charles Wilkins in Riverside, USA, have to be named as the first pioneers of on-line LC–NMR. Major applications employing LC–NMR have been performed by Tony Keller, Manfred Spraul, Ian Wilson, John Lindon, Jeremy Nicholson and Alfred Preiss. The nanoliter approach has been pioneered by Jonathan Sweedler, Andrew Webb and Timothy Peck.

Currently, on-line LC–NMR is just at the starting point for dramatic miniaturization and parallelization. Prototypes of NMR probes with remarkable sensitivity values are currently available in 2002, although the ease of operation of the combined separation–detection system is not guaranteed. Because the design of the current analytical LC–NMR probes is so simple, they can be used in many real-world applications. However, current research areas such as proteomics and metabolomics necessitate the utmost sensitivity, using very small amounts of sample, which can only be achieved by employing capillary probes.

Future successful real-world application of the on-going developments might likely be described in the second edition of this current text. Nevertheless, the

On-Line LC–NMR and Related Techniques
Edited by K. Albert.

wealth of practical applications of on-line LC–NMR on the analytical scale described in this present volume should also convince the 'innocent' chromatographer and NMR spectroscopist not familiar with LC–NMR coupling to apply this time- and money-saving technique for the solution of his/her demanding structure elucidation problems.

INDEX

Page numbers in italic, e.g. *19*, refer to figures. Page numbers in bold, e.g. **124**, signify entries in tables.